Logic: Mathematics, Language, Computer Science and Philosophy

VOLUME II

Logic and Computer Science

LOGIC:

Mathematics, Language, Computer Science and Philosophy

Volume II

Logic and Computer Science

by

H. C. M. DE SWART

J. H. Geuvers and R. P. Nederpelt, J.-J. Ch. Meyer, R. Kuiper and W. Penczek, E. A. de Kogel and W. M. J. Ophelders

PETER LANG

Frankfurt am Main · Berlin · Bern · New York · Paris · Wien

Die Deutsche Bibliothek - CIP-Einheitsaufnahme

Swart, Henricus C. M. de:

Logic : mathematics, language, computer science, and philosophy / H. C. M. de Swart. - Frankfurt am Main ; Berlin ; Bern ; New York ; Paris ; Wien : Lang, 1994

Vol. 2 Logic and computer science. - 1994
ISBN 3-631-45435-X

ISBN 3-631-45435-X

Printed in Germany 1 2 3 5 6 7

ACKNOWLEDGEMENTS

An author who experiences as much cooperation as I did is a very happy one.

First of all there are my co-authors: Rob Nederpelt and Herman Geuvers, who were responsible for sections 32 and 33 on untyped and typed Lambda-calculus; Jules Meyer who wrote section 34 on epistemic logic for distributed systems; Ruurd Kuiper and W. Penczek who dealt with verification by means of temporal logic in sections 35 and 36; Eric de Kogel and Willy Ophelders who presented in the Appendix an automated theorem prover for classical predicate logic (without equality) based on the tableaux method extended with unification. It was a pleasure working with them.

Then there are José van Grimbergen and Brigitte van Gisbergen who over a period of many years typed the manuscript in WordMarc Composer with great precision and much enthusiasm.

Marianne Sanders, Mark Vitullo and Vickie Wightman improved our use of the English language, very patiently and competently, over a period of about two years, checking all thirty-six sections.

In the final stage the manuscript was converted into LaTeX. Huub Mulders developed a conversion program and provided the know-how. Unfortunately, he was not able to finish the project. He passed away on January 23, 1993. In the meantime, Sjaak Rens and Jozef Pijnenburg were able to continue the project. They did so with great skill and in a very pleasant way.

Last, but not least, there is the Department of Philosophy of Tilburg University which enabled me to execute this enterprise.

My dealings with Verlag Peter Lang which resulted in the publication of the manuscript in two volumes, were, owing to their cooperation and professionalism, most pleasant.

All the people mentioned behaved as friends aiming at one and the same goal: finishing this book successfully. It was their friendship that made this enterprise a most enjoyable one!

Tilburg, May, 1993 H.C.M. de Swart

CONTENTS

CONTENTS OF VOLUME I

PREFACE

In Volume I we studied propositional logic in Chapter I, set theory in Chapter II and predicate logic in Chapter III, paying special attention to the intimate relation of logic with mathematics, language and philosophy. In this book, Volume II, consisting of Chapter IV, we dwell upon the intimate relation of logic with computer science, both from a theoretical and from a practical point of view.

In sections 25 and 26 we explain how the language of logic can be used as a declarative programming language. By building a logical inference mechanism into the program, the computer can simulate reasoning and hence can be said to possess Artificial Intelligence. In section 27 we present relational databases from a conceptual point of view and make clear how the language of set theory is used as a query language.

In section 28 a mathematical model of real life computers is presented in terms of a Turing machine. In section 29 we prove that the Halting Problem is unsolvable, i.e., there is no algorithm that decides for any Turing machine T, given any input w, whether T will halt on input w. In order to be able to prove this, one has to specify mathematically what one means by an algorithm. A. Turing defined this notion as a Turing machine that halts for any input, answering Yes or No. The Church-Turing Thesis is the statement that the intuitive notion of an algorithm for problem P coincides with the mathematically defined notion of a Turing machine that decides P. If the Halting Problem is reduced to the validity problem for predicate logic, it follows that the latter problem is also unsolvable. In other words, predicate logic is undecidable.

There are (practical) problems which can be solved in theory but not in practice because they would require too much computing time. It seems reasonable to define a realistic or practical algorithm as a Turing machine that given any input computes the answer in a time that polynomially depends on the length of the given input. In section 30 we prove that the satisfiability problem for propositional logic is **NP**-complete, meaning roughly that 1. this problem belongs to the class **NP**, and 2. if this problem could be realistically solved, then any problem in the class **NP** could be solved realistically. We stress that the proof of the **NP**-completeness of the satisfiability problem for propositional logic is very close to the proof of the undecidability of the validity problem for predicate logic.

In section 31 we present the notion of grammar introduced by Chomsky in order to explain the linguistic competence of people. We show that the class of languages generated by a grammar is precisely the class of languages recognized by a Turing machine.

The lambda calculus is the subject of sections 32 and 33. The untyped lambda calculus, as originally introduced by A. Church and H.B. Curry, is treated in section 32. Nowadays this calculus is the basis for functional programming languages such as Miranda. The typed lambda calculus, the subject of section 33, is closely linked to logic and proof systems via the formulas-as-types embedding.

Knowledge, common knowledge and implicit knowledge are studied in the context of distributed systems in section 34. The story about the dean and the hatted professors illustrates the reasoning about common knowledge and the co-ordinated attack problem illustrates that if communication is not guaranteed, it is impossible in principle to attain common knowledge. It is shown that common knowledge is constant in any non-simultaneous execution (of the Kripke model associated with a distributed system). A list of references is given at the end of this section.

Temporal logic is presented as a special case of modal logic: necessity corresponds to always-in-the-future and possibility to sometime-in-the-future. Accordingly, the accessibility relation in Kripke models reflects the passage of time. Two different intuitions about time are used: linear and branching. The former is used to verify systems by hand using proof rules, the latter for automatic verification using model checking.

The appendix contains a tableaux-based automated theorem prover for classical predicate logic (without equality). This theorem prover systematically searches for a tableaux deduction of a given formula B from given premisses $A_1, \ldots, A_m$. In order to make this search efficient, the tableaux method, presented in Volume I (sections 11 and 22), is extended with unification.

Most sections can be read independently. Exceptions are section 30 on Complexity, which presupposes section 29 on Undecidability, which in its turn presupposes section 28 on Computability. Section 33 on typed lambda calculus also requires knowledge of section 32 on untyped lambda calculus. And section 36 presupposes section 35.

Each section ends with a number of exercises; the answers can be found at the end of this book.

Chapter IV
LOGIC AND COMPUTER SCIENCE

§25 LOGIC PROGRAMMING AND PROLOG:

AN EXAMPLE, ANSWERING QUESTIONS; RECURSION; DECLARATIVE VERSUS PROCEDURAL PROGRAMMING; SYNTAX; MATCHING VERSUS UNIFICATION; LISTS, ARITHMETIC; CUT; NEGATION AS FAILURE; APPLICATIONS; PITFALLS

The language of logic can be used as a declarative programming language, i.e., the programmer has to describe *what* the problem is, not *how* it should be solved. We introduce logic programming by means of an example and explain how the system answers questions given a certain program. The possibility of recursive definitions is one of the cornerstones of logic programming. Prolog is a particular form of logic programming; it has been implemented in a certain way. As a consequence, although declarative in principle, Prolog also has certain procedural aspects. The syntax of logic programming in general and of Prolog in particular is very simple. Although the reasoning mechanism should use unification, many systems work with a simpler form, called matching, for reasons of efficiency. Lists are important terms in logic programming. Cut is a procedural device needed to keep programs efficient. Negation is implemented by means of cut and hence differs from logical negation. Logic programming has many applications in (deductive) databases and in Artificial Intelligence. We discuss the most important pitfalls.

AN EXAMPLE; ANSWERING QUESTIONS

The best way to introduce the subject of logic programming seems to be to give an example of a concrete logic program. The following example is from I. Bratko, *PROLOG, Programming for Artificial Intelligence.*

EXAMPLE 25.1.

parent(pam, bob). (1)
parent(tom, bob). (2)
parent(tom, liz). (3)

parent(bob, ann). (4)
parent(bob, pat). (5)

parent(pat, jim). (6)
grandparent(X, Z) :-
parent(X, Y), parent(Y, Z). (7)

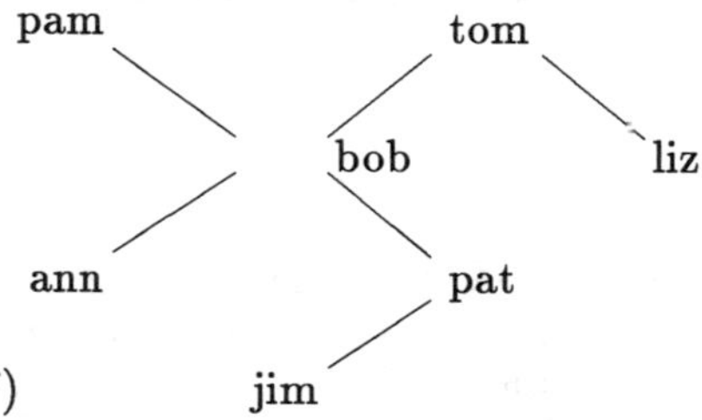

This logic program consists of seven *clauses*. The first six of them are called *facts* ; they express that pam is a parent of bob, tom is a parent of bob, etc. The last clause is called a *rule*; it expresses that X is a grandparent of Z if X is a parent of Y and Y is a parent of Z. The symbol ":-" is to be read as "if" and the comma between "parent(X, Y)" and "parent(Y, Z)" is to be read as "and". "X", "Y" and "Z" are called *variables*; it is allowed to replace them by the names of arbitrary individual objects. In the rule L :- L_1, L_2 "L" is called the *head* of the rule and "L_1, L_2" is called the *body*.

Given a logic program P, the user may ask questions. For instance, given the logic program just presented one might ask who are the parents of bob; in other words, for which X is it true that X is a parent of bob? This question is formulated as follows:

:- parent(X, bob).
or
?- parent(X, bob).

And the question whether bob and liz have a common parent is formulated as follows:

:- parent(X, bob), parent(X, liz).
or
?- parent(X, bob), parent(X, liz).

In *Prolog* , which stands for PROgramming in LOGic and which is just a particular form of logic programming, the answers to these questions are found as follows.

Given the logic program mentioned above and given the question

?- parent(X, bob).,

the Prolog system tries to *match* or *unify* the clause "parent(X, bob)" with the first fact "parent(pam, bob)" in the given program. This matching or unifying succeeds by replacing "X" by "pam". So, "X = pam" is the first answer to this question. Next, the Prolog system tries to match or unify the clause "parent(X, bob)" with the second fact "parent(tom, bob)" in the given program, yielding the second answer "X = tom". Since the Prolog system cannot succeed in unifying or matching the clause "parent(X, bob)" with the other facts in the program, there are no more answers to this question. The following picture describes graphically how the Prolog system finds the answers to the question "?- parent(X, bob)." given the program above.

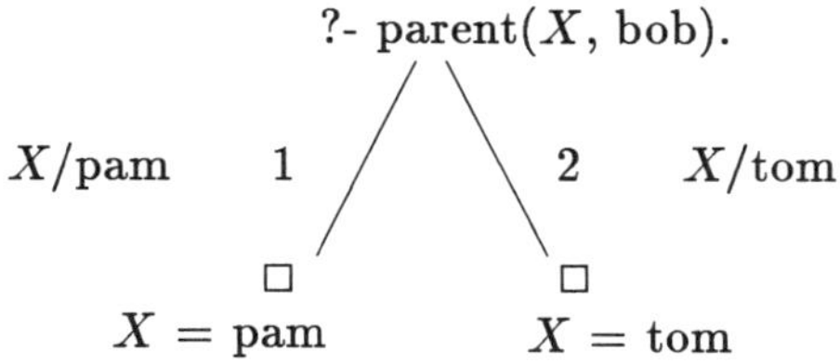

This picture is called the *search tree* for the question "?- parent(X, bob)." given the program above. The numbers 1 and 2 refer to the first and second facts in the program. "X/pam" indicates that "X" is substituted by "pam" in order to match the clause "parent(X, bob)" with fact (1) in the program. The symbol "□" indicates that no other questions remain to be answered.

The search tree for the question

?- parent(X, bob), parent(X, liz).,

given the logic program above, looks as follows.

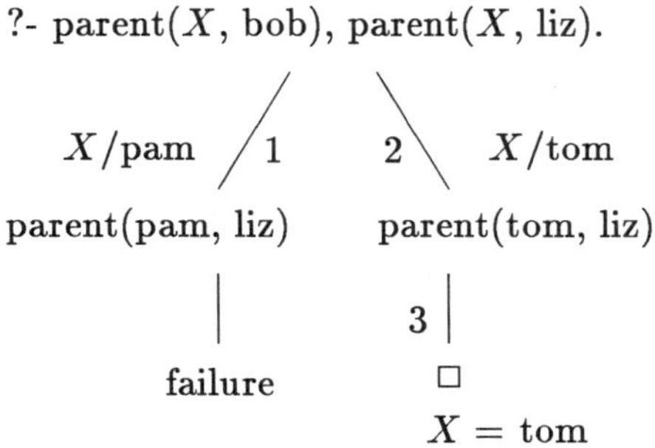

In this example there are two simultaneous questions or *goals*: parent(X, bob), parent(X, liz). The Prolog system selects the *left-most* goal first and tries to match it with the facts in the program. The first possibility to do so is by matching it with fact (1) in the program via the *substitution* X/pam. Then only the goal "parent(X, liz)" remains with "X" replaced by "pam". However, this goal cannot be realized, in the sense that there is no such fact in the program. Hence, this trial to realize the two goals fails. Then the Prolog system *backtracks* and tries to realize the first goal in another way. This can be done by matching it with fact (2) in the program via the substitution X/tom. The second goal "parent(X, liz)" remains with "X" replaced by "tom". Since "parent(tom, liz)" occurs as fact (3) in the program, this goal can be realized by the program and no other goals remain.

Given the logic program mentioned above, the search tree for the question

?- grandparent(X, pat).,

i.e., "for which X is it true that X is a grandparent of pat", in other words, "who are pat's grandparents", looks as follows.

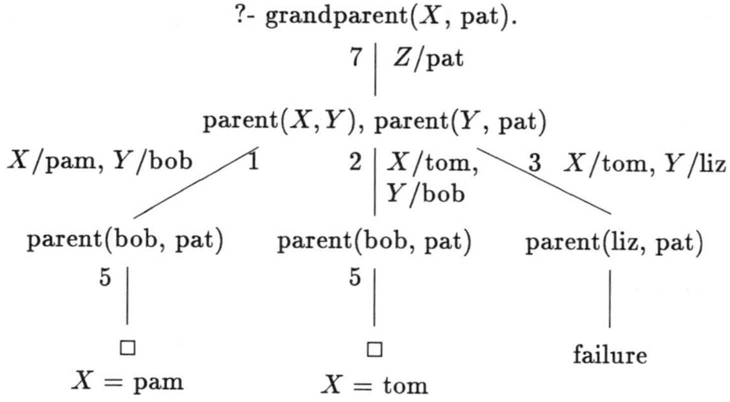

Given the question or goal "?- grandparent(X, pat).", the Prolog system looks for facts of this form in the given program, but does not find any. It also tries to match or unify the goal with the head of a rule in the program. This succeeds: the goal "grandparent(X, pat)" can be unified with the head of clause (7) in the program via the substitution Z/pat. Then the original goal is replaced by two new goals:

$$\text{parent}(X, Y),\ \text{parent}(Y,\ \text{pat}).$$

The *left-most* goal is selected first. Looking at the program, we see that there are six different ways to realize this goal. The first possibility is to use the first clause in the program via the substitution X/pam, Y/bob. The second goal "parent(Y, pat)" remains with "Y" substituted by "bob". Since "parent(bob, pat)" is the 5th clause in the program, this goal is realized and no other goals remain. The Prolog system answers "X = pam". Next the system *backtracks* and tries to realize the left-most goal "parent(X, Y)" in another way. This can be done by using the second clause in the program via the substitution X/tom, Y/bob. Then the second goal "parent(Y, pat)" remains with "Y" replaced by "bob". Since this is the 5th fact in the program the second branch in the search tree is completed successfully and the system answers "X = tom". Again, backtracking takes place and Prolog realizes the left-most goal "parent(X, Y)" by using the third fact in the program via the substitution X/tom, Y/liz. Then the goal "parent(liz, pat)" remains. Looking at the program, the Prolog system discovers that this goal cannot be realized. It backtracks and tries to realize the left-most goal "parent(X, Y)" in a fourth way, and so on.

TO SUMMARIZE, in logic programming a program consists of *facts* and *rules*. A logic program is a kind of database. A question is a finite sequence of one or more goals. Given a logic program, questions are answered by trying exhaustively to realize the goals by *matching* (or *unifying*) them with the facts and/or the heads of the rules in the program, possibly via *substitution* of the variables. A logic programming system accepts facts and rules as a set of (non-logical) axioms and a question as a putative theorem or conclusion. The logic programming system tries to deduce this putative conclusion logically from the axioms.

Typical features of Prolog are that it has a *left-most selection rule*, that the search is *depth-first* (not breadth-first) and that after successful or unsuccessful termination of a branch in the search tree, *backtracking* takes place in order to find alternative solutions.

The reader is advised to do Exercise 25.1.

RECURSION

The use of recursive definitions (of predicates or relations) is typical in logic programming. As an example we might add the following two clauses to the logic program (Example 25.1.) of the preceding subsection.

pred(X, Z) :- parent(X, Z).
pred(X, Z) :- parent(X, Y), pred(Y, Z).

These two clauses define the predecessor relation, abbreviated by "pred". The first clause expresses that X is a predecessor of Z if X is a parent of Z. And the second one expresses that X is a predecessor of Z if (for some Y) X is a parent of Y and Y is a predecessor of Z. So, in the second clause the relation "pred" recurs in the definition of "pred(X, Z)". For this reason one speaks of a *recursive definition.*

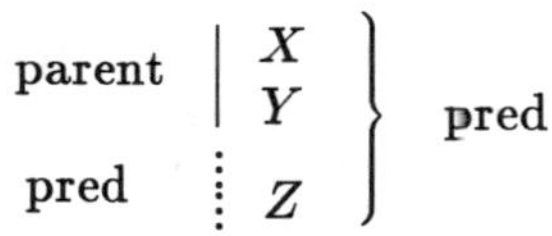

In order to understand the role of recursive definitions in logic programming, let us ask the question

?- pred(tom, pat).,

given the program which results from adding the two clauses for "pred" to the program in Example 25.1. This program looks as follows.

parent(pam, bob). (1)
parent(tom, bob). (2)
parent(tom, liz). (3)
parent(bob, ann). (4)
parent(bob, pat). (5)
parent(pat, jim). (6)
grandparent(X, Z) :- parent(X, Y), parent(Y, Z). (7)
pred(X, Z) :- parent(X, Z). (8)
pred(X, Z) :- parent(X, Y), pred(Y, Z). (9)

Given this program, the search tree for the question (or goal)

?- pred(tom, pat).

looks as follows.

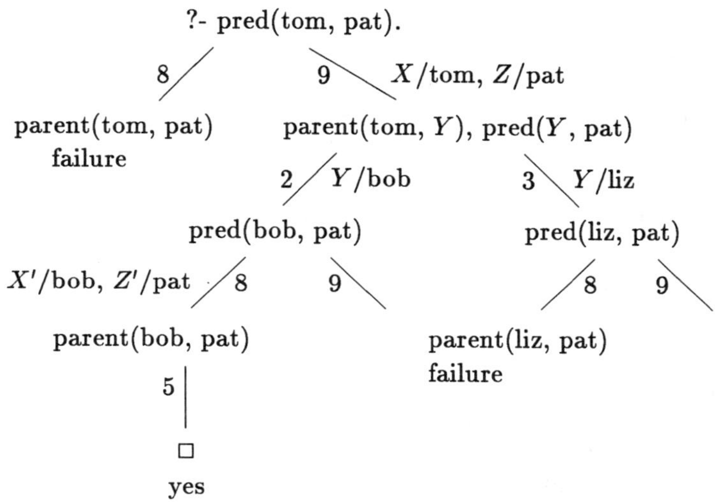

In order to answer the question "pred(tom, pat)" the Prolog system tries to unify this goal with a fact or the head of a rule in the program. The first possibility is to unify "pred(tom, pat)" with the head of clause (8) via the substitution X/tom, Z/pat. The goal "parent(tom, pat)" is the result. Since there is no such fact in the given program, the left-most branch in the search tree terminates unsuccessfully and backtracking takes place. The original goal can also be unified with the head of clause (9) in the program via the substitution X/tom, Z/pat. Then the original goal is replaced by two new goals: parent(tom, Y), pred(Y, pat). Prolog selects the left-most goal first. It can be unified with clause (2) in the program via the substitution Y/bob. Then the goal "pred(Y, pat)" remains with "Y" substituted by "bob". In order to realize this latter goal, Prolog first matches it with the head of clause (8). Since "X" and "Z" have already been substituted by "tom" and "pat" respectively, the Prolog system has replaced the variables X and Z in clause (8) by X' and Z' respectively. Clause (8) now looks as follows:

$$\text{pred}(X', Z') \text{ :- parent}(X', Z'). \qquad (8')$$

The goal "pred(bob, pat)" can now be unified with the head of clause ($8'$). This yields the new goal "parent(bob, pat)". Because of the 5th clause in the program, the second branch (from the left) in the search tree terminates successfully, and the answer to the original question is "yes". Backtracking takes place in order to see whether the goal "pred(bob, pat)" can be realized in other ways. In our search tree we have not worked this out. But it turns out that the goal "pred(bob, pat)" cannot be realized in another way. Further backtracking takes place in order to see whether the left-most goal in "parent(tom, Y), pred(Y, pat)" can be realized in another way. This may indeed be the case. Applying clause (3) in the program and substituting "liz" for "Y", the goal "pred(Y, pat)" remains with "Y" replaced by "liz", and so on.

DECLARATIVE VERSUS PROCEDURAL PROGRAMMING

Logic programming is in principle *declarative*: the programmer only has to describe the problem, in other words, he must formulate *what* the problem is; but he does not have to specify *how* the problem has to be solved. The logic programmer is more concerned with *knowledge* than with *algorithms*.

In order to answer certain questions, the logic programmer has to formulate all relevant information, consisting of facts and rules, in a logic program. Given a certain program, the logic programming system will try systematically to deduce the answer to any question from the facts and rules in the program. This is done by an exhaustive search, which can be represented in a *search tree*.

Given the program of the preceding subsection, the answer to the question "?- pred(tom, pat)." is "yes". This means that pred(tom, pat) logically follows from (is a valid consequence of) the facts and the rules in the program. The answer to the question "?- pred(pam, liz)." will be "no", meaning that pred(pam, liz) is not a logical consequence of the (facts and rules in the) program. This does NOT mean that "not pred(pam, liz)" is a logical consequence of the program in question! And the answer to the question

?- parent(X, bob), parent(X, liz).

was "X = tom". This means that both parent(tom, bob) and parent(tom, liz) logically follow from the given program.

One can prove what is called *soundness*: given any logic program P and question or goal G every computed answer logically follows from P. The converse problem is *completeness* : given any logic program P, is the logic programming system able to compute any goal which logically follows from P? This problem is more difficult and cannot be answered with a simple "yes" or "no". We will return to this in Section 26.

Programming languages like Pascal, Algol and C are *procedural* languages: the programmer has to specify *how* the problem has to be solved. Although logic programming is in principle declarative and not procedural, the logic programmer has to take into account certain procedural aspects. We have already noticed above that Prolog, being a particular – but most popular – logic programming system, has a *left-most selection rule* and a *depth-first search* strategy. This strategy first developes the left-most branch in the search tree. As long as this branch has not been terminated, no other branches are developed. It also searches for facts and rules in the program in the order they have been programmed. The programmer, who writes a logic program in Prolog, has to take the procedural aspects of the Prolog system into account. The order of the facts and rules in his program may be important and even the order of the goals in the body of a rule may be important. In order to make this clear, look at the following four definitions of the predecessor relation.

pred(X, Z) :- parent(X, Z).
pred(X, Z) :- parent(X, Y), pred(Y, Z).

pred2(X, Z) :- parent(X, Y), pred2(Y, Z).
pred2(X, Z) :- parent(X, Z).

pred3(X, Z) :- parent(X, Z).
pred3(X, Z) :- pred3(X, Y), parent(Y, Z).

pred4(X, Z) :- pred4(X, Y), parent(Y, Z). (I)
pred4(X, Z) :- parent(X, Z). (II)

In the definition of "pred2" and "pred4" the order of the clauses is reversed with respect to the definition of "pred". And in the definition of "pred3" and "pred4" the goals in the body of the recursion clause are reversed with respect to the definition of "pred". This may have disastrous consequences. Consider the search tree for ?- pred4(tom, pat).

?- pred4(tom, pat).

I | $X/\text{tom}, Z/\text{pat}$

pred4(tom, Y), parent(Y, pat)

I | $X'/\text{tom}, Z'/Y$

pred4(tom, Y'), parent(Y',Y), parent(Y, pat)

I |

.
.
.

ad infinitum

The left-most branch in this search tree will be infinitely long since clause I will be applied again and again. And the other branches in the search tree will not be developed. So, the Prolog system will give no answer to the question "?- pred4(tom, pat).". And although the Prolog system does answer the questions "?- pred2(tom, pat)." and "?- pred3(tom, pat)." positively, exercise 25.2. makes clear that from a procedural point of view, pred2 and pred3 do not give an adequate description of the predecessor relation.

To summarize, although the definitions of "pred", "pred2", "pred3" and "pred4" are equivalent from a logical or declarative point of view, they are quite different from a procedural point of view. Consequently, the logic programmer has to take into account the particular procedural aspects of the logic programming system he or she is working with. For recursive definitions there is a simple rule.

i) The more simple clause is put first (supposing that the logic programming system searches through the program from top to bottom). In the case of the predecessor relation this is the clause

$$\text{pred}(X, Z) \text{ :- } \text{parent}(X, Z).$$

ii) The goals in the body of the recursion clause should be ordered from simple to more complex (supposing a left-most selection rule in the logic programming system). So, the recursion clause for the predecessor relation should be

$$\text{pred}(X, Z) \text{ :- } \text{parent}(X, Y), \text{pred}(Y, Z).,$$

since "parent" is a more simple relation than "pred".
Another procedural aspect of Prolog is the cut, denoted by "!". The cut prunes part of the search tree. This enhances efficiency, but may be dangerous if the pruned part contains successful branches. We will discuss this feature of Prolog further on in a special subsection.

SYNTAX

The syntax of a logic programming language is that of first-order predicate logic (see § 18 and § 24) with possibly some modifications in notation. The *alphabet* consists of:

i) *Individual variables* , such as $X1, X2, \ldots, X, Y$, Person, ...
ii) *Individual constants* , such as pam, bob, liz, ... ; 0, 1, 2, ...
iii) *Function symbols* , such as father_of, mother_of, ... ; +, ∗, ...
iv) *Predicate symbols* , such as parent, male, female; =, <, ...

The programmer is free to choose his own symbols in the alphabet. However, in Prolog any expression starting with a capital is a variable. For instance, the expressions "Pam" and "Bob" are variables in Prolog, while "pam" and "bob" are individual constants. Each function and predicate symbol is k-ary for some k; the arity is chosen by the programmer.

Terms are defined as in Section 24:

i) Each individual variable and each individual constant is a term.
ii) If f is a k-ary function symbol and $t_1, \ldots, t_k$ are terms, then $f(t_1, \ldots, t_k)$ is a term.

Examples of terms:

1) Person, tom, bob; 1, 2.
2) father_of(Person), mother_of(bob); +(1, 2), usually written as $1 + 2$.
3) mother_of(father_of(Person)); ∗(1, +(1, 2)), usually written as $1 * (1 + 2)$.

Atomic formulas are defined as in Section 18: if p is an n-ary predicate symbol and $t_1, \ldots, t_n$ are terms, then $p(t_1, \ldots, t_n)$ is an atomic formula.
Examples of atomic formulas: parent(tom, bob), parent(X, liz), female(X), male(bob); = (X, 1), usually written as $X = 1$ and <(2, 3), usually written as $2 < 3$.

DEFINITION 25.2. (Cf. Definition 4.8.):

i) A *definite program clause* is any expression of the form B :- $A_1, \ldots, A_m$ $(m \geq 0)$, where B and $A_1, \ldots, A_m$ are atomic formulas. If $m = 0$, one writes simply "B" instead of "B :- ". "B :- $A_1, \ldots, A_m$" is to be read as: B if A_1 and ... and A_m; in the notation of § 18: $B \subset A_1$ & ... & A_m. B is called the *head* and $A_1, \ldots, A_m$ is called the *body* of the clause B :- $A_1, \ldots, A_m$. If $m = 0$, one calls the definite program clause a *fact*, otherwise a *rule*.

ii) A *definite goal* is any expression of the form :- $A_1, \ldots, A_m$ $(m \geq 0)$, where $A_1, \ldots, A_m$ are atomic formulas. Each A_i is called a *subgoal* of the goal. If $m = 0$, one speaks of the *empty goal* or *empty clause*, denoted with "$\Box$".

iii) A *definite program* is a finite set of definite program clauses. See, for instance, Example 25.1.

iv) A *definite clause* or *Horn clause* is either a definite program clause or a definite goal.

DEFINITION 25.3.

i) A *literal* is an atomic formula A or the negation "not A" of an atomic formula A. In the notation of § 2 and § 18 "not A" was written as "$\neg A$".

ii) A *normal program clause* is any expression of the form B :- $L_1, \ldots, L_m$, where B is an atomic formula and $L_1, \ldots, L_m$ are literals. *Example*: sister(X, Y) :- parent(Z, X), parent(Z, Y), female(X), not $X = Y$.

iii) A *normal goal* is any expression of the form :- $L_1, \ldots, L_m$, where $L_1, \ldots, L_m$ are literals.

iv) A *normal program* is a finite set of normal program clauses.

From a theoretical point of view, in particular with respect to completeness, definite logic programs are to be preferred to normal logic programs. However, for practical purposes the latter ones are often needed. We return to the problems connected with negation later in this section.

Definite and normal program clauses are formulas of a special kind. In order to see this, let us repeat the definition of *formulas* (Definition 18.1).

i) Any atomic formula is a formula.

ii) If A and B are formulas, then so are $(A \sim B)$, $(A \supset B), (A \,\&\, B)$, $(A \vee B)$ and $(\neg A)$, to be read as "A if and only if B", "if A, then B", "A and B", "A or B" and "not A", respectively.

iii) If A is a formula and x is an individual variable, then $\forall x[A]$ and $\exists x[A]$ are formulas, to be read as "for all x, A" and "there is at least one x such that A", respectively. (See Definition 18.1. for more details.)

In Theorem 4.7. we have shown that any formula not containing the quantifiers $\forall$ and $\exists$ is equivalent to a finite conjunction of clauses, where a *clause* is a formula of the form

$$A_1 \,\&\, \ldots \,\&\, A_m \supset B_1 \vee \ldots \vee B_n$$

or, in different notation,

$$B_1 \vee \ldots \vee B_n \text{ :- } A_1, \ldots, A_m,$$

the A's and B's being *atomic* formulas. (See the subsection on Knowledge Representation and Prolog in section 4; note that $B_1 \vee \ldots \vee B_n$:- $A_1, \ldots, A_m$ is equivalent to $B_1 \vee \ldots \vee B_n \vee \neg A_1 \vee \ldots \vee \neg A_m$.)

However, in Prolog only *definite clauses* are allowed, i.e., clauses with $n \leq 1$, for reasons of efficiency. For instance, suppose we had a program containing the clause $p(1) \vee q(2)$. Now the question ?- $p(X) \vee q(X)$ should be answered as follows: $X = 1$ if $p(1)$ and $X = 2$ if $q(2)$. It is hard to implement a system that is able to give such conditional answers.

In the subsection on Skolemization and Clausal Form in § 24 we have defined the *clausal form* $C(A)$ of any formula A, such that

i) $C(A)$ is a conjunction of clauses, and

ii) A is satisfiable iff $C(A)$ is satisfiable.

In the subsection on Automated Theorem Proving and Logic Programs in § 24 we have shown that any definite logic program is a formula in clausal form.

MATCHING VERSUS UNIFICATION

In the preceding examples we have seen that a Prolog system, given a certain program and answering a certain question, makes use of what is called matching or unification. In this subsection we want to describe matching and unification more precisely and to point out the difference between them.

Matching is a process that takes as input two terms or atomic formulas and checks whether they match. The rules governing this process are the following:

1. Two individual constants match only if they are syntactically the same.
2. If X is a variable and t a term, then they match and X is instantiated to, or substituted by, t.
3. Two terms match only if they have the same principal function symbol and all their arguments match.
4. Two atomic formulas match only if they have the same principal predicate symbol and all their arguments match.

EXAMPLE 25.4.:

i) The pair {parent(X, bob), parent(pam, bob)} can be matched.

ii) The pair {parent(X, bob), parent(pat, jim)} cannot be matched, because the second arguments cannot be matched.

iii) The pair $\{p(f(X), Z),\ p(Y, c)\}$ can be matched via the substitution $Y/f(X), Z/c$.

iv) The pair $\{p(f(X), c),\ p(Y, f(Z))\}$ cannot be matched, since the second arguments cannot be matched.

v) The pair $\{p(X, X),\ p(Y, f(Y))\}$ can be matched. The first arguments can be matched via the substitution X/Y. The resulting pair is $\{p(Y, Y),\ p(Y, f(Y))\}$. The second arguments can be matched via the substitution $Y/f(Y)$.

The following logic program shows that in some cases matching yields undesirable results. Let P be the following logic program:

$$\text{parent}(Y, \text{child\_of}(Y)). \tag{1}$$

Note that "parent(X, X)" is similar to "$p(X, X)$" and that "parent(Y, child_of(Y))" is similar to "$p(Y, f(Y))$" in Example 25.4. v). The search tree for the question "?- parent(X, X).", given this program, looks as follows.

?- parent(X, X).

(1) | X/Y, Y/child_of(Y)

□

yes

However, this result is undesired, since "parent(X,X)" does *not* logically follow from the program P. In the intended interpretation the only formula (1) in P expresses a true proposition, while "parent(X,X)" expresses a false proposition for any value of X.

For reasons of efficiency, most logic programming systems make use of matching and take for granted that in some cases this may yield the wrong results. What they should do, however, is to replace matching by unification and take for granted that in some cases this may be inefficient. *Unification* can be characterized by the following slogan:

Unification = matching + occur check.

The *occur check* involves checking whether in the substitution of a term t for a variable X (clause 2 in the definition of matching), the variable X does not occur in t. If X does occur in t, then unification fails, while matching may succeed.

In Example 25.4. v) we have seen that the pair $\{p(X,X),\ p(Y,f(Y))\}$ can be matched. However, this pair cannot be unified. After the substitution X/Y the resulting pair is $\{p(Y,Y),\ p(Y,f(Y))\}$. And although the second arguments in this pair can be matched, they cannot be unified since the variable Y does occur in $f(Y)$.

The *unification algorithm* is like the matching algorithm given above, except that the occur check is added to clause 2. Below, we demonstrate how the unification algorithm works in a few examples.

EXAMPLE 25.5. (Lloyd):

1. Is it possible to unify $p(f(c),g(X))$ and $p(Y,Y)$?
   i) The predicate symbols are identical.
   ii) The left-most arguments that differ are $f(c)$ and Y.
   Occur check: Y does not occur in $f(c)$. So, replace Y by $f(c)$.
   Result: $p(f(c),g(X))$ and $p(f(c),f(c))$.
   iii) The left-most arguments that differ are $g(X)$ and $f(c)$. These terms have different principal function symbols and hence cannot be unified.

   Conclusion: $p(f(c),g(X))$ and $p(Y,Y)$ cannot be unified. Nor can they be matched.
2. Can $p(c,X,h(g(Z)))$ and $p(Z,h(Y),h(Y))$ be unified?
   i) The predicate symbols are identical.
   ii) The left-most arguments that differ are c and Z.
   Occur check: Z does not occur in c. So, replace Z by c.

Result: $p(c, X, h(g(c)))$ and $p(c, h(Y), h(Y))$.

iii) The left-most arguments that differ now are X and $h(Y)$.
Occur check: X does not occur in $h(Y)$. So, replace X by $h(Y)$. Result: $p(c, h(Y), h(g(c)))$ and $p(c, h(Y), h(Y))$.

iv) The left-most arguments that differ now are $h(g(c))$ and $h(Y)$.
a) The principal function symbols are identical.
b) The arguments $g(c)$ and Y are different.
Occur check: Y does not occur in $g(c)$. So, replace Y by $g(c)$.
Result: $h(g(c))$ and $h(g(c))$.

Conclusion: $p(c, X, h(g(Z)))$ and $p(Z, h(Y), h(Y))$ are unifiable via the substitutions $Z/c, X/h(Y)$ and $Y/g(c)$.

3. As we have seen above, $p(X, X)$ and $p(Y, f(Y))$ cannot be unified because of the occur check, although they can be matched.

For more details about substitution and unification the reader is referred to Section 26. See also Exercise 25.3.

Note that both the matching and the unification algorithms result in a *most general unifier* (substitution) in the sense that no more is substituted than strictly necessary. For instance, unifying date(D, M, 1983) and date($D1$, may, Y) results in the substitution $D/D1$, M/may and $Y/1983$. The substitutions $D/1$, $D1/1$, M/may and $Y/1983$ also unify the two terms, but are less general.

One can show that given a logic program P any answer to a question is correct in the sense that the computed answer is a logical consequence of the given program, provided the system uses unification instead of matching. For instance, given the program of Example 25.1. the Prolog system computes two answers to the question ?- grandparent(X, pat): X = pam; X = tom. The theorem just mentioned, called the *soundness* theorem, then says that "grandparent(pam, pat)" and "grandparent(tom, pat)" are logical consequences of the given program. For more details we refer the reader to Section 26 where we also discuss the converse problem whether any logical consequence of a given program can be computed by the Prolog system (*completeness*).

LISTS; ARITHMETIC

Lists are very important terms in the practice of logic programming.

For instance, if one wants to represent information about families in a logic program, one has the problem that different families have different numbers of children. By putting the children of any family in a list, one can represent any family in a uniform way:

family(Father, Mother, List_of_Children).

Here, "family" is a predicate symbol taking three arguments, no matter how many children there are.

Lists are terms of a special kind, defined as follows.

DEFINITION 25.6. (*Lists*):
i) [ ] is a list, called the *empty list.*
ii) If t is any term and L is a list, then $[t \mid L]$ is a list.
In $[t \mid L]$ t is called the *head* and L is called the *tail* of the list $[t \mid L]$.

Examples of lists:
i) [ ].
ii) $[c \mid [\,]]$, usually rendered as $[c]$.
iii) $[b \mid [c]]$, usually rendered as $[b, c]$.
iv) $[a \mid [b, c]]$, usually rendered as $[a, b, c]$.

Working with lists, one needs a program that determines the elements or members of a given list. Reading "member(X, L)" as "X is a member of list L", the membership relation is defined recursively as follows.

$$\text{member}(X, [X \mid L]). \tag{1}$$
$$\text{member}(X, [Y \mid L]) \text{ :- } \text{member}(X, L). \tag{2}$$

In words: X is a member of a given list if either (1) X is the head of the list, or (2) X is a member of the tail of the list. Given this program, the search tree for the question "?- member$(X, [b, c])$." (What are the members of the list $[b, c]$?) looks as follows.

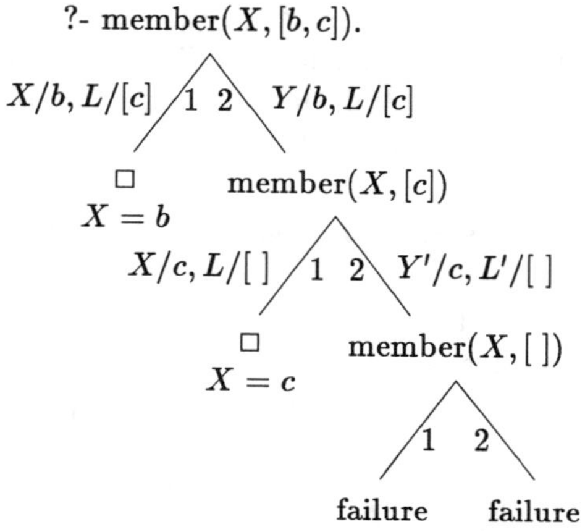

In Exercise 25.4. the reader is invited to define a concatenation relation for lists and in Exercise 25.5. to define a relation for deleting members from a given list.

Prolog contains some built-in arithmetic operations which can be used in the infix notation, i.e., $2+3$ instead of $+(2,3)$, etc. Among them are
$+$ for addition , $*$ for multiplication ,
$-$ for subtraction , $/$ for division .
When doing arithmetic in Prolog it is important to realize that "=" is a built-in matching operator, while "is" is a built-in operator that forces the evaluation of the term in question. In order to make the difference clear, consider the following examples.

?- $X = 2+3$.	?- X is $2+3$.	?- $2+3 = 3+2$.
$X = 2+3$	$X = 5$	no

Other built-in operators which force the evaluation of the terms in question are the following:

>	is greater than;
>=	is greater than or equal to;
=:=	the values of the left and right terms are equal;
=\=	the values of the left and right terms are not equal.

Example: ?- $2+3 =:= 3+2$. yes ?- $2*3 > 5$. yes

It is important to realize that all arguments must be instantiated to numbers at the time that the evaluation is carried out. Examples:

?- X is 2*3, $X > 5$. ?- $X > 5$, X is 2*3.

$X = 6$ control error

In Exercise 25.6. the reader is invited to define a relation for counting the number of members of a given list.

CUT

max(X, Y, Z), to be read as "Z is the maximum of X and Y", can be defined as follows.

max(X, Y, X) :- $X >= Y$. (1)
max(X, Y, Y) :- $Y > X$. (2)

Now the programmer, but not the logic programming system, knows that if the goal $X >= Y$ succeeds, then the goal $Y > X$ is bound to fail. So, given this program and the question ?- max(3, 2, Z), it is a waste of time and energy to try to apply the second clause via backtracking, once the left-most branch in the search tree has terminated successfully.

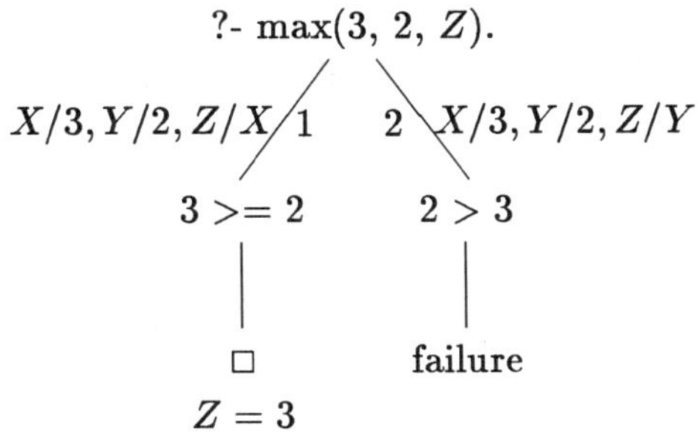

It is attractive to have a control facility that prunes that part of a search tree that only contains unsuccessful branches. Prolog has such a control facility, called *cut* and denoted by "!". ! can be conceived of as a true atomic formula or as a goal that always succeeds. However, while the declarative or logical meaning of "!" is "true", the procedural meaning of "!" is the pruning of the search tree. Given the program

max1(X, Y, X) :- $X >= Y$, !. (1)
max1(X, Y, Y)., (2)

the search trees for the questions ?- max1(3, 2, Z) and ?- max1(2, 3, Z) look as follows:

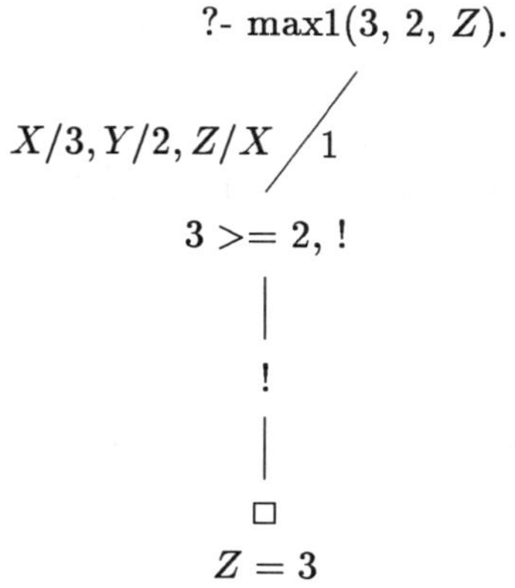

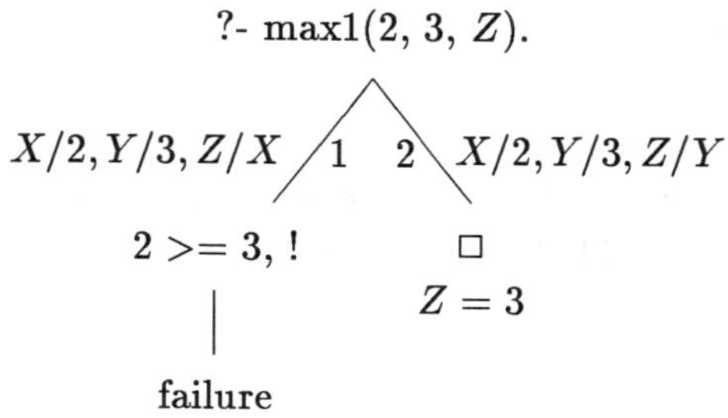

The right-most branch in the search tree for ?- max1(3, 2, Z) is pruned because first the goal 3 >= 2 succeeds and next the cut is passed. The goal 2 >= 3 in the left-most branch in the search tree for ?- max1(2, 3, Z) fails, hence the cut is not passed and backtracking takes place as usual.

From a procedural point of view, the programs for max(X, Y, Z) and for max1(X, Y, Z) yield the same results. However, from a declarative or logical point of view the program for max1(X, Y, Z) is not an adequate description of the maximum relation. Since the declarative meaning of "!" is "true", from a declarative point of view the program for max1 is equivalent to the following one.

max2(X, Y, X) :- X >= Y. (1)
max2(X, Y, Y). (2)

But the question ?- max2(3, 2, Z) yields two answers: $Z = 3$ and $Z = 2$.

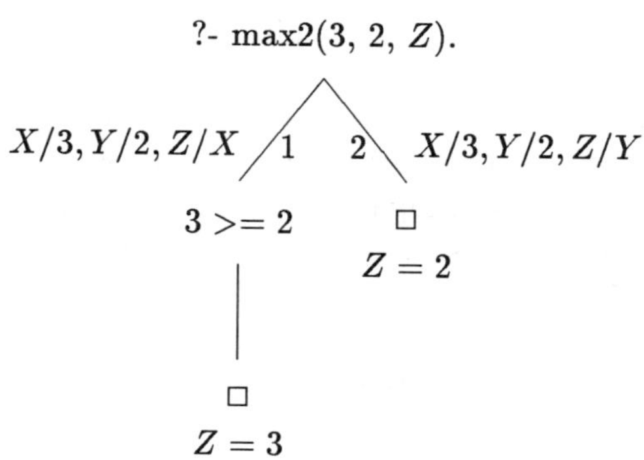

So, if one wants a program for the maximum relation that is both correct from a declarative point of view and efficient from a procedural point of view, the following program is to be preferred.

max3(X, Y, X) :- $X >= Y$, ! . (1)
max3(X, Y, Y) :- $Y > X$. (2)

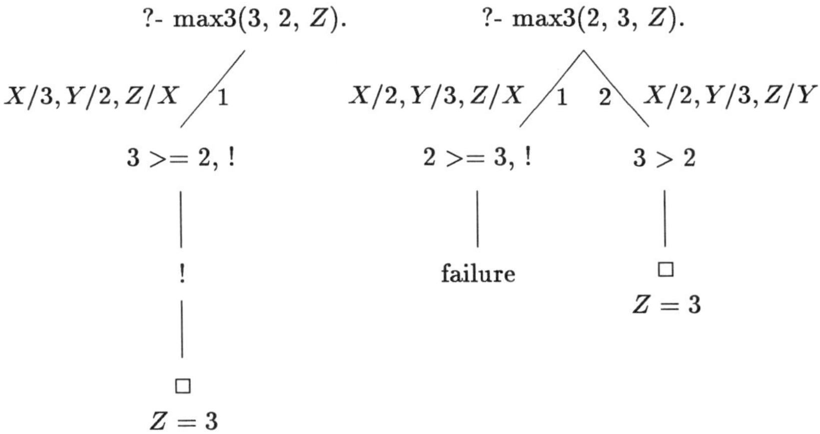

So, a cut prunes the search tree. This is safe if the pruned part contains no successful branches. In that case the cut merely enhances efficiency; it saves time. However, if the pruned part contains successful branches, the use of cut may have disastrous consequences. For that reason the programmer should be very careful in using this control facility. Unfortunately, more complicated programs often require the use of cuts in order to keep the program efficient.

What part of the search tree is pruned by using cut? In order to answer this question more precisely, consider the following program P.

$p(X)$:- $q(X), r(X)$.
⋮
$q(X)$:- $s(X), t(X)$, !, $u(X)$.
$q(X)$:- $v(X)$.
⋮
$r(1)$.
$s(1)$.
$t(1)$.

$u(X)$:- $X = 5$.

$u(X)$:- $X > 2$.

The following picture shows the effect of cut on the search tree for the question "?- $p(X)$.", given program P above.

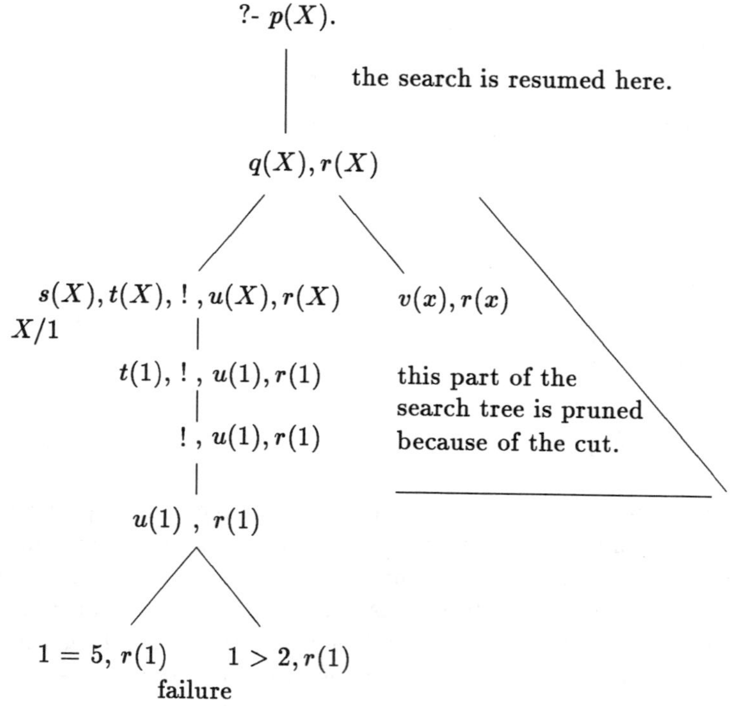

Given the program P above, the goal "?- $p(X)$." will be answered with "no", even if we add the fact $v(1)$ to P. In that case there is a successful branch in the search tree, namely, the branch with $v(X), r(X)$. However, this branch will be pruned because of the cut. If we add a second rule to P, "$p(X)$:- $X = 2$.", the goal "?- $p(X)$." will have "$X = 2$" as its only solution.

In order to formulate precisely what the effect of cut on the search tree is, we have to define the notion of *parent* goal. The *parent* goal is the goal that causes the clause containing the cut to be activated. In our example this is $q(X)$. The cut commits the system to all choices made between the time the parent goal was involved and the time the cut was encountered. The remaining alternatives between the parent goal and the cut are discarded.

Exercises 25.7. and 25.8. give some other programs containing cuts.

NEGATION AS FAILURE

Prolog has a built-in operator "not", which has been defined, using cut, as follows.

not(A) :- A, !, fail.

not(A) :- true.

In order to understand this definition, the reader should know that "fail" and "true" are built-in expressions which always fail or succeed respectively, when they are invoked. From this definition it follows immediately that

i) the goal "not(A)" fails if the search tree for "?- A." is finite and has a successful branch, and

ii) the goal "not(A)" succeeds if the search tree for "?- A." is finite and has no successful branches.

Note that if the search tree for "?- A." contains no successful branches and has at least one infinite branch, then the Prolog system cannot answer the question "?- not(A).".

In order to see how Prolog handles negation, let us consider the following program P.

student(tom).

student(jane).

teacher(mary).

It is important to note that neither "student(mary)" nor "not student(mary)" are logical consequences of P.

Given this program, the question "?- not student(mary)." is answered by the Prolog system as follows.

```
?- not student(mary).   ------->   ?- student(mary).
          :                              |
          :                             fail
         yes            <-------         no
```

The Prolog system uses what is called the *Negation as Failure* (*NF*-)rule: if the search tree for A, given a certain program, is finite and has no successful branches, then conclude "not A".

The Negation as Failure rule is *non-monotonic*, i.e., adding new facts and/or rules to the given program may eliminate some former conclusions. For instance, if we add the fact "student(mary)" to the program P above, the conclusion "not student(mary)" can no longer be drawn. More information may lead to different (and other) conclusions.

Program P above is equivalent to the following one.

student(X) :- X = tom.
student(X) :- X = jane.
teacher(X) :- X = mary.

In most cases what the programmer has in mind is not the program P itself, but what is called the *completion* of P:

student(X) iff X = tom or X = jane.
teacher(X) iff X = mary.

The completion of P is obtained by replacing the if's in program P by iff's. And although "not student(mary)" is not a logical consequence of P, it is a logical consequence of the completion of P. Both the Negation as Failure rule and the process of completion capture the idea that information not given by the program is taken to be false.

Exercises 25.9. and 25.10. make clear that for programs which contain negation, the use of cut may affect the soundness of the system.

APPLICATIONS: DEDUCTIVE DATABASES AND ARTIFICIAL INTELLIGENCE

In Example 25.1. we have given a very simple application of logic programming to databases. In this example a database is given containing facts or data concerning who is a parent of whom. This database has been extended with rules stating under what conditions the grandparent relation applies. We have seen that one can add other rules such as rules for the predecessor relation, the mother relation, etc. (see also Exercise 25.1.). These rules enable the user to derive conclusions from the database which are not explicitly present in the database (as static facts), but which can be logically deduced from the facts in the database by means of application of the rules. For this reason one speaks of *deductive databases.* A logic program can be viewed as a (deductive) database, consisting of facts and rules. *Relational databases* correspond to logic programs consisting only of facts.

Prolog contains a number of facilities for updating databases. For instance, the goal "assert(C)" will always succeed and will result in adding the program clause C to the database. The goal "asserta(C)" adds C at the beginning of the database and the goal "assertz(C)" adds C at the end of the database. The goal "retract(C)" deletes a program clause that matches C.

The following non-trivial example of a deductive database is from I. Bratko, *Prolog; Programming for Artificial Intelligence*, § 4.1.

EXAMPLE 25.7.: The database or logic program contains facts of the following form:

```
family(
     person(tom, fox, date(7, may, 1950), works(bbc, 15200)),
     person(ann, fox, date(9, may, 1951), unemployed)        ,
    [person(pat, fox, date(5, may, 1973), unemployed),
     person(jim, fox, date(5, may, 1973), unemployed)]
                                                              ).
```

These atomic formulas are built from a ternary predicate symbol "family", a 4-ary function symbol "person", a ternary function symbol "date", a binary function symbol "works" and a number of individual constants. The overall structure of these facts is: family (Father, Mother, List_of_Children). Now, given a database of the type above, the question "give name and surname of all married woman who have at least two children" can be formulated in Prolog as follows:

```
?- family(_, person(Name, Surname, _, _), [_, _| _]).
```

In order to understand this formulation the reader should know that "_" is a so-called *anonymous variable*, i.e., a variable whose value is not given when Prolog answers the question. Among the answers to this question would be:

Name = ann,
Surname = fox.

In Exercise 25.11. the reader is invited to add a number of rules to the database such that many other questions can be asked in a straightforward manner.

Since any logic programming system is equipped with a reasoning mechanism (see also Section 26), one might say that any such system is able to simulate reasoning and hence disposes of *Artificial Intelligence* (*A.I.*). This makes logic programming a very appropriate tool for solving many problems, which are generally considered to belong to the field of Artificial Intelligence. Many puzzles can be solved by appropriate logic programs. A nice example is *cryptarithmetic puzzles*, such as

$$\begin{array}{r} \text{S E N D} \\ \underline{\text{M O R E}}\;+ \\ \text{M O N E Y} \end{array}$$

where the problem is to assign decimal digits to the letters of the alphabet such that the above sum is correct. I. Bratko's book *Prolog* contains a Prolog program for solving cryptarithmetic puzzles (Section 7.1.).

Below we give a simple logic program for colouring a given map, such that the colour in each region is different from the colours in all its adjacent regions.

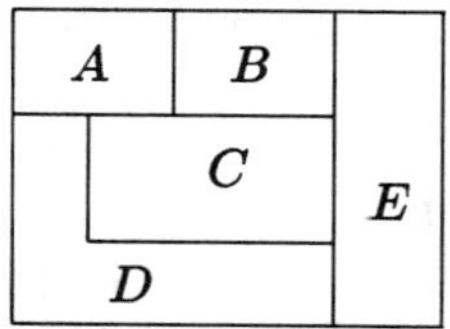


```
color(X) :- X = red.
color(X) :- X = blue.
color(X) :- X = green.
color(X) :- X = black.
next(X, Y) :- color(X), color(Y), not(X = Y).
colormap([A, B, C, D, E]) :- next(A, B), next(A, C), next(A, D),
      next(B, C), next(B, E), next(C, D), next(C, E), next(D, E).
```

Given this program, the appropriate question to ask is

```
?- colormap(Z).
```

Another example of the use of logic programming in the domain of Artificial Intelligence is for *parsing* sentences. The following program is for parsing sentences in a very simple and small fragment of English.

```
tv([loves]).      % "loves" is a transitive verb.
tv([hates]).
cn([man]).        % "man" is a common noun.
cn([woman]).
cn([dog]).
np([john]).       % "john" is a noun phrase.
np([mary]).
np([bill]).
```

```
det([a]).          % "a" is a determiner.
det([the]).
vp([walks]).       % "walks" is a verb phrase.
vp([talks]).
np(L) :- conc(L1, L2, L), det(L1), cn(L2).
vp(L) :- conc(L1, L2, L), tv(L1), np(L2).
s(L) :- conc(L1, L2, L), np(L1), vp(L2).
conc([ ], L, L).
conc([X | L1], L2, [X | L3]) :- conc(L1, L2, L3).
```

In this program "conc($L1, L2, L$)" should be read as "L is the concatenation of $L1$ and $L2$", and "$s(L)$" should be read as "L is a sentence". Given this program, questions one might ask are:

```
?- s([john, hates, the, dog]).
yes
?- s([john, hates, the, walks]).
no
```

The question "?- s(S)." will generate all syntactically correct sentences in the given fragment of English.

When a logic program behaves like an expert in some specific domain such as medical diagnosis or system break-down diagnosis, the logic program is called an *expert system* or a *knowledge-based-system.* By "behaving like an expert" we mean that 1) the logic program contains some expertise information concerning a specific domain, 2) that the program must be able to ask certain questions to the user and 3) that the program must be able to indicate in a user friendly manner how it has derived the answer(s) to a given question. Logic programs which satisfy these conditions become rather complex. Relatively simple examples of such expert systems can be found, among others, in I. Bratko, *Prolog*, Chapter 14.

PITFALLS

There are at least four pitfalls the logic programmer should be aware of.

1. We have already mentioned that most actual logic programming systems use matching instead of unification for reasons of efficiency. However, as indicated in the subsection on Matching versus Unification, it may happen as a consequence that some goal is answered affirmatively,

while the goal does not logically follow from the given program. In other words, the lack of the occur check destroys the soundness of the system.

2. The occurrence of a cut in a *definite* program does not affect the soundness of the system, although it may affect the completeness of the system by pruning successful branches. However, Exercise 25.10. makes clear that the use of cut in a *normal* program may even destroy the soundness of the system.

3. Consider the following program P (from J.W. Lloyd, *Foundations of Logic Programming*, § 10).

$$
\begin{array}{ll}
p(a,b). & (1)\\
p(c,b). & (2)\\
p(X,Z) \text{ :- } p(X,Y),\ p(Y,Z). & (3)\\
p(X,Y) \text{ :- } p(Y,X). & (4)
\end{array}
$$

Now it is easy to see that $p(a,c)$ is a logical consequence of P. From (2) and (4) it follows that $p(b,c)$ (5). And from (1), (5) and (3) it follows that $p(a,c)$.

However, given this program, the question "?- $p(a,c)$." will not be answered by any of the existing Prolog systems. In order to see why, let us consider the search tree for this question.

Any logic programming system that uses a depth-first search, combined with a fixed order for trying clauses given by their ordering in the program, will never find the success branch, because the left-most branch in the search tree is infinite. We have seen that all the clauses (1), (2), (3) and (4) were used in concluding $p(a,c)$ from P. However, in the left-most branch of the search tree for "?- $p(a,c)$.", clause (4) will never be applied. Interchanging clauses (3) and (4) in the program P would result in a left-most branch in which clause (3) is never applied, while all the clauses in P are necessary to deduce $p(a,c)$.

The solution to this problem would be a logic programming system with a breadth-first search rule. However, it is unlikely that such a system can be implemented efficiently.

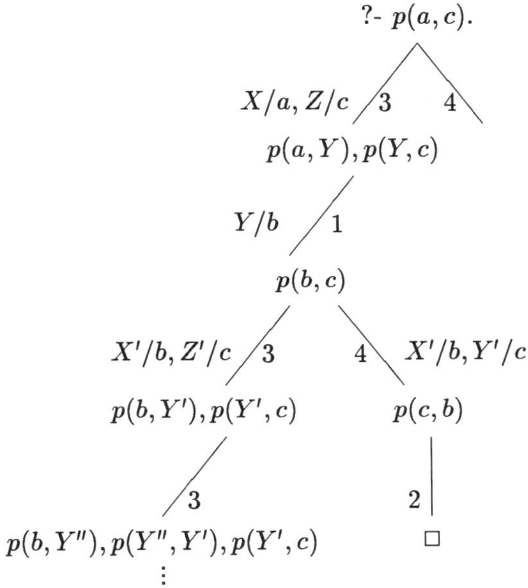

4. Many logic programming systems do not satisfy the *safeness condition*: negative literals are only allowed to be selected if they do not contain any variables. The safeness condition can be implemented by delaying the treatment of negative subgoals until any variable in the subgoal has been substituted by a term not containing variables.

Violation of the safeness condition affects the soundness of the system. Consider, for instance, the following program P.

bachelor(X) :- not married(X), man(X). (1)
man(bob). (2)
married(alice). (3)

What the programmer actually has in mind is not P itself, but the *completion of* P, consisting of the following formulas, interpreting "$\sim$ " as "if *and only if*".

bachelor(X) $\sim$ not married(X), man(X).
man(X) $\sim$ X = bob.
married(X) $\sim$ X = alice.

From the completion of P it logically follows that for some X, bachelor(X), namely X = bob. However, a logic programming system that does not satisfy the safeness condition will answer the question "?- bachelor(X)." with "no".

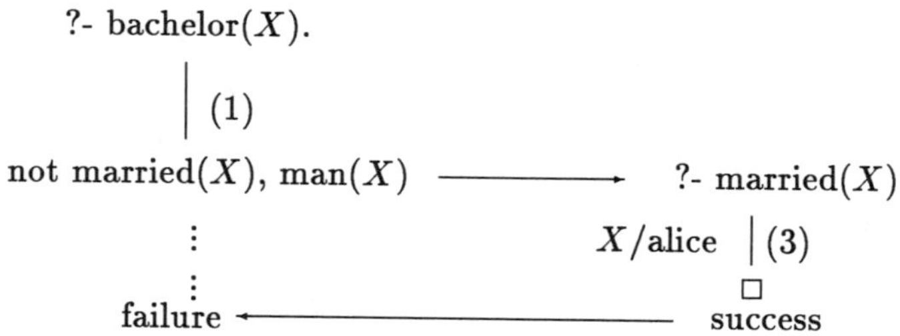

A logic programming system that delays the treatment of a negative subgoal, until all variables in the subgoal have been replaced by terms not containing variables, will answer the question "?- bachelor(X)." with "X = bob".

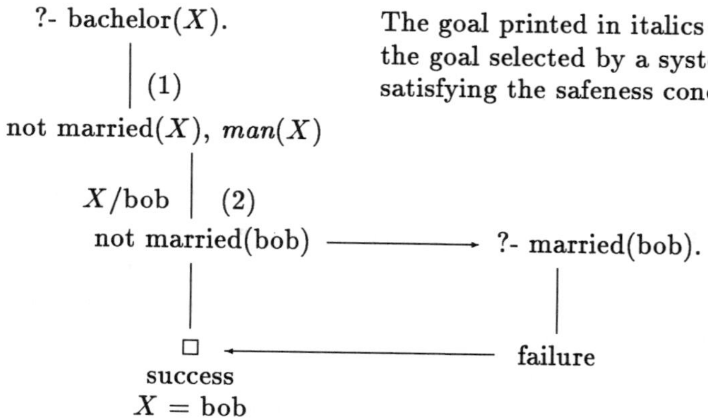

The goal printed in italics is the goal selected by a system satisfying the safeness condition.

READING LIST

Bratko, I., *Prolog; Programming for Artificial Intelligence.*

Lloyd, J.W., *Foundations of Logic Programming.*

Sterling, L. and Shapiro, E., *The Art of Prolog: Advanced Programming Techniques.*

EXERCISES

25.1. Extend the program concerning the parent relation in Example 25.1. with rules which define the offspring, the father, the mother, the sister and the brother relation. It will be necessary to introduce unary predicate symbols "male" and "female" and to add some facts about the sex of the persons whose names occur in the program. Given the extended program, construct the search trees for the following questions.

?- mother(tom, liz). ?- mother(X, bob).
?- sister(ann, pat). ?- father(bob, Y).

25.2. Let the predecessor relation be added to the program in Example 25.1. (concerning the parent relation) in the following ways.

pred2(X, Z) :- parent(X, Y), pred2(Y, Z).
pred2(X, Z) :- parent(X, Z).

pred3(X, Z) :- parent(X, Z).
pred3(X, Z) :- pred3(X, Y), parent(Y, Z).

Construct the search trees for the following questions.

?- pred2(tom, pat).
?- pred3(tom, pat).
?- pred3(liz, jim).

Conclude that from a procedural point of view, pred2 and pred3 do not describe the predecessor relation in an adequate way. (The examples are from I. Bratko, *Prolog*, § 2.6.2.)

25.3. Determine whether the following pairs can be matched or unified respectively.
i) $p(f(X), Z)$ and $p(Y, c)$.
ii) X and $f(X)$.
iii) $p(f(X), c)$ and $p(Y, f(Z))$.

25.4. Give a recursive definition of the concatenation relation, reading "conc($L1, L2, L3$)" as "$L3$ is the concatenation of the lists $L1$ and $L2$".

25.5. Give a recursive definition of the deletion relation, reading "del($X, L, L1$)" as "$L1$ results from the list L by deleting one occurrence of X".

25.6. Give a recursive definition for establishing the length of a list, reading "length(L, N)" as "N is the number of elements in the list L".

25.7. (Bratko) Let P be the following program.

```
p(1).
p(2) :- ! .
p(3).
```

Construct the search trees for the following goals.

i) ?- $p(X)$.
ii) ?- $p(X), p(Y)$.
iii) ?- $p(X)$, !, $p(Y)$.

25.8. Let P be the following program.

```
p(X, 0) :- X < 1.
p(X, 1) :- X >= 1, X < 2.
p(X, 2) :- X >= 2.
```

Using cuts, change P into a program which is declaratively equivalent, but procedurally more efficient.

25.9. (Lloyd) Consider the following program P for the subset relation, representing sets by lists.

```
subset(X,Y) :- not p(X,Y).                         (1)
p(X,Y) :- member(Z,X), not member(Z,Y).            (2)
member(X, [X | L]).                                (3)
member(X, [Y | L]) :- member(X,L).                 (4)
```

$p(X, Y)$ expresses that X is not a subset of Y.
Make clear how the Prolog system answers the question

?- subset([1, 2], [1, 2, 3]).

25.10. If we replace clause (3) in the program of the preceding exercise by clause (3′)

```
member(X, [X | L]) :- ! .                          (3') .
```

then the membership program will generate just one solution and not all possible solutions.
Verify that if we do so, then the question

?- subset([1, 2, 3], [1]).

will be answered affirmatively, while "not subset([1, 2, 3], [1])" is a logical consequence of P. So, the use of cut in combination with negation may affect the soundness of the system!

25.11. Extend the database in Example 25.7. with appropriate rules, such that the following questions can be formulated in Prolog in an adequate way. (Confer I. Bratko, *Prolog*, § 4.1.)

i) Give the names and surnames of all people in the database.
ii) Give all children born in 1973.
iii) Give the names and surnames of all employed wives.
iv) Give the names and surnames of all unemployed people who were born before 1960.
v) Give all people born before 1960 whose salary is more than 10000.
vi) Give the surnames of all families with at least two children.
vii) Give the surnames of all families without children.

25.12. Instead of $f(t1, t2)$, Prolog also allows the *infix notation* $t1ft2$. For that purpose it is necessary to define f as an operator with a given precedence. The *precedence* of an arbitrary term is then defined as follows.

1. The precedence of individual variables and individual constants is 0.
2. The precedence of $f(t_1, \ldots, t_n)$ is the precedence of f.
3. The precedence of $(t), t$ a term, is 0.

In order to ensure that $a+b*c$ is interpreted as $a+(b*c)$ and not as $(a+b)*c$, the operators $+$ and $*$ may be defined as follows.

op(500, yfx, +). (1)
op(400, yfx, $*$). (2)

In (1) the operator + is defined as an infix operator (i.e., occurring between its arguments) with precedence 500. "y" represents an argument whose precedence must be lower than or equal to that of the operator, and "x" represents an argument whose precedence must be strictly lower than that of the operator.

i) Check that under the definitions (1) and (2) $a + b * c$ is understood as $a + (b * c)$ and not as $(a + b) * c$.
ii) Defining "$-$" by "op(500, yfx, $-$).", check that $a - b - c$ is read as $(a - b) - c$ and not as $a - (b - c)$.
iii) Defining "has" by "op(600, xfx, has).", check that instead of "has(peter, information)." the programmer can write "peter has information.".

§26 THEORETICAL ASPECTS OF LOGIC PROGRAMMING:

HERBRAND MODELS; SLD-RESOLUTION; RESOLUTION

Having given a practical introduction to logic programming in Section 25, we now spend some time on the theoretical foundations. We prove that the least Herbrand model of a definite logic program P is precisely the set of ground atoms which are a logical consequence of P. Next we study SLD-resolution, the reasoning mechanism of Prolog, and prove its soundness. Finally, we make clear that SLD-resolution is a special form of resolution in general.

HERBRAND MODELS

DEFINITION 26.1.: A *clause* is a formula of the form
$\forall x_1 \ldots \forall x_n[B_1 \vee \ldots \vee B_k \vee \neg A_1 \vee \ldots \vee \neg A_m]$, or, equivalently,
$\forall x_1 \ldots \forall x_n[B_1 \vee \ldots \vee B_k \subset A_1 \;\&\; \ldots \;\&\; A_m]$, where all the B's and A's are *atomic* formulas and $x_1, \ldots, x_n$ are all the variables occurring in these atomic formulas.

Notation: The universal quantifiers are often suppressed for reasons of efficiency. Instead of $B_1 \vee \ldots \vee B_k \vee \neg A_1 \vee \ldots \vee \neg A_m$ one frequently writes $\{B_1, \ldots, B_k, \neg A_1, \ldots, \neg A_m\}$.

Note that a *definite program clause* (Definition 25.2.) is a clause with $k = 1$, that a *definite goal* is a clause with $k = 0$, that the *empty clause* is a clause with $k = 0$ and $m = 0$ and that a *definite clause* or *Horn clause* is a clause with $k \leq 1$. (Confer Definition 25.2.)
If $k = 0$, $B_1 \vee \ldots \vee B_k$ is equivalent to $\perp$ (contradiction) and
if $m = 0$, $A_1 \;\&\; \ldots \;\&\; A_m$ is equivalent to $\top$ (an always true formula).
So, a *definite goal* ($k = 0$) is a formula of the form

$\forall x_1 \ldots \forall x_n[\perp \subset A_1 \;\&\; \ldots \;\&\; A_m]$ or, equivalently,
$\forall x_1 \ldots \forall x_n[\neg(A_1 \;\&\; \ldots \;\&\; A_m)]$ or, equivalently,
$\forall x_1 \ldots \forall x_n[\neg A_1 \vee \ldots \vee \neg A_m]$ or, equivalently,
$\neg\exists x_1 \ldots \exists x_n[A_1 \;\&\; \ldots \;\&\; A_m]$.

And the *empty clause* ($k = 0$, $m = 0$), $\square$, is equivalent to

$$\forall x_1 \ldots \forall x_n[\perp \subset \top], \text{ i.e., to } \perp \text{ (contradiction)}.$$

The reasoning mechanism of Prolog is based on *resolution* which is a refutation mechanism. That is, given a logic program P and a putative conclusion $\exists x_1 \ldots \exists x_n[A_1 \ \& \ \ldots \ \& \ A_m]$, the negation :- $A_1, \ldots, A_m$ (which is equivalent to $\neg\exists x_1 \ldots \exists x_n[A_1 \ \& \ \ldots \ \& \ A_m]$) is added to the program and the mechanism tries to derive a contradiction (the empty clause). For instance, given the logic program P of Example 25.1., if we want to know whether $\exists x$[parent(bob, x)], we add to P the clause :- parent(bob, x) which is short for $\forall x[\neg$parent(bob, x)] or, equivalently, $\neg\exists x$[parent(bob, x)], after which the reasoning mechanism will try to deduce a contradiction, or, more precisely, the empty clause. If this succeeds, it follows that $P \cup \{\neg\exists x$[parent(bob, x)]$\}$ is unsatisfiable (i.e., has no model), or, equivalently, that $P \models \exists x$[parent(bob, x)].

We will show below that if $P \cup \{\neg E\}$ is a set of clauses, $P \cup \{\neg E\}$ is satisfiable (i.e., has at least one model) if and only if $P \cup \{\neg E\}$ has a particular kind of model, called *Herbrand model.*

DEFINITION 26.2.:

i) A *ground term* is a term (see § 25) that does not contain any variables.

ii) A *ground atom* is an atomic formula (see § 25) that does not contain any variables.

Let Γ be a set of clauses.

iii) The *Herbrand universe* for Γ, notation U_Γ, is the set of all ground terms that can be constructed from the individual constants and function symbols which occur in Γ. In case no individual constants occur in Γ, we add one, say "c".

iv) The *Herbrand base* for Γ, notation B_Γ, is defined as follows: if q is an n-ary predicate symbol occurring in Γ and $t_1, \ldots, t_n \in U_\Gamma$, then $q(t_1, \ldots, t_n) \in B_\Gamma$.

EXAMPLE 26.3.: Let P be the following logic program.

$$p(X) \text{ :- } q(X, f(X)).$$
$$q(g(c), c).$$

$U_P = \{c, f(c), g(c), f(f(c)), f(g(c)), g(f(c)), g(g(c)), \ldots\}$.

$B_P = \{p(c), q(c,c), p(f(c)), p(g(c)), q(c, f(c)), q(f(c), c), \ldots\}$.

DEFINITION 26.4.: Let Γ be a set of clauses containing the predicate symbols $p_1, \ldots, p_m$ and the function symbols $f_1, \ldots, f_n$.

i) $M = \langle D_M; p_1^M, \ldots, p_m^M; f_1^M, \ldots, f_n^M \rangle$ is a *Herbrand interpretation* for Γ := M is an interpretation of (the predicate and function symbols in) Γ (see Definition 24.1.) such that
   (a) $D_M = U_\Gamma$, and
   (b) for each k-ary function symbol f_i occurring in Γ, $f_i^M : U_\Gamma^k - U_\Gamma$ is defined by $f_i^M(t_1, \ldots, t_k) = f_i(t_1, \ldots t_k)$.

   In particular, for each individual constant c occurring in Γ, $c^M = c$. Note that we do not require any special conditions for the p_i^M. According to Definition 24.1., if p_i is k-ary, then $p_i^M \subseteq U_\Gamma^k$ or, equivalently, $p_i^M : U_\Gamma^k \to \{0,1\}$. Note that one may identify a Herbrand interpretation for Γ with a subset of B_Γ.

ii) M is a *Herbrand model* of Γ := M is a Herbrand interpretation for Γ such that M is a model of (all clauses in) Γ.

EXAMPLE 26.3. (continued): Given program P above, let $p^M(t) := t = c$ and $q^M(t_1, t_2) := t_1 = g(c)$ and $t_2 = c$. Then $M = \langle U_P; p^M, q^M; f^M, g^M \rangle$ with $f^M(t) := f(t)$ and $g^M(t) := g(t)$, is a Herbrand model of P. M can be identified with the subset $\{p(c), q(g(c), c)\}$ of B_P.

THEOREM 26.5.: Let Γ be a set of clauses.

Γ has a model if and only if Γ has a Herbrand model.

PROOF: If Γ has a Herbrand model, then Γ has a model. Conversely, let Γ be a set of clauses and let M be a model of Γ. Define the Herbrand interpretation M' as follows. For each k-ary predicate symbol p occurring in Γ and for all $t_1, \ldots, t_k \in U_\Gamma$, $p^{M'}(t_1, \ldots, t_k) := M \models p(t_1, \ldots, t_k)$. Then for all $t_1, \ldots, t_k \in U_\Gamma$,

$$M' \models p(X_1, \ldots, X_k)[t_1, \ldots, t_k] \text{ iff } M \models p(t_1, \ldots, t_k). \qquad (*)$$

Remember that Γ is a set of clauses. So, Γ is a set of formulas of the form $\forall x_1 \ldots \forall x_k[L_1 \vee \ldots \vee L_m]$, where each L_i is a literal (i.e., an atomic formula or the negation of an atomic formula). Now let $\forall x_1 \ldots \forall x_k[L_1 \vee \ldots \vee L_m]$ be a clause in Γ. In order to show that $M' \models \forall x_1 \ldots \forall x_k[L_1 \vee \ldots \vee L_m]$ we have to prove that for all ground terms $t_1, \ldots, t_k$ in U_Γ, $M' \models L_1 \vee \ldots \vee L_m[t_1, \ldots, t_k]$. Because of $(*)$ this is equivalent to $M \models L_1(t_1, \ldots, t_k) \vee \ldots \vee L_m(t_1, \ldots, t_k)$, which holds since M is a model of Γ.□

Note that $(*)$ in the proof of Theorem 26.5. cannot be generalized to formulas of the form $\exists x[q(x)]$. Theorem 26.5. does not hold if Γ is a set of arbitrary formulas. Counterexample: let $\Gamma = \{p(c),\ \exists x[\neg p(x)]\}$. Then $M = \langle\{0,1\};$ is even; $0\rangle$ is a model of Γ. But Γ does not have a Herbrand model, since the Herbrand universe U_Γ has only one element; $U_\Gamma = \{c\}$.

Theorem 26.5. says in particular that given any logic program P and goal $\neg E$ (or, equivalently, :- E), $P \cup \{\neg E\}$ is unsatisfiable if and only if $P \cup \{\neg E\}$ does not have a Herbrand model.

THEOREM 26.6.: Let P be a *definite* logic program and $\{M_i\}_{i\in I}$ a non-empty set of Herbrand models of P. Then $\bigcap_{i\in I} M_i$ is again a Herbrand model of P.

PROOF: That the intersection of a non-empty set of Herbrand interpretations is again a Herbrand interpretation is evident. Suppose P is a *definite* logic program and for each $i \in I$, M_i is a Herbrand model of P. We have to show that $\bigcap_{i\in I} M_i$ is a model for all definite program clauses B :- $A_1, \ldots, A_m$ in P.
Case 1: for all $i \in I$, $M_i \models B$. Then $\bigcap_{i\in I} M_i \models B$ and hence

$$\bigcap_{i\in I} M_i \models B \text{ :- } A_1, \ldots, A_m.$$

Case 2: for some $i \in I$, $M_i \not\models B$. Then, since $M_i \models B$:- $A_1, \ldots, A_m$, $M_i \not\models A_j$ for some j, $1 \leq j \leq m$. So, $\bigcap_{i\in I} M_i \not\models A_j$ and consequently, $\bigcap_{i\in I} M_i \models B$:- $A_1, \ldots, A_m$. □

In the proof of Theorem 26.6. we have used the hypothesis that the program clauses in P are definite. This is essential. Theorem 26.6. does not hold if P contains clauses $B_1 \vee \ldots \vee B_k \subset A_1\ \&\ \ldots\ \&\ A_m$ with $k \geq 2$. For instance, let $p(c) \vee q(c)$ be the only clause in P ($k = 2, m = 0$). The Herbrand interpretations $M_1 = \{p(c)\}$ and $M_2 = \{q(c)\}$ are models of P. But the Herbrand interpretation $M_1 \cap M_2 = \emptyset$ is not a model of P.

Let P be a definite logic program. Since the Herbrand base B_P for P is a model of (the clauses in) P, it follows that the set of all Herbrand models of P is non-empty.

DEFINITION 26.7.: Let P be a definite logic program.

$$M_P := \cap\{M \mid M \text{ is a Herbrand model of } P\}.$$

M_P is the *least Herbrand model* of P (see Theorem 26.6.) and can be conceived of as the canonical model of P.

In the following theorem we prove that the least Herbrand model of a definite logic program P is precisely the set of ground atoms which are a logical consequence of P.

THEOREM 26.8.: Let P be a definite logic program. Then for any $A \in B_P$, $A \in M_P$ iff $P \models A$.

PROOF: Let P be a definite logic program and $A \in B_P$.
$A \in M_P$ iff for all Herbrand models M of P, $M \models A$
iff there is no Herbrand model of $P \cup \{\neg A\}$
(26.5.) iff there is no model of $P \cup \{\neg A\}$
iff $P \models A$. □

EXAMPLE: Let P be the following definite logic program.

$p(c)$.
$q(b)$.
$q(f(X))$:- $q(X), p(c)$.

$M_P = \{p(c), q(b), q(f(b)), q(f(f(b))), \ldots\}$ is the least Herbrand model of P. See Exercise 26.1. for another example.

For a fixpoint characterization of M_P the reader is referred to J.W. Lloyd, *Foundations of Logic Programming*, § 6.

SLD-RESOLUTION

For a precise definition of *SLD*-resolution we need substitutions and composition of substitutions.

DEFINITION 26.9.:

i) A *substitution* θ is a finite set of the form $\{X_1/t_1, \ldots, X_n/t_n\}$, where each X_i is a variable and each t_i is a term different from X_i.
ii) Let $\theta = \{X_1/t_1, \ldots, X_n/t_n\}$ be a substitution, t a term and L a literal. Then $t\theta$ and $L\theta$ are obtained from t and L respectively by simultaneously replacing each occurrence of the variable X_i in t and

L by the term t_i ($i = 1, \ldots, n$). $t\theta$ and $L\theta$ are called the *instance* of t and L respectively by θ. If $S = \{E_1, \ldots, E_n\}$ is a finite set of terms and/or literals and θ is a substitution, then $S\theta := \{E_1\theta, \ldots, E_n\theta\}$.

iii) Let $\theta = \{X_1/s_1, \ldots, X_m/s_m\}$ and $\sigma = \{Y_1/t_1, \ldots, Y_n/t_n\}$ be substitutions. The *composition* $\theta\sigma$ of θ and σ is the substitution obtained from the set $\{X_1/s_1\sigma, \ldots, X_m/s_m\sigma,\ Y_1/t_1, \ldots, Y_n/t_n\}$ by deleting any binding $X_i/s_i\sigma$ for which $X_i = s_i\sigma$ and deleting any binding Y_j/t_j for which $Y_j \in \{X_1, \ldots, X_m\}$.

EXAMPLES: Let $\theta = \{X/f(Y), Y/Z\}$ and $\sigma = \{X/a, Z/b\}$.
Let $L_1 = p(X, Y, g(Z))$ and $L_2 = \neg q(X, Y, f(a))$.
Then $L_1\theta = p(f(Y), Z, g(Z))$ and $L_2\sigma = \neg q(a, Y, f(a))$.
$\theta\sigma = \{X/f(Y), Y/b, Z/b\}$ and $\sigma\theta = \{X/a, Z/b, Y/Z\}$. Note that $(L_1\theta)\sigma = L_1(\theta\sigma) = p(f(Y), b, g(b))$. See Exercise 26.2.

DEFINITION 26.10.: Let S be a finite set of literals or terms and θ a substitution.

i) θ is a *unifier* for S := $S\theta$ is a singleton, i.e., $S\theta$ has only one element.
ii) S is *unifiable* := there is a substitution θ such that θ is a unifier for S.
iii) θ is a *most general unifier* (*mgu*) for S :=
a) θ is a unifier for S, and
b) for each unifier σ for S there is a substitution γ such that $\sigma = \theta\gamma$.

EXAMPLES: Let $S = \{p(f(Y), X, g(Z)),\ p(X1, X1, X2)\}$ and $\theta = \{X1/f(Y), X/f(Y), X2/g(Z)\}$. Then θ is a most general unifier for S. $\sigma = \{Y/c, X1/f(c), X/f(c), X2/g(b), Z/b\}$ is a unifier for S. Note that $\sigma = \theta\gamma$ for $\gamma = \{Y/c, Z/b\}$.

In the subsection on Matching versus Unification in § 25 we have described informally a *unification algorithm* which takes as input a finite set S of terms or literals; this algorithm always terminates and after termination yields either a most general unifier for S, if one exists, or reports that S is not unifiable. For a detailed description of this algorithm and for a proof that the algorithm has the properties just mentioned, see J.W. Lloyd, *Foundations of Logic Programming*, § 4.

For the moment we will not discuss resolution in its most general form, but restrict ourselves to *SLD-resolution* which is the reasoning mechanism used in pure Prolog, i.e., Prolog with definite clauses only. The "S" stands for the Selection rule which is used in Prolog; in most Prolog systems the left-most goal in :- $A_1, \ldots, A_m$ is selected. The "L" stands for "Linear". A picture of a sequence of SLD-resolution steps shows a linear chain. And the "D" indicates that we restrict ourselves to Definite clauses only.

In what follows we suppose there is some fixed selection rule. The goal selected by this rule will be underlined.

DEFINITION 26.11.: Let P be a definite logic program and let C be the program clause B :- $A_1, \ldots, A_m$ in P. Let G be the definite goal clause :- $H_1, \ldots, \underline{H_i}, \ldots, H_n$, the underlined goal being selected by the given selection rule.

Let σ be a substitution such that $C\sigma$ (i.e., $B\sigma$:- $A_1\sigma, \ldots, A_m\sigma$) and G have no variables in common. And let θ be a most general unifier for $\{H_i, B\sigma\}$.

Let G_1 be the goal :- $(H_1, \ldots, H_{i-1}, A_1\sigma, \ldots, A_m\sigma, H_{i+1}, \ldots, H_n)\theta$. Then we say that G_1 is *derived from* G using C and the most general unifier θ via (a one step) *SLD-resolution.* So, an SLD-resolution step looks as follows.

$$\begin{array}{ll} G & :\text{-}\ H_1, \ldots, \underline{H_i}, \ldots, H_n. \\ \Big\downarrow & C\sigma \text{ (i.e., } B\sigma :\text{-}\ A_1\sigma, \ldots, A_m\sigma),\ \theta \text{ (a mgu of } H_i \text{ and } B\sigma). \\ G_1 & :\text{-}\ (H_1, \ldots, H_{i-1}, A_1\sigma, \ldots, A_m\sigma, H_{i+1}, \ldots, H_n)\theta. \end{array}$$

Notation: $G \xrightarrow[C,\theta]{} G_1.$

The clause C is called the *input clause* of the resolution step

$$G \xrightarrow[C,\theta]{} G_1.$$

In order to prove that SLD-resolution is sound (Theorem 26.16), we need the following lemma.

LEMMA 26.12.: Under the conditions and notation as in Definition 26.11., if (1)

$P \models \forall[(H_1 \,\&\, \ldots \,\&\, H_{i-1} \,\&\, A_1\sigma \,\&\, \ldots \,\&\, A_m\sigma \,\&\, H_{i+1} \,\&\, \ldots \,\&\, H_n)\theta]$,
then $P \models \forall[(H_1 \,\&\, \ldots \,\&\, H_{i-1} \,\&\, H_i \,\&\, H_{i+1} \,\&\, \ldots \,\&\, H_n)\theta]$.

PROOF: Suppose (1). Note that $P \models \forall[A_1\sigma \,\&\, \ldots \,\&\, A_m\sigma \supset B\sigma]$ because B :- $A_1, \ldots, A_m$ is a clause in P.
Therefore, $P \models \forall[(H_1 \,\&\, \ldots \,\&\, H_{i-1} \,\&\, B\sigma \,\&\, H_{i+1} \,\&\, \ldots \,\&\, H_n)\theta]$.
But θ is a most general unifier of H_i and $B\sigma$, i.e., $H_i\theta = (B\sigma)\theta$. Hence, $P \models \forall[(H_1 \,\&\, \ldots \,\&\, H_{i-1} \,\&\, H_i \,\&\, H_{i+1} \,\&\, \ldots \,\&\, H_n)\theta]$. □

DEFINITION 26.13.: Let P be a definite logic program and G a definite goal. i) An *SLD-refutation* of $P \cup \{G\}$ is a finite sequence

$$G \xrightarrow[C_1,\theta_1]{} G_1 \xrightarrow[C_2,\theta_2]{} \cdots \xrightarrow[C_k,\theta_k]{} G_k \xrightarrow[C_{k+1},\theta_{k+1}]{} \Box$$

of SLD-resolution steps ending with the empty goal, □.

A finite sequence of SLD-resolution steps starting with G and ending in a non-empty goal, the selected atom of which does not unify with the head of any program clause in P, is called a *failed SLD-derivation* for $P \cup \{G\}$.
ii) Given an SLD-refutation of $P \cup \{G\}$, the composition $\theta_1\theta_2\ldots\theta_{k+1}$, restricted to the variables in G, is called a *computed answer substitution for* $P \cup \{G\}$, where $\theta_1, \ldots, \theta_{k+1}$ are the most general unifiers used in the given SLD-refutation of $P \cup \{G\}$.

An SLD-refutation of $P \cup \{G\}$ corresponds to a successful branch in the search tree for G, given program P.

EXAMPLE 26.14. (J.W. Klop): Let P be the following definite logic program.

$q(Y)$:- $r(X)$.
$p(1)$.
$r(2)$.

And let G be the definite goal :- $q(Y), p(X)$. Then the following sequence of SLD-resolution steps is an SLD-refutation of $P \cup \{G\}$.

G $\quad$:- $q(Y), p(X)$.

$\downarrow$ $\quad (q(Y)$:- $r(X))\{Y/Y1, X/X1\},\ \theta_1 = \{Y/Y1\}$

G_1 $\quad$:- $r(X1), p(X)$.

$\downarrow$ $\quad r(2)\ ,\ \theta_2 = \{X1/2\}$

G_2 $\quad$:- $p(X)$.

$\downarrow$ $\quad p(1)\ ,\ \theta_3 = \{X/1\}$

$\square$ $\quad$:- $\top$

The composition $\theta_1\theta_2\theta_3 = \{Y/Y1, X1/2, X/1\}$, restricted to the variables in G, i.e., $\{Y/Y1, X/1\}$, is a computed answer substitution for $P \cup \{G\}$. This example also demonstrates that in an application of an SLD-resolution step it is necessary to rename the variables of the input clause C via a substitution σ, such that the resulting clause $C\sigma$ and the goal in question have no variables in common. Leaving out this renaming of variables would result in the following failed SLD-derivation for $P \cup \{G\}$, while $q(1)$ & $p(1)$ is a logical or valid consequence of P.

G $\quad$:- $q(Y), p(X)$.

$\downarrow$ $\quad q(Y)$:- $r(X)\ ,\ \theta_1 = \varepsilon$

G_1 $\quad$:- $r(X), p(X)$.

$\downarrow$ $\quad r(2)\ ,\ \theta_2 = \{X/2\}$

G_2 $\quad$:- $p(2)$.

DEFINITION 26.15.: Let P be a definite logic program and G a definite goal :- $A_1, \ldots, A_m$. Let θ be a substitution for the variables in G. θ is a *correct answer substitution* for $P \cup \{G\} := \forall[(A_1\ \&\ \ldots\ \&\ A_m)\theta]$ is a logical consequence of P.

Next we prove that SLD-resolution is sound: all answers computed via SLD-resolution logically follow from the given program.

THEOREM 26.16. (*Soundness* of *SLD*-resolution): Let P be a definite logic program and G a definite goal. Every computed answer substitution for $P \cup \{G\}$ is a correct answer substitution for $P \cup \{G\}$.

PROOF: Let P be a definite logic program and let G be a definite goal. Suppose $\theta = \theta_1 \ldots \theta_k\theta_{k+1}$ is a computed answer substitution for $P \cup \{G\}$. Let

$$G \xrightarrow[C_1,\theta_1]{} G_1 \xrightarrow[C_2,\theta_2]{} \cdots \xrightarrow[C_k,\theta_k]{} G_k \xrightarrow[C_{k+1},\theta_{k+1}]{} \Box$$

be the corresponding *SLD*-refutation of $P \cup \{G\}$. We prove by induction that for any goal G_i, say :- $A_1^i, \ldots, A_{m_i}^i$, in the given *SLD*-refutation of $P \cup \{G\}$, $\forall x[(A_1^i \,\&\, \ldots \,\&\, A_{m_i}^i)\theta]$ is a logical consequence of P.
Induction basis: $G_{k+1} = \Box$, i.e., :- $\top$. Clearly, $P \models \top$.
Induction step: This is Lemma 26.12.
It follows that θ is a correct answer substitution for $P \cup \{G\}$. □

A natural question to ask is whether conversely every valid consequence of a given logic program can be computed. For definite logic programs the answer is positive in the sense that every correct answer substitution for $P \cup \{G\}$ can be obtained from a computed answer substitution θ for $P \cup \{G\}$ by specialization, i.e., by composition of θ with some substitution γ. The answer substitutions computed by *SLD*-resolution are the most general ones: all other correct answer substitutions are specializations of them. For instance, consider Example 26.14. P was the program

$q(Y)$:- $r(X)$.
$p(1)$.
$r(2)$.

And G was the goal :- $q(Y), p(X)$.
We have seen that $\theta = \{Y/Y1, X/1\}$ is a *computed* answer substitution for $P \cup \{G\}$. $\sigma_1 = \{Y/1, X/1\}$ and $\sigma_2 = \{Y/2, X/1\}$ are *correct* answer substitutions for $P \cup \{G\}$. And $\sigma_1 = \theta\{Y1/1\}$ and $\sigma_2 = \theta\{Y1/2\}$, if we restrict the composition to the variables X and Y.

Below we formulate, but do not prove, the completeness of *SLD*-resolution for definite programs and definite goals. For a proof of this theorem the reader is referred to J.W. Lloyd, *Foundations of Logic Programming*, § 7. Herbrand models play an important role in this proof.

THEOREM 26.17. (*Completeness* of SLD-resolution): Let P be a definite logic program and G a definite goal. For every correct answer substitution σ for $P \cup \{G\}$ there exists a computed answer substitution θ for $P \cup \{G\}$ such that $\sigma = \theta\gamma$ for some substitution γ. In addition, this result is independent of the selection rule used.

It is important to note that this result does not hold for normal programs and/or normal goals. See Exercise 26.4.! For these programs different, but similar, results hold. See J.W. Lloyd, *Foundations of Logic Programming*, Chapter 3. The reader should also realize that although SLD-resolution is sound and complete in the case of definite programs and definite goals, the use of cut as a procedural device destroys completeness because it may cut off successful branches. And leaving out the occur check in the unification algorithm, as is done in most Prolog systems, even destroys the soundness of SLD-resolution (see the subsection on Matching versus Unification in § 25).

RESOLUTION

SLD-resolution is a special case of resolution: resolution is applied to a definite clause and a definite goal, which are of the form B :- $A_1, ..\ A_m$ and :- $H_1, \ldots, H_n$ respectively or, equivalently, $B \vee \neg A_1 \vee \ldots \vee \neg A_m$ and $\neg H_1 \vee \ldots \vee \neg H_n$, with at most one positive literal. Arbitrary clauses may have more than one positive literal: $B_1 \vee \ldots \vee B_k \vee \neg A_1 \vee \ldots \vee \neg A_m$, frequently written as $\{B_1, \ldots, B_k, \neg A_1, \ldots, \neg A_m\}$. Remember that clauses are universally quantified. So, $\{B_1, \ldots, B_k, \neg A_1, \ldots, \neg A_m\}$ is short for $\forall x_1 \ldots \forall x_n[B_1 \vee \ldots \vee B_k \vee \neg A_1 \vee \ldots \vee \neg A_m]$ or, equivalently, $\forall x_1 \ldots \forall x_n[B_1 \vee \ldots \vee B_k \subset A_1\ \&\ \ldots\ \&\ A_m]$, where $x_1, \ldots, x_n$ are all the variables occurring in the atomic formulas $B_1, \ldots, B_k, A_1, \ldots, A_m$.

DEFINITION 26.18.: Let A be a formula.
A is in *clausal form* := A is a conjunction of clauses with different variables.

EXAMPLE: $\neg\exists y\forall x[p(x) \supset q(y)]$ is not in clausal form, neither is $\forall y[p(g(y))\ \&\ \neg q(y)]$. However, $\forall y[p(g(y))]\ \&\ \forall z[\neg q(z)]$ is in clausal form. A frequently used notation for this form is $\{p(g(Y))\}, \{\neg q(Z)\}$.

In Section 24 we have given an algorithm to construct for any formula A a clausal form $C(A)$ of A such that
i) A is a valid consequence of $C(A)$, and
ii) A is satisfiable if and only if $C(A)$ is satisfiable.

A *resolution* step in its more general form looks as follows. Let C and C' be two clauses that have no variables in common (if not, one can rename the variables without objection because of the universal quantifiers). Let L and L' be literals in C and C' respectively and let θ be a substitution such that $L\theta$ and $L'\theta$ are complementary, i.e., a pair consisting of a positive literal and its negation. Then from C and C' one can obtain via (one) *resolution* (step) the *resolvent*, $res_{L,\theta}(C, C')$, of C and C', defined by taking the union of the clauses $C\theta$ and $C'\theta$ (conceived as sets of literals) leaving out $L\theta$ and $L'\theta$. In other words,

$$res_{L,\theta}(C, C') := (C - \{L\})\theta \cup (C' - \{L'\})\theta.$$

If it is clear from the context what L and θ are, then one may write $res(C, C')$.

EXAMPLES:

i) Modus Ponens is a special case of resolution. Let C be $\forall x[p(x)]$ or $\{p(X)\}$, and let $C' := \forall y[p(y) \supset q(y)]$ or $\forall y[\neg p(y) \vee q(y)]$ or $\{\neg p(Y), q(Y)\}$. Then $res(C, C') = \{q(Y)\}$ or $\forall y[q(y)]$.

ii) Let C be $\{\neg r(c), p(X), q(Y)\}$, i.e., $\forall x \forall y[r(c) \supset p(x) \vee q(y)]$.
Let C' be $\{r(Z), \neg p(b), \neg q(c)\}$, i.e., $\forall z[p(b) \,\&\, q(c) \supset r(z)]$.
Then there are three different resolvents of C and C'.
Resolving $\neg r(c)$ and $r(Z)$ with $\theta_1 = \{Z/c\}$:
$$\{p(X), q(Y), \neg p(b), \neg q(c)\}.$$
Resolving $p(X)$ and $\neg p(b)$ with $\theta_2 = \{X/b\}$:
$$\{\neg r(c), q(Y), r(Z), \neg q(c)\}.$$
Resolving $q(Y)$ and $\neg q(c)$ with $\theta_3 = \{Y/c\}$:
$$\{\neg r(c), p(X), r(Z), \neg p(b)\}.$$

iii) Let G be a definite goal clause $\{\neg H_1, \ldots, \neg H_n\}$, i.e., :- $H_1, \ldots, H_n$. Let C_1 be a definite program clause $\{B, \neg A_1, \ldots, \neg A_m\}$, i.e., B :- $A_1, \ldots, A_m$. Then there can be only one resolvent of G and C_1, since there is only one positive literal. In addition, the resolvent G_1 will be a new goal clause. So, a second resolution step will require another (definite) program clause C_2. (Confer Definition 26.11. and Example 26.14.) If only definite program clauses are available, as is the case in pure Prolog, a *linear chain* of resolution steps results. This explains the "L" ("linear") in "*SLD-resolution*".

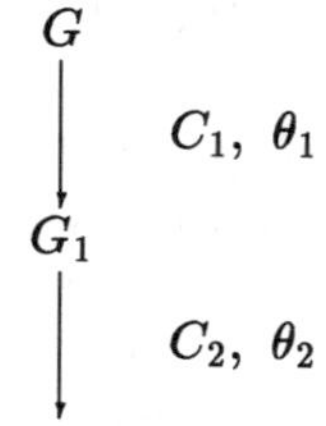

The *soundness* of resolution is easily seen. If $res(C, C')$ is a resolvent of the clauses C and C', then $C,\ C' \models res(C, C')$. Consequently, if there is a resolution deduction of the empty clause, $\square$, from clauses $C_1, \ldots, C_n$, then there can be no model of $C_1, \ldots, C_n$.

If one wants a resolution deduction of B from $A_1, \ldots, A_m$, one first has to construct the clausal forms $C(A_1), \ldots, C(A_m)$ and $C(\neg B)$ of $A_1, \ldots, A_m$ and $\neg B$. Section 24 gives an algorithm for doing so. A resolution deduction is then sought of the empty clause, $\square$, from $C(A_1), \ldots, C(A_m)$ and $C(\neg B)$. If that search is successful, it can be concluded that there is no model of $C(A_1), \ldots, C(A_m), C(\neg B)$ and hence (see § 24) there can be no model of $A_1, \ldots, A_m, \neg B$, in other words, $A_1, \ldots, A_m \models B$.

EXAMPLE: We want to find a resolution deduction of $P \supset R$ from $P \supset Q$ and $Q \supset R$. We first have to determine the clausal forms of the premisses and of the negation of the conclusion.

$C(P \supset Q) = \neg P \vee Q$, i.e., $\{\neg P, Q\}$.
$C(Q \supset R) = \neg Q \vee R$, i.e., $\{\neg Q, R\}$.
$C(\neg(P \supset R)) = P\ \&\ \neg R$, i.e., $\{P\}$ and $\{\neg R\}$.

This example (see Figure 1) shows that a sequence of resolution steps is not always a linear chain. In this particular case there is also a linear resolution deduction of the empty clause from the given clauses (see figure 2). However, Exercise 26.3. i) gives an example where no linear resolution deduction of the empty clause from given clauses exists, although a non-linear resolution deduction does exist.

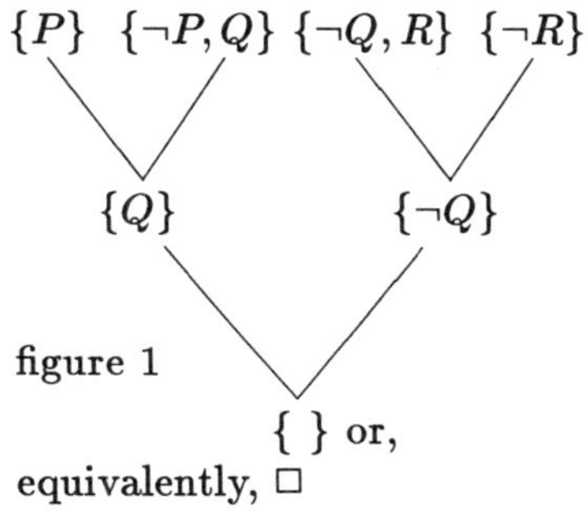

figure 1

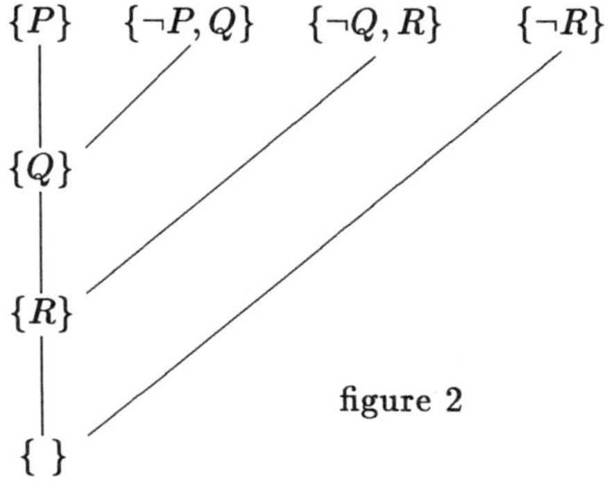

figure 2

EXAMPLE: Let $A := \forall x \exists y[p(x) \supset q(y)]$ and $B := \exists y \forall x[p(x) \supset q(y)]$. We want to find a resolution deduction of B from A.
$C(A) = \forall x[p(x) \supset q(f(x))]$, i.e., $\{\neg p(X), q(f(X))\}$.
$\neg B = \neg \exists y \forall x[p(x) \supset q(y)]$ or, equivalently, $\forall y \exists x[p(x) \,\&\, \neg q(y)]$.
$C(\neg B) = \forall y[p(g(y))] \,\&\, \forall z[\neg q(z)]$, i.e., $\{p(g(Y))\}$ and $\{\neg q(Z)\}$.
For more examples, see Exercise 26.3.

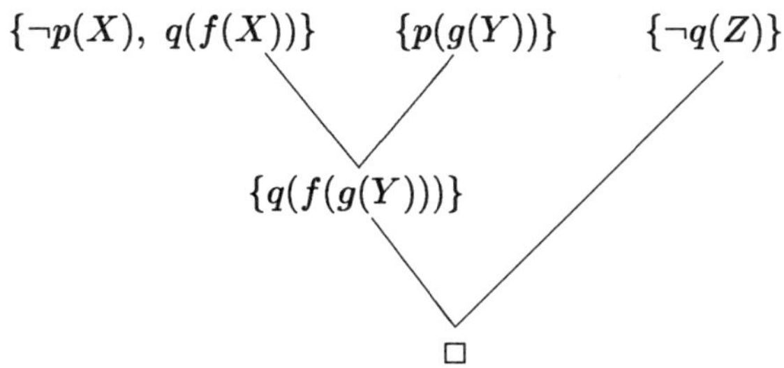

Completeness of resolution can be shown in the following sense: if a set Δ of clauses is unsatisfiable, then there is a resolution deduction of the empty clause from Δ. See, for instance, M.R. Genesereth and N.J. Nilsson, *Logical Foundations of Artificial Intelligence.*

READING LIST

Lloyd, J.W., *Foundations of Logic Programming.*

Genesereth, M.R., and Nilsson, N.J., *Logical Foundations of Artificial Intelligence.*

EXERCISES

26.1. Let P be the following definite logic program.

$p(a, b)$.
$p(c, b)$.
$p(X, Z)$:- $p(X, Y), p(Y, Z)$.
$p(X, Y)$:- $p(Y, X)$.

Determine the Herbrand universe U_P, the Herbrand base B_P and the least Herbrand model M_P of P.

26.2. Let ε be the identity substitution given by the empty set. Let θ, σ and γ be substitutions. Prove:

(a) $\theta\varepsilon = \varepsilon\theta = \theta$.
(b) $(L\theta)\sigma = L(\theta\sigma)$ for every literal L.
$(t\theta)\sigma = t(\theta\sigma)$ for every term t.
(c) $(\theta\sigma)\gamma = \theta(\sigma\gamma)$.

26.3. Give a resolution deduction of

i) $p \,\&\, q$ from $p \vee q$, $p \supset q$ and $q \supset p$.
ii) $p \supset (q \supset p)$.
iii) $p \vee q \supset r$ from $p \supset r$ and $q \supset r$.
iv) $\forall x \exists y[p(x, y)]$ from $\exists y \forall x[p(x, y)]$.

26.4. (Lloyd) Let P be the following normal program.

r :- p.
r :- *not* p.
p :- p.

i) Verify that r is a logical consequence of P.
ii) Construct the search tree for the question "?- r.", given program P.

Conclude that SLD-resolution extended with the Negation as Failure rule is not complete for arbitrary normal programs.

§27 RELATIONAL DATABASES AND SQL:

OBJECT-CHARACTERISATION; CONSTRAINTS; KEY; DATABASE-CHARACTERISATION; DATABASE-TYPE; BOYCE-CODD NORMAL FORM; JOIN; SQL

In this section we shall concentrate on the conceptual schema, i.e., the description of a database on a logical level. Only the relational model of databases will be discussed; neither the network-model nor the hierarchical model will be treated here, since these are less interesting from a logical and set-theoretical point of view.

The description of the logical structure of relational databases in set theoretic terms shows that a Query Language such as SQL is a very natural one. Tuple-, table- and database *constraints* are discussed. The notion of *key* is introduced and we also discuss the *Boyce-Codd Normal Form*, the *projection* of a table and the (natural) *join* of two tables.
The material presented in this section is based on F. Remmen's book *Databases* and on E.O. de Brock's book *De grondslagen van semantische databases*, both in Dutch.

By a database we mean a class of permanent data, which is available to all users of an information system. These data relate to the objects which are relevant to the information system and to the attributes which are relevant to these objects. For instance, the permanent data of a hospital organisation include, among other things, the name, address and residence of each patient.

These permanent data should be available to all users of an information system. This availability for many users has important consequences as different groups of users will be interested in the data in different manners. For instance, an administrator in a hospital organisation will be in need of financial data about persons and rates, while a specialist needs to have at his disposal all medical data of persons and of all treatments to be applied.

Which object with what properties are relevant to the information system can only be determined by the users. The design and implementation of a database will be a compromise between the different and partly clashing desires of the different users. It is the task of the Data Base Administrator (DBA) to bring about such a compromise.

In current database terminology the difference between species and individual is usually indicated by the difference between *type* and *occurrence*. So one can speak of the (object) type patient, and of the (object) occurrence of a patient in a hospital-organisation. In this example we have one type (species) with – in general – many occurrences (individuals).

Each user communicates with the database via the Data Base Management System (DBMS). In fact, a DBMS can be considered as a special expansion of the operating system.

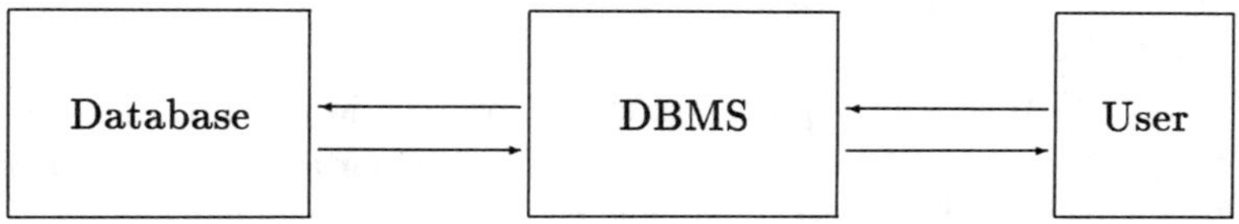

The users of an information-system are interested in information about individual objects and information about an individual object can only be provided in the form of values of one or more attributes. In general, many possible values will be available for each attribute. For instance, the attribute 'pnr' (short for 'patientnumber') of the object 'patient' may have a value between 1 and 100000, and the attribute 'pnm' (patient-name) may have a value consisting of a combination of at most 25 characters. In general, we demand that the values of an attribute form a set.

The set of attributes of an object together with the sets of values belonging to them is called the *object-characterisation* of that object. In the following examples it is made clear how we shall render an object-characterisation.

obchar patient =

attrib	pnr : $\{1, \ldots, 100000\}$	,	patientnummer
	pnm : chs25	,	name
	padr : chs20	,	address
	pres : chs20	,	residence
	db : $\{8800101, \ldots, 9991231\}$	,	date of birth
	sex : $\{m, f\}$	,	sex

endobchar

By chs25 (character string 25) we mean the set of all strings of at least one and at most 25 signs (letters, figures). In the following example –

an object-characterisation of the object 'admission' (into a hospital) – we use the abbreviation 'dat' for the set $\{500101, \ldots, 991231\}$, i.e., the set of all natural numbers between 500101 and 991231.

obchar admission =

attrib	pnr	: $\{1,\ldots,100000\}$,	
	pnm	: chs25 ,	
	padr	: chs20 ,	
	pres	: chs20 ,	
	indat	: dat ,	date of admission
	outdat	: dat ,	date of discharge
	reas	: chs25 ,	reason of admission
	snr	: $\{1,\ldots,100000\}$,	number of specialist
	snm	: chs25 ,	name of specialist
	rnr	: $\{1,\ldots,1000\}$,	number of nursing-room
	wnr	: $\{1,\ldots,15\}$,	number of ward

endobchar

Lastly, we give as an example an object-characterisation of the object 'specialist'.

obchar specialist =

attrib	snr	: $\{1,\ldots,100000\}$,	registration-number
	snm	: chs25 ,	name
	sadr	: chs20 ,	address
	sres	: chs20 ,	residence
	wnr	: $\{1,\ldots,15\}$,	number of ward
	nbd	: $\{1,\ldots,100\}$,	number of beds

endobchar

Let O be an object with attributes $A_1, \ldots, A_m$ and let $W_1, \ldots, W_m$ be the sets of values belonging to $A_1, \ldots, A_m$ respectively. Then we define

$$F_O := \{(A_1, W_1), \ldots, (A_m, W_m)\},$$

and call it the *object-characterisation* of O.

Next we define $\pi(F_O) :=$

$$\{t \mid t = \{(A_1, w_1), \ldots, (A_m, w_m)\} \text{ for some } w_1 \in W_1, \ldots, w_m \in W_m\}.$$

The elements of $\pi(F_O)$ are called *tuples* for O. A tuple for O is a function (and hence a relation) with domain $\{A_1, \ldots, A_m\}$. If t is a tuple for O and $(A_i, w_i) \in t$, then we write $t(A_i)$ for w_i.

So a *tuple* t for O is a set $\{(A_1, w_1), \ldots, (A_m, w_m)\}$ with $w_1 \in W_1$,

$\dots, w_m \in W_m$. We write $w_1 = t(A_1), \dots, w_m = t(A_m)$. Each tuple for O represents one object-occurrence. By mentioning the attributes at the head of columns, we can list the tuples for a given object O in a table, each row in the table corresponding to a tuple for O. For instance, below we give a partial table for the object 'patient'.

pnr	pnm	padr	pres	db	sex
537	Blunt	36 Evans Drive	Cranbury	9080527	m
498	Kiviat	67 Main Street	Newark	9090730	f

In general it holds that not all possible combinations of values of the attributes $A_1, \dots, A_m$ will be allowed. In the literature on databases these restrictions are called *constraints*. We distinguish

constraints on tuples,
constraints on tables, and
constraints on databases.

Below we give some examples of tuple-constraints, i.e., constraints on tuples.

$C_1(t)$: if $t(\text{pnr}) < 200$, then $t(\text{db}) < 9000101$; t being a tuple for the object 'patient'.

$C_2(t)$: $t(\text{indat}) < t(\text{outdat})$; t being a tuple for the object 'admission' (into hospital).

$C_3(t)$: if $t(\text{wnr}) = 9$, then $t(\text{sres}) = \text{Princeton}$; and if $t(\text{wnr}) = 7$, then $t(\text{nbd}) \leq 2$; t being a tuple for the object 'specialist'.

So a *tuple-constraint* is a condition on tuples for a given object O, such that it can be determined whether the condition holds for a given tuple t or not completely independent of the other tuples.

Given an object O and a tuple-constraint C

$$T\text{-}O := \{t \in \pi(F_O) \mid C(t)\}.$$

$T\text{-}O$ is called the *tuple-type* for O (determined by the constraint C) and is the set of all tuples t for O satisfying the condition C.

If the number of tuples t for O satisfying a condition C is finite (and in practice not too large), then the tuple-type $T\text{-}O$ for O can be rendered by an exhaustive list of all object-occurrences satisfying condition C.

A *table for* O is by definition a subset of $T\text{-}O$. There may also be constraints on such tables. We give some examples below.

$TC_1(D) : \forall t_1, t_2 \in D[t_1(\text{pnr}) = t_2(\text{pnr}) \supset t_1 = t_2]$; D being a table for the object 'patient'. $\forall t_1, t_2 \in D$ stands for 'for all t_1 and t_2 in D'.

This table-constraint TC_1 is also formulated as follows:

{pnr} is *uniquely identifying*, or
{pnr} uni, for short.

$TC_2(D) : \forall t_1, t_2 \in D[t_1(\text{pnr}) = t_2(\text{pnr}) \ \& \ t_1(\text{indat}) = t_2(\text{indat}) \supset t_1 = t_2]$; D being a table for the object 'admission'.

This table-constraint TC_2 is also formulated as follows:

{pnr, indat} is *uniquely identifying*, or
{pnr, indat} uni, for short.

$TC_3(D)$: {snr} uni, and
{snm, sadr, sres} uni, and
the number of specialists at ward 9 is at least 2;
D being a table for the object 'specialist'.

So, a *table-constraint* indicates which subsets of a tuple-type are allowed. The set of all tables allowed, given an object O, is called a table-type for O. More precisely, let O be an object, T a tuple-type for O and TC a table-constraint for O. Then

$$TT\text{-}O := \{D \in P(T) \mid TC(D)\}$$

is called a *table-type* for O.

If $D \in TT\text{-}O$, we say D is a *table of type* O.

Let O be an object with attributes $A_1, \ldots, A_m$, and let D be a table of type O. Let $V, W \subseteq \{A_1, \ldots, A_m\}$.

$V \to W$ in $D := \forall t_1, t_2 \in D[t_1\lceil V = t_2\lceil V \supset t_1\lceil W = t_2\lceil W]$, where $t\lceil V$ is the restriction of t to V. In words: *V functionally determines W in D*, or *W is functionally dependent on V in D*.

V is *uniquely identifying within* $D := V \to \{A_1, \ldots, A_m\}$ in D, i.e., $\forall t_1, t_2 \in D[t_1\lceil V = t_2\lceil V \supset t_1 = t_2]$.

$V \to W$ for $O :=$ for every table D of type O, $V \to W$ in D.

V is a *key for* $O := V$ is uniquely identifying within every table of type O. Examples:

{pnr} is a key for 'patient'; {pnr, indat} is a key for 'admission';
{pnm, padr, pres} is a key for 'patient'; {snr} is a key for 'specialist'.

Within the framework of an information-system one usually will be interested in more than only one table-type. For instance, in a hospital-organisation one may be interested in patients, admissions (into the hospital) and specialists and hence also in the table-types TT-patient, TT-admission and TT-specialist belonging to them. At a certain moment the situation of the hospital, at least with respect to patients, admissions and specialists, can be summed up by three tables, one of type TT-patient, one of type TT-admission and one of type TT-specialist. Such a triple of tables is called a *relational database.*

A set of objects together with table-types belonging to them is called a *database-characterisation.* More precisely, let $O_1, \ldots, O_n$ be objects, together with table-types $TT\text{-}O_1, \ldots, TT\text{-}O_n$ belonging to them. Then

$$F_{DB} := \{(O_1, TT\text{-}O_1), \ldots, (O_n, TT\text{-}O_n)\}$$

is a database-characterisation. In the following example it is made clear how we shall render a database-characterisation.

Example: The database-characterisation for the combination of the objects patient, admission and specialist looks as follows:

dbchar hospital =

obj	pat	:	TT-patient ,
	adm	:	TT-admission,
	spec	:	TT-specialist

enddbchar

Note the analogy between an object-characterisation and a database-characterisation: the attributes are replaced by objects and the sets of values by table-types.

$$\pi(F_{DB}) := \{\{(O_1, D_1), \ldots, (O_n, D_n)\} \mid D_1 \in TT\text{-}O_1, \ldots, D_n \in TT\text{-}O_n\}.$$

The elements of $\pi(F_{DB})$ are called *(relational) databases.*

Given a database-characterisation with objects $O_1, O_2, \ldots, O_n$ and table-types $TT\text{-}O_1$, $TT\text{-}O_2, \ldots, TT\text{-}O_n$ belonging to them, in general not all databases in $\pi(F_{DB})$ will be allowed. This brings us to the last class of constraints, the so-called *database-constraints.* The set of all databases satisfying a certain database-constraint is called a *database-type.*

An important subclass of database-constraints is formed by the so-called *subset-requirements.* We make this notion clear by means of the following two database-constraints.

$DC_1(D_1, D_2, D_3) := \forall t_2 \in D_2\ \exists t_1 \in D_1[t_2(\text{pnr}) = t_1(\text{pnr})]$, in words: for each admission-tuple t_2 there is a patient-tuple t_1 such that the value of pnr in t_2 is equal to the value of pnr in t_1.

$DC_2(D_1, D_2, D_3) := \forall t_2 \in D_2\ \exists t_3 \in D_3[t_2(\text{snr}) = t_3(\text{snr})]$, in words: for each admission-tuple t_2 there is a specialist-tuple t_3 such that the value of snr in t_2 is equal to the value of snr in t_3.

The constraint DC_1 means that for any database allowed the set of pnr-values in the admission-table is a subset of the set of pnr-values in the patient-table. A similar remark is to be made for DC_2.

For these subset-requirements the following notation is used: for DC_1, ssr(adm.pnr, pat.pnr); for DC_2, ssr(adm.snr, spec.snr).

Below we give an example of a database-type. The symbols *tatp, tutp, obchar, attrib* stand for table-type, tuple-type, object-characterisation and attribute respectively. The symbols *tuc, tac, dbc* stand for tuple-constraint, table-constraint and database-constraint respectively.

```
Type nr    : {1,...,100000}
     hoev  : {1,...,100}
     dat   : {700101,...,991231}
  tatp TT-patient =
      tutp T-patient =
          obchar patient =
              attrib pnr   : nr,
                     pnm   : chs25,
                     padr  : chs20,
                     pres  : chs20,
                     db    : {8800101, ...,9991231},
                     sex   : {m,f}
          endobchar;
          tuc   pnr < 200 ⊃ db < 9000101
      endtutp;
      tac       {pnr} uni,
                {pnm, padr, pres} uni
  endtatp,
```

```
tatp TT-admission =
    tutp T-admission =
        obchar admission =
            attrib pnr     : nr              ,
                   pnm     : chs25           ,
                   padr    : chs20           ,
                   pres    : chs20           ,
                   indat   : dat             ,
                   outdat  : dat             ,
                   reas    : chs25           ,
                   snr     : nr              ,
                   snm     : chs25           ,
                   rnr     : {1,...,1000}    ,
                   wnr     : {1,...,15}
        endobchar;
        tuc  indat < outdat,
             reas = 'informaritis' ⊃ rnr = 5
    endtutp ;
    tac      {pnr, indat} key
endtatp,

tatp TT-specialist =
    tutp T-specialist =
        obchar specialist =
            attrib snr   : nr            ,
                   snm   : chs25         ,
                   sadr  : chs20         ,
                   sres  : chs20         ,
                   wnr   : {1,...,15}    ,
                   nbd   : hoev
        endobchar;
        tuc  wnr = 9 ⊃ sres = 'Princeton',
             wnr = 7 ⊃ nbd ≤ 2
    endtutp;
    tac      keys {{snr}, {snm, sadr, sres}},
             "at least two specialists at ward 9"
endtatp,
```

```
    dbtype DT-hospital =
        dbchar hospital =
            obj pat   : TT-patient    ,
                adm   : TT-admission  ,
                spec  : TT-specialist
        enddbchar;
        dbc       ssr (adm.pnr, pat.pnr),
                  ssr (adm.snr, spec.snr)
    enddbtype ,
endtype
```

Looking at the database-type given above, we see that some attributes are *redundant*: pnm, padr and pres in the table for 'admission' are uniquely determined by pnr and already occur in the table for 'patient'; snm in the table for 'admission' is uniquely determined by snr and also occurs in the table for 'specialist'; and wnr in the table for 'admission' is uniquely determined by rnr, although a table in which both rnr and wnr already occur is not (yet) available. For these reasons we say that the table-type *TT*-admission given above is not *normal*. In concrete cases this means that in case of a change of address of a patient not only the table for 'patient' has to be updated, but also the table for 'admission'; otherwise, an inconsistent database would result.

Let O be an object with attributes $A_1, \ldots, A_m$ and let TT-O be a table-type for O. TT-O is in *Boyce-Codd Normal Form* (BCNF) := for all $V \subseteq \{A_1, \ldots, A_m\}$ and for all $A \in \{A_1, \ldots, A_m\}$: if $V \rightarrow \{A\}$ for O and $A \notin V$, then V is a key for O. Informally: TT-O is in BCNF if every set V of attributes which determines an attribute outside of V is a key for O.

Since {pnr} → {pnm} for 'admission', pnm ∉ {pnr}, but {pnr} is not a key for 'admission', it follows that *TT*-admission is not in Boyce-Codd Normal Form. (Remember that {pnr, indat} is a key for 'admission'.)

In the literature one also finds various other normal forms including the first, second, third and fourth normal forms. If TT-O is in BCNF, then TT-O is also in 3NF (Third Normal Form).

A *normal database-type* is a database-type in which each table-type is normal. We can convert the database-type given above into a normal database-type by applying the following two operations to the given database-type.

1. In the table-type TT-admission leave out the attributes pnm, padr, pres, snm and wnr.
2. Add a table-type TT-room as follows:

```
tatp TT-room =
    tutp T-room =
        obchar room =
            attrib rnr  : {1,...,1000},
                   wnr  : {1,...,15}
        endobchar;
    endtutp;
    tac          {rnr} uni
endtatp.
```

The result is a normal database-type, in which redundancies are avoided, while all information has been saved. (However, in practice, redundant *storage* of data may be necessary, for instance, because of the required time of response.)

Of course, a database-type for any actual hospital organisation will be much more complex than the simple example considered here.

Let O be an object with attributes $A_1, \ldots, A_m$, $V \subseteq \{A_1, \ldots, A_m\}$ and let D be a table of type O. $D \parallel V$, the *projection* of D on V, is by definition $\{t\lceil V;\ t \in D\}$. For instance, for the following table D_1:

nr	name	sal	sex	dept
8	Johnson	2200	male	1
7	Johnson	3100	female	2
9	Kiviat	2900	male	1

$D_1 \parallel$ {sex, dept} is the table

sex	dept
male	1
female	2

.

Let O_1 be an object with attributes $A_1, \ldots, A_m$ and O_2 an object with attributes $B_1, \ldots, B_n$. Let t_1 be a tuple for O_1 and t_2 a tuple for O_2.
t_1 and t_2 are *compatible* :=

$$t_1\lceil\{A_1, \ldots, A_m\} \cap \{B_1, \ldots, B_n\} = t_2\lceil\{A_1, \ldots, A_m\} \cap \{B_1, \ldots, B_n\}.$$

Let D_1 be a table of type O_1 and D_2 a table of type O_2.
$D_1 \bowtie D_2 := \{t_1 \cup t_2 \mid t_1 \in D_1$ and $t_2 \in D_2$ and t_1 and t_2 are compatible$\}$.
$D_1 \bowtie D_2$ is called the (natural) *join* of D_1 and D_2.
For instance, let D_2 be the table:

anr	name	man
2	planning	7
1	production	9

and let D_3 result from D_1 above by replacing 'dept' by 'anr' and 'name' by 'wnm'. Then $D_3 \bowtie D_2$ is the table:

nr	wnm	sal	sex	anr	name	man
8	Johnson	2200	male	1	production	9
7	Johnson	3100	female	2	planning	7
9	Kiviat	2900	male	1	production	9

SQL

The purpose of a *query-language* is to enable the user to make use of the data stored in the database in an user-friendly manner. In order to give a more concrete idea of a query-language, we shall treat some elements of the query-language SQL (Structured Query Language; 1980) on the basis of some examples. Having understood the logical structure of a relational database, query-languages such as SQL become very perspicuous.

The terminology of SQL is familiar to the terminology of set theory, as will become clear from the examples below. In these examples P, ADM and SP stand for the set (or table) of all patients, the set of all admissions and the set of all specialists, respectively. The examples all refer to the objects described in the *normalized database given above.*

EXAMPLE 1:

a) Describe the set of numbers, names and addresses of all patients who live in Princeton and were born before 1960. *Answer*:
$\{t\lceil\{\text{pnr, pnm, padr}\} \mid t \in \text{P}\mid t(\text{pres}) = \text{'Princeton'} \ \& \ t(\text{db}) < 9600101\}$.
b) Now, in SQL the query 'give number, name and address of all patients who live in Princeton and were born before 1960' is formulated as follows.

```
SELECT  t.pnr, t.pnm, t.padr
FROM    P t
WHERE   t.pres = 'Princeton'
AND     t.db < 9600101
```

Here t.pnr corresponds to t(pnr).

EXAMPLE 2:

a) Describe the set of numbers, names and addresses of all patients who were admitted into hospital in the period between May 26 and July 11, 1981. *Answer*: $\{t\lceil\{\text{pnr, pnm, padr}\} \mid t \in \text{P} \mid \exists s \in \text{ADM}$
$[s(\text{pnr}) = t(\text{pnr})\ \ \&\ s(\text{indat}) \geq 810526\ \&\ s(\text{indat}) \leq 810711]\}$
or, equivalently,
$\{t\lceil\{\text{pnr, pnm, padr}\} \mid t \in \text{P} \mid t(\text{pnr}) \in \{s(\text{pnr}) \mid s \in \text{ADM} \mid$
$s(\text{indat}) \geq 810526\ \&\ s(\text{indat}) \leq 810711\}\}$.

b) Now, in SQL the query 'give number, name and address of all patients who were admitted into hospital in the period between May 26 and July 11, 1981' is formulated as follows.

```
SELECT  t.pnr, t.pnm, t.padr
FROM    P t
WHERE   t.pnr IN
            (SELECT  s.pnr
             FROM    ADM s
             WHERE   s.indat ≥ 810526
             AND     s.indat ≤ 810711)
```

EXAMPLE 3:

a) Describe the set of names, addresses and residences of all specialists who were responsible for an admission in August 1977 of a patient from Princeton for reason 034.
Answer: $\{t\lceil\{\text{snm, sadr, sres}\} \mid t \in \text{SP} \mid \exists s \in \text{ADM}[s(\text{snr}) = t(\text{snr})$
$\&\ s(\text{reas}) = 034\ \&\ 770801 \leq s(\text{indat}) \leq 770831\ \&$
$\exists u \in \text{P}[u(\text{pnr}) = s(\text{pnr})\ \ \&\ u(\text{pres}) = \text{'Princeton'}]]\}$
or, equivalently,
$\{t\lceil\{\text{snm, sadr, sres}\} \mid t \in \text{SP} \mid t(\text{snr}) \in \{s(\text{snr}) \mid s \in \text{ADM} \mid s(\text{reas}) =$
$034\ \ \&\ 770801 \leq s(\text{indat}) \leq 770831\ \&$
$s(\text{pnr}) \in \{u(\text{pnr}) \mid u \in \text{P} \mid u(\text{pres}) = \text{'Princeton'}\}\}\}$.

b) Now, in SQL the query 'give name, address and residence of all specialists who were responsible for an admission in August 1977 of a patient from Princeton for reason 034' is formulated as follows.

```
SELECT  t.snm, t.sadr, t.sres
FROM    SP t
WHERE   t.snr IN
        (SELECT  s.snr
         FROM    ADM s
         WHERE   s.reas = 034
         AND     s.indat ≤ 770831
         AND     s.indat ≥ 770801
         AND     s.pnr IN
                 (SELECT  u.pnr
                  FROM    P u
                  WHERE   u.pres = 'Princeton'))
```

EXAMPLE 4:

a) Describe the set of numbers and names of all patients, reason of admission and number of nursing-room, who were admitted into hospital between September 1 and 5, 1977, in ward number 9.

Answer: $\{t\lceil\{\text{pnr, pnm, reas, rnr}\} \mid t \in \text{P} \bowtie \text{ADM} \mid 770901 \leq t(\text{indat}) \leq 770905 \;\&\; \exists s \in \text{ROOM}[s(\text{rnr}) = t(\text{rnr}) \;\&\; s(\text{wnr}) = 9]\}$

or, equivalently,

$\{t\lceil\{\text{pnr, pnm, reas, rnr}\} \mid t \in \text{P} \bowtie \text{ADM} \mid 770901 \leq t(\text{indat}) \leq 770905 \;\&\; t(\text{rnr}) \in \{s(\text{rnr}) \mid s \in \text{R} \mid s(\text{wnr}) = 9\}\}$,

where $P \bowtie$ ADM is the join of P and ADM.

b) Now, in SQL the query 'give number and name of all patients, reason of admission and number of nursing-room, who were admitted into hospital in the period between September 1 and 5, 1977, in ward number 9' is formulated as follows.

```
SELECT  t1.pnr, t1.pnm, t2.reas, t2.rnr
FROM    P t1, ADM t2
WHERE   t1.pnr = t2.pnr
AND     t2.indat ≥ 770901
AND     t2.indat ≤ 770905
AND     t2.rnr IN
```

```
(SELECT  s.rnr
FROM     R s
WHERE    s.wnr = 9)
```

EXAMPLE 5:

a) Describe the set of all room-numbers, in which no patients from Cranbury were hospitalized in the period between August 11 and 17, 1977.
Answer: $\{s(\text{rnr})|\ s \in \text{R} \mid \neg\exists t \in \text{ADM}[t(\text{rnr}) = s(\text{rnr})\ \ \&\ 770811 \leq t(\text{indat}) \leq 770817\ \&\ A(t)]\}$ where $A(t) :=$
i) $\exists u \in \text{P}[u(\text{pnr}) = t(\text{pnr})\ \ \&\ u(\text{pres}) = \text{'Cranbury'}]$ or, equivalently,
ii) $t(\text{pnr}) \in \{u(\text{pnr})|\ u \in \text{P} \mid u(\text{pres}) = \text{'Cranbury'}\}$.
Note that $\neg\exists t \in \text{ADM}[t(\text{rnr}) = s(\text{rnr})\ \ \&\ \ldots\ \&\ A(t)]$ is equivalent to $s(\text{rnr}) \notin \{t(\text{rnr})|\ t \in \text{ADM} \mid \ldots\ \&\ A(t)\}$.

b) Now, in SQL the query 'give the numbers of all rooms, in which no patients from Cranbury were hospitalized in the period between August 11 and 17, 1977' can be formulated as follows.

```
SELECT  s.rnr
FROM    R s
WHERE   s.rnr NOT IN
          (SELECT  t.rnr
          FROM     ADM t
          WHERE    t.indat ≤ 770817
          AND      t.indat ≥ 770811
          AND      t.pnr IN
                    (SELECT  u.pnr
                    FROM     P u
                    WHERE    u.pres = 'Cranbury'))
```

For further reading, the reader is referred to E. O. de Brock, *The Foundations of Semantic Databases.*

EXERCISES

27.1. The following queries all refer to the normalized database given above. Formulate these queries into SQL.

i) Give name, address and residence of all specialists from ward number 9, having more than two beds.

ii) Give number of all rooms in which no patients from Princeton were hospitalized in the period between May 9 and 18, 1980.

iii) Give number, name, address and residence of all patients who were hospitalized by a specialist of ward number 9.

iv) Give number and name of all specialists who were responsible for admission on March 3, 1980, because of informaritis.

27.2. Let D_1 be the table

nr	name	sal	sex	dept
8	Johnson	2200	male	1
7	Johnson	3100	female	2
9	Kiviat	2900	male	1

and let D_2 be the table

anr	name	man
2	planning	7
1	production	9

.

Determine $D_1 \bowtie D_2$.
Let D_3 result from D_1 by replacing 'dept' by 'anr' and 'name' by 'wnm'. Determine $D_3 \bowtie D_2$.
Let D_4 result from D_2 by replacing 'man' by 'nr' and 'name' by 'anm'. Determine $D_1 \bowtie D_4$ and $D_3 \bowtie D_4$.

27.3. Make clear why the following set does not describe the set of all room-numbers in which no patients from Cranbury were hospitalized in the period between August 11 and 17, 1977 (compare Example 5).
$\{s(\text{rnr}) \mid s \in \text{R} \mid \exists t \in \text{ADM}[t(\text{rnr}) = s(\text{rnr}) \;\; \& \; 770811 \leq t(\text{indat}) \leq 770817 \; \& \; \neg\exists u \in \text{P}[u(\text{pnr}) = t(\text{pnr}) \;\; \& \; u(\text{pres}) = \text{'Cranbury'}]]\}$.
Hint: Consider the following tables.

R	rnr	wnr
s1	11	5
s2	12	5
s3	13	6

ADM	pnr	indat	rnr
t1	400	770812	11
t2	500	770813	11
t3	600	770814	12

P	pnr	pres
u1	400	Princeton
u2	500	Cranbury
u3	600	Cranbury

§28 COMPUTABILITY:

PROCEDURES AND ALGORITHMS; ALPHABETS AND LANGUAGES; TURING MACHINES; TURING-ACCEPTABLE AND TURING-DECIDABLE LANGUAGES; TURING-COMPUTABLE FUNCTIONS; CHURCH-TURING THESIS

In this section we will start with an informal discussion of procedures and algorithms, giving several examples of both. An algorithm is a computational procedure that always terminates.

An alphabet Σ is defined as a finite set of symbols. A string or word over Σ is a finite sequence of symbols from Σ. And a language over Σ is any set of strings over Σ.

In order to make the notions of procedure and algorithm mathematically precise, we will introduce the notion of the Turing machine. A Turing machine consists of a program, to be specified below, and a two-way infinite tape, which serves both as input and as output tape.

A language is called *Turing-acceptable* if there is a Turing machine which serves as a procedure for recognizing words in the language; such a Turing machine does not halt for words not in the language. A language is called *Turing-decidable* if there is a Turing machine which provides us with an algorithm to decide whether a given input string belongs to the language; such a Turing machine halts for every input, answering either Yes or No.

In addition, Turing machines can be used for computing functions. The *Church-Turing thesis* identifies the somewhat vague intuitive concept of an effectively calculable function with the mathematically defined notion of a *Turing-computable* function. Evidently, it is not possible to prove this thesis mathematically.

PROCEDURES AND ALGORITHMS

In section 2 we presented an *algorithm* or *effective computational procedure* to decide whether an arbitrary formula A of the propositional calculus is valid or not. The algorithm in question consists simply of making the truth table of a given formula A and checking whether this table has 1 in each line. See also section 5 for an algorithm that decides, given any finite set of formulas B, $A_1, \ldots, A_m$ of the propositional calculus, whether B is a valid consequence of $A_1, \ldots, A_m$ and § 11, 12 for an algorithm to decide whether B is deducible from $A_1, \ldots, A_m$ in the propositional calculus.

Making a truth table of A and checking if it contains only 1's is recognized by us as being an algorithm (effective computational procedure), although so far we have not specified precisely what we mean by the word "algorithm". Whatever definition is suggested, it should certainly be such that the truth table method counts as an algorithm. For the time being, an algorithm for a certain class of questions (for example, "is a given formula A of the propositional calculus valid") will mean intuitively any set of instructions that provides a mechanical procedure by which the answer to any question of the class (for example, "is $P \vee Q \supset (\neg P \supset Q)$ valid") can be obtained. It is assumed that these instructions do not require any "creative" thought in their execution, making it possible for them to be carried out by a machine. Familiar examples of algorithms used in daily life are the algorithm to calculate the net salary of a given employee and the algorithm to add two given natural numbers. Another example of an algorithm is given in exercise 28.1.

It is interesting to note that the algorithm to decide the validity of a given formula A (of the propositional calculus) may in practice turn out to be a rather theoretical one. If a formula A is built, for instance, from 100 atomic propositional formulas, then a truth table for A will contain 2^{100} lines (see § 2). Since $2^{10} = 1024$ is greater than 10^3 and $2^{100} = (2^{10})^{10}$, such a truth table has more than $(10^3)^{10} = 10^{30}$ lines. Even if we could examine 10^{10} lines per second – a pace far beyond the capabilities of existing or projected computers – the required time for completing this calculation would be 10^{20} seconds, which – one year is 3.1536 times 10^7 seconds – is more than 10^{10} human lifetimes! Thus, the fact that a problem (such as "is a given formula A valid") is solvable in theory does not immediately imply that it can be *realistically* solved. (See section 3 and Theorem 30.17.)

If we extend the propositional language of section 1 to more complex languages like the predicate language mentioned in section 18, which contains the quantifiers $\forall x$ (for all x) and $\exists x$ (for some x), the truth table for a given formula A itself may become infinitely long (since in order to calculate the value of $\forall x[P(x)]$, for instance, one may have to calculate the values of infinitely many instances $P(a_1)$, $P(a_2),\ldots$). As mentioned in section 22, no algorithm is known for deciding whether a given formula A of the (first-order) predicate language is valid or not. It is true that in section 22 a *procedure* is given such that A is valid (or equivalently, formally provable) if and only if the procedure (of searching for a formal proof of A) stops with a formal proof of A as a result. But this procedure does not always terminate: if there is no formal proof of A, then the procedure may not stop. In other words, if there is a formal proof of A, the procedure described in § 22 will find one. However, if there is no formal proof of A (in the predicate calculus), the procedure may run forever. And as long as the procedure described in § 22 continues to run, we have no way of knowing if it will eventually halt. Since an *algorithm* is an *effective* computational procedure, i.e., one which always terminates, the procedure in section 22 is not an algorithm. (See exercise 28.2. for another example of a non-effective procedure.) Moreover, in § 29 we shall see that not only is no algorithm known, but *none is possible* (Theorem 29.12.). However, in order to show that no algorithm to decide validity for formulas of the predicate calculus can exist, it is necessary to give an exact mathematical definition of the notion of "algorithm".

Before trying to define "algorithm", we want to discuss Richard's paradox (see also exercise 20.1.), from which we learn – still on the intuitive level – that there can be no algorithm for deciding whether an alleged algorithm is indeed an algorithm.

The expressions in English language are denumerable, i.e., they are in one-to-one correspondence with the natural numbers (see Chapter II); let $E_0, E_1, E_2, \ldots$ be an enumeration. Some of these expressions describe an algorithm which defines a function $f : \mathbb{N} \to \mathbb{N}$. By leaving out in the enumeration $E_0, E_1, E_2, \ldots$ those expressions which do not describe such an algorithm, we find an enumeration $E_{n_0}, E_{n_1}, E_{n_2}, \ldots$ of algorithms which define functions $f_0, f_1, f_2, \ldots$ from $\mathbb{N}$ to $\mathbb{N}$ respectively. Now let $f : \mathbb{N} \to \mathbb{N}$ be defined as follows: for all $n \in \mathbb{N}, f(n) := f_n(n) + 1$. Then for each natural number n, $f \neq f_n$, since $f(n) \neq f_n(n)$. So, there is no algorithm which defines f. On the other hand it seems we just have given an algorithm for computing f. Contradiction (*Richard's paradox*, 1905).

The solution of the paradox is that something is wrong with the step "leaving out those expressions which do not describe such an algorithm".

At this point we assume that we have at our disposal an algorithm which enables us to decide whether an English expression describes an algorithm for computing the values of a function $f : \mathbb{N} \to \mathbb{N}$. Since this assumption yields a contradiction, it follows that this assumption is false. So we have shown, still on the intuitive level, that the problem of verifying that an alleged algorithm is indeed an algorithm is more complex than one might expect. In fact, having defined the notion of algorithm more precisely, we will show that this problem is unsolvable (see Theorem 29.5.).

ALPHABETS AND LANGUAGES

By an *alphabet* we mean a finite set of *symbols*. Familiar examples are the Roman alphabet {a, b, ..., z}, the alphabet of the propositional language (see section 1) $\{P, 1, \sim, \supset, \&, \vee, \neg, (\ ,)\}$ and the binary alphabet $\{0,1\}$. Another familiar example is the alphabet of symbols on the keyboard of your favorite computer.

A *string* or *word* or *sentence* over an alphabet is a finite sequence of symbols from the alphabet. Thus *orchid* is a string over the alphabet {a, ... , z }, $P_1 \supset P_{11}$ is a string over the alphabet of the propositional language and 0 1 1 0 is a string over the alphabet $\{0,1\}$. The *empty string*, denoted by e, contains no symbols at all.

Given an alphabet Σ, Σ^* denotes the set of all strings over Σ, including the empty string. Given a string w over an alphabet Σ, the length $|w|$ of w is its length as a sequence. Thus, the *length* of the string *orchid* is 6, in other words, $|\text{orchid}| = 6$; $|0110| = 4$ and $|e| = 0$. If w is a string, then $w(j)$, where $1 \leq j \leq |w|$, is the symbol in the jth position of w. For example, if w = *orchid*, then $w(1) = o$, $w(2) = r$, $w(3) = c$, $w(4) = h$, $w(5) = i$ and $w(6) = d$.

The *concatenation* of two strings x and y, written $x_\circ y$ or simply xy, is the string x followed by the string y. For example, $01_\circ 10 = 0110$ and $hand_\circ book = handbook$. Of course, $w_\circ e = e_\circ w = w$ for any string w.

A string v is a *substring* of a string w if and only if there are strings x and y, possibly empty, such that $w = x\ v\ y$. The *reversal* of a string w, denoted by w^R, is the string "spelled backwards". For instance, $dog^R = god$ and $01^R = 10$.

A *language* over an alphabet Σ is by definition any set of strings over Σ, i.e., any subset of Σ^*. Thus, Σ^* and $\emptyset$ are languages. We can specify a finite language by listing all its strings. For example, {orchid, mother, abca} is a finite language over the alphabet {a, b, ..., z} And {0, 01, 001, 0001,...} is an infinite language over the alphabet {0, 1}. Examples of other infinite languages are $\{w \in \{0,1\}^* \mid w$ has an equal number of 0's and 1's$\}$ and $\{w \in \{0,1\}^* \mid w = w^R\}$.

TURING MACHINES

The Turing machine was proposed by Alan Turing in 1936-37 as a mathematical model for describing procedures. The *Church-Turing thesis* states that our intuitive notion of a procedure as a finite sequence of instructions which can be mechanically carried out is equivalent to the precise notion of a procedure computable by a Turing machine. This thesis cannot be proved, since only one of the two concepts it mentions is mathematically defined. However, from the definition of a Turing machine, it will be readily apparent that any computation that can be described by means of a Turing machine can be carried out mechanically. Thus, the definition is not too broad. It can also be shown that any computation that can be performed on a computer can be described by means of a Turing machine. Thus, if one ever found a procedure that fits the intuitive notions, but could not be described by means of a Turing machine, it would indeed be of an unusual nature since it could not possibly be programmed for any existing computer. Many other formalizations of a procedure have been proposed, and they have been shown to be equivalent to the Turing machine formalization. This supports the belief that the Turing machine is general enough to encompass the intuitive notion of a procedure. For these other formalizations, as well as for many historical details, the reader is referred to S.C. Kleene's books *Introduction to Metamathematics and Mathematical Logic.*

A Turing machine may be conceived as a description of a computing device from a purely conceptual point of view. A *Turing machine* consists of a finite control and a two-way infinite tape.

The *finite control* contains the instructions of the program, in other words, it contains a finite description of a procedure. The control unit admits a fixed finite number of possible "states", denoted by $q_0, \ldots, q_k$. These states might be denoted by arbitrary words as in computer languages, thus permitting them to be given names that remind us of their functions, e.g., "start", "loop", etc. The internal state of a physical computing machine is the actual arrangement of its parts.

The infinite *tape* consists of cells (or squares) which may be thought of as being indexed by the integers. Each cell of the tape holds exactly one symbol from a given alphabet Σ. This alphabet is supposed to contain a special symbol, #, called the *blank*.

Communication between the finite control and the tape is provided by a read/write *head*. This head can read only one cell at a time. It can move along the tape in both directions. It can also be used for writing on that tape, replacing one symbol by another one. By allowing the tape to be infinite, the Turing machine has an unlimited capacity for storing information, giving it a potentially infinite memory. In this sense the Turing machine is an idealization of any physical computing device.

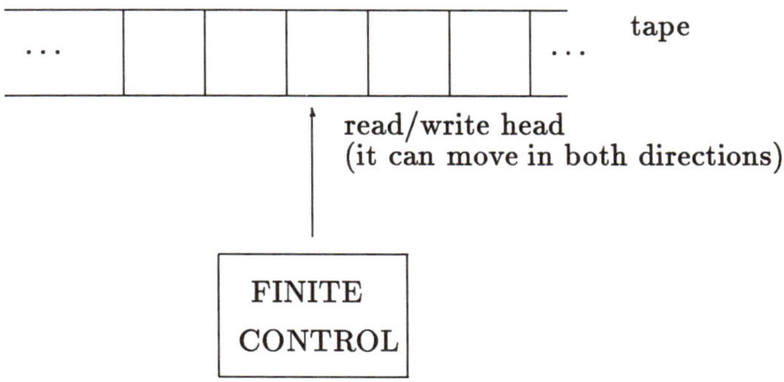

Depending on the current state of the finite control and on the tape symbol currently scanned by the read/write head, the Turing machine can execute one of its instructions; this involves the following actions:

1. the control unit will be put into a new state;
2. the read/write head will write a symbol in the tape cell currently scanned, replacing the one already there;
3. the head will move one cell to the left or to the right or remain in its former position.

Consequently, the instructions or the program of a Turing machine can be represented by quintuples (p, a, q, b, m), to be read as follows: when in state p and scanning symbol a, enter state q, rewrite a as b and move the head one cell to the left, to the right or do not move the head for $m = L$, R or O respectively.

A Turing machine is supplied with input by inscribing that input string on consecutive tape cells. The infinitely many remaining cells are supposed to contain the blank symbol #. The Turing machine is supposed to start in its *initial state* q_0 with its read/write head in a position just before the beginning of the input. There is a special *halt state* , denoted by h, also called the *accepting* state, that signals the end of a computation. Because a Turing machine can write on its tape, it can leave an answer on the tape at the end of a computation. We agree that the output consists of all non-blank tape symbols to the left of the head in the halting configuration.

We can now present the mathematical definition of a Turing machine. By allowing more tapes, more heads on a tape, etc., a large variety of Turing machines results. However, all these machines are equivalent in the sense that anything which can be computed with one type of Turing machine can also be computed by another type of Turing machine.

DEFINITION 28.1.: A (deterministic) *Turing machine* T is a quadruple $\langle K,\ \Sigma,\ q_0,\ \delta\rangle$, where

i) K is a finite set, not containing h. The elements of K are called *states; h is called the halt state* (or *accepting state*).
ii) Σ is an alphabet, i.e., a finite set of symbols, containing the *blank* symbol #, but not containing the symbols L, R and O.
iii) $q_0 \in K$ (q_0 is an element of K). q_0 is called the *initial state.*
iv) δ is a partial function from $K \times \Sigma$ to $(K \cup \{h\}) \times \Sigma \times \{L,\ R,\ O\}$. $\delta(p, a) = (q, b, m)$ is to be read as: if the machine T is in state p and reads symbol a, then it should enter state q, replace the symbol a by b and move its head one cell to the left, to the right or do not move at all, depending on whether $m = L$, $m = R$ or $m = O$, respectively.

Since δ assigns at most one triple (q, b, m) to each pair (p, a), the operation of T is *deterministic.* In section 29 we will also consider non-deterministic Turing machines (Definition 29.16.). They are like (deterministic) Turing machines with the important difference that for a given state and tape symbol scanned by the tape head, the machine may have several choices for the next move.

A Turing machine T will *halt* only when it enters the halt state h; and T will *stop* only when it halts or when it arrives at a state p scanning the symbol a, while $\delta(p, a)$ has not been defined.

The only function of the internal state is to specify the next act of the computer, given knowledge of the symbol that appears on the scanned square. The counterpart of the internal state of a Turing machine in a real life computer is the actual arrangement of its parts.

The overall *program* of instructions that a Turing machine $T = \langle K,\ \Sigma,\ q_0,\ \delta\rangle$ is to follow can be specified in various ways, e.g., by a *machine table* or by a *flow graph* or by a *set of quintuples*. The three sorts of descriptions are illustrated in the figure below for the case of a machine which writes three symbols "1" on a blank tape and then halts, scanning the first cell after the rightmost of the three.

EXAMPLE 28.2.: $T = \langle\{q_0, q_1, q_2\}, \{\#, 1\},\ q_0,\ \delta\rangle$, where

$$\delta(q_0, \#) = (q_1, 1, R),$$
$$\delta(q_1, \#) = (q_2, 1, R),$$
$$\delta(q_2, \#) = (h, 1, R),$$

is a Turing machine, which will write a 1 in the square it is initially scanning (we suppose we start with a blank tape), move right one square, write a 1 there, move right one more square, write a third 1 there, move right again and halt.

This Turing machine can also be specified by

(a) the *machine table*

		scanned symbol	
		$\#$	1
present state	q_0	q_1 1 R	–
	q_1	q_2 1 R	–
	q_2	h 1 R	–

(b) the *flow graph*

$$q_0 \xrightarrow{\#\ :\ 1,\ R} q_1 \xrightarrow{\#\ :\ 1,\ R} q_2 \xrightarrow{\#\ :\ 1,\ R} h$$

(c) the *set of quintuples* $(q_0, \#, q_1, 1, R)$, $(q_1, \#, q_2, 1, R)$ and $(q_2, \#, h, 1, R)$.

A virtue of the flow graph as a way of representing the machine program is that if the starting state is somehow indicated (e.g., if it is understood that the leftmost node represents the starting state unless there is an indication to the contrary), we can dispense with the names of the states: it doesn't matter what you call them.

For other examples of Turing machines, we refer the reader to the exercises. We can indicate how a Turing machine operates by writing down its sequence of *configurations*. Each configuration contains information on what is on the tape at some stage of the computation, what state the machine is in at that stage, and which square is being scanned.

DEFINITION 28.3.: A *configuration* of the Turing machine $T = \langle K,\ \Sigma,\ q_0,\ \delta\rangle$ is a triple (q, α, i) such that 1) q is in $K \cup \{h\}$, designating the current state of T; 2) α is a string in Σ^*, designating the non-blank portion of the tape; and 3) i is an integer, designating the distance of the tape head of T from the left end of α. A configuration whose state component is h will be called a *halted* or *accepting configuration.*
Notation: If $\alpha = a_1 \ldots a_n$, $1 \le i \le n$, we can write the configuration (q, α, i) as $(q, a_1 \ldots \underline{a_i} \ldots a_n)$, the underlined symbol indicating the head position. $(q, a_1 \ldots a_n,\ n+1)$ can be written as $(q, a_1 \ldots a_n\underline{\#})$.

The following configurations occur in a computation of the Turing machine in Example 28.2.:

$$(q_0, \#\underline{\#}\#\#\#),\ (q_1, \#1\underline{\#}\#\#),\ (q_2, \#11\underline{\#}\#),\ (h, \#111\underline{\#}).$$

Other notations for a configuration $(q, a_1 \ldots \underline{a_i} \ldots a_n)$ are
$a_1 \ldots a_{i-1}\ q\ a_i \ldots a_n$ and $a_1 \ldots \underset{q}{a_i} \ldots a_n$.

DEFINITION 28.4.: Let $T = \langle K,\ \Sigma,\ q_0,\ \delta\rangle$ be a Turing machine. We define a *move* of T as follows:

i) If $\delta(p, a) = (q, b, O)$, then by definition for any $\alpha, \beta \in \Sigma^*$

$$(p,\ \alpha\ \underline{a}\ \beta) \vdash_T (q,\ \alpha\ \underline{b}\ \beta).$$

That is, T, when in state p and scanning symbol a, will enter state q, rewrite a as b and leave the head in the same position.

ii) If $\delta(p, a) = (q, b, R)$, then by definition for any $\alpha, \beta \in \Sigma^*$, $c \in \Sigma$

$$(p,\ \alpha\ \underline{a}\ c\ \beta) \vdash_T (q,\ \alpha\ b\ \underline{c}\ \beta).$$

That is, T, when in state p and scanning symbol a, will enter state q, rewrite a as b and move its head one square to the right.

iii) If $\delta(p, a) = (q, b, L)$, then by definition for any $\alpha, \beta \in \Sigma^*$, $c \in \Sigma$

$$(p,\ \alpha\ c\ \underline{a}\ \beta) \vdash_T (q,\ \alpha\ \underline{c}\ b\ \beta).$$

That is, T, when in state p and scanning symbol a, will enter state q, rewrite a as b and move its head one square to the left.

If two configurations are related by $\vdash_T$, we say that the first *yields* the second (*in one step*) or that the second *results* from the first *by one move.*

If one configuration results from another by some finite number of moves, including zero moves, they are related by the relation $\vdash_T^*$. We say that configuration C_1 *yields* configuration C_2 if $C_1 \vdash_T^* C_2$.

DEFINITION 28.5.: Let T be a Turing machine. A *computation* by T is a sequence of configurations $C_0, C_1, \ldots, C_n$ for some $n \geq 0$ such that $C_0 \vdash_T C_1$, $C_1 \vdash_T C_2, \ldots,$ $C_{n-1} \vdash_T C_n$. We say that the computation is of *length* n or has n *steps*.

EXAMPLE 28.6.: Consider the Turing machine
$T = \langle \{q_0, q_1\},\ \{a, \#\},\ q_0,\ \delta \rangle$, where δ is defined as follows:

$\delta(q_0, a) = (q_0, \#, R)$,

$\delta(q_0, \#) = (h, \#, O)$.

If T is started in configuration $(q_0,\ \underline{a}\ a\ a)$, its computation would be represented as follows.

$$\begin{aligned}(q_0,\ \underline{a}\ a\ a)\ &\vdash_T (q_0, \#\underline{a}\ a) ,\\ &\vdash_T (q_0, \#\#\underline{a}) ,\\ &\vdash_T (q_0, \#\#\#\underline{\#}) ,\\ &\vdash_T (h, \#\#\#\underline{\#}) .\end{aligned}$$

This computation has four steps.

Changing a non-blank symbol to the blank symbol will be called *erasing* the non-blank symbol.

TURING-ACCEPTABLE AND TURING-DECIDABLE LANGUAGES

Several conventions can be followed for presenting *input* to Turing machines. Here we adopt the following policy:

- The input string is surrounded by a blank on each side; the infinitely many remaining cells of the tape must also contain the blank symbol.
- The head is positioned at the tape square containing the blank that marks the left end of the input;
- The machine starts operating in its initial state q_0.

DEFINITION 28.7.: Let $T = \langle K,\ \Sigma,\ q_0,\ \delta\rangle$ be a (deterministic) Turing machine and let $w \in \Sigma^*$ (w is the input).

T accepts w (or, *T halts* on input w) := $(q_0, \underline{\#}w\#)$ yields some halted (or accepting) configuration, i.e., for some $\alpha \in \Sigma^*$ and $i \in \mathbf{N}$,

$$(q_0, \underline{\#}w\#) \vdash_T^* (h, \alpha, i).$$

In the definition of "T accepts w" one might demand that $i = |\alpha|+1$, i.e., that in the halting (or accepting) configuration the head is positioned at the tape square containing the blank that marks the right end of α.

If $(q_0, \underline{\#}w\#) \vdash_T^* (h, \#w'\underline{\#})$, then one might consider w' as the *output* of Turing machine T given input w.

DEFINITION 28.8.: Let Σ_0 be some alphabet not containing $\#$, $w \in \Sigma_0^*$. Let $T = \langle K,\ \Sigma,\ q_0,\ \delta\rangle$ be a Turing machine with $\Sigma_0 \subseteq \Sigma$, i.e., Σ_0 is a subset of Σ.

i) Let L be a language over the alphabet Σ_0, i.e., L is a subset of Σ_0^*.

T accepts (or *recognizes*) L := for all $w \in \Sigma_0^*$, $w \in L$ if and only if T accepts w; in other words, $L = \{w \in \Sigma_0^* \mid T \text{ accepts } w\}$.

L is *Turing-acceptable* := there is a Turing machine T such that T accepts L.

ii) Let P be a unary predicate over Σ_0^*.

T accepts (or *recognizes*) P := T accepts $\{w \in \Sigma_0^* \mid P(w)\}$, i.e., for all $w \in \Sigma_0^*$, $P(w)$ iff T halts on input w.

EXAMPLE 28.9.:
Let $\Sigma_0 = \{0,1\}$ and let $L = \{w \in \Sigma_0^* \mid w \text{ contains at least one } 0\}$. Then L is accepted by the Turing machine $T = \langle\{q_0\},\ \{0,1,\#\},\ q_0,\ \delta\rangle$, where δ is defined as follows: $\delta(q_0,\#) = (q_0,\#,R)$, $\delta(q_0,1) = (q_0,1,R)$, and $\delta(q_0,0) = (h,0,R)$.

This machine, when started in configuration $(q_0, \underline{\#}w\#)$ for some $w \in \{0,1\}^*$, simply scans right until a 0 is encountered and then halts. If no 0 is found, the machine does not halt. So, L is exactly the set of strings w in $\{0,1\}^*$ such that T halts on input w, and therefore T accepts L. Hence, L is Turing-acceptable.

DEFINITION 28.10.: Let Σ_0 be some alphabet not containing $\#$. Let $T = \langle K,\ \Sigma,\ q_0,\ \delta\rangle$ be a Turing machine with $\Sigma_0 \subseteq \Sigma$ and such that Y (Yes) and N (No) are in Σ.

i) Let L be a language over Σ_0. T *decides* L, or T is a *decision procedure* for L, iff for every $w \in \Sigma_0^*$,

if $w \in L$, then $(q_0, \underline{\#}w\#) \vdash_T^* (h, \#Y\underline{\#})$, and
if $w \notin L$, then $(q_0, \underline{\#}w\#) \vdash_T^* (h, \#N\underline{\#})$.

L is *Turing-decidable* := there is some Turing machine T such that T decides L.

ii) Let P be a unary predicate over Σ_0^*. T *decides* P (or T is a *decision procedure* for P) := T decides $\{w \in \Sigma_0^* \mid P(w)\}$, i.e., for all $w \in \Sigma_0^*$,

if $P(w)$, then $(q_0, \underline{\#}w\#) \vdash_T^* (h, \#Y\underline{\#})$, and
if not $P(w)$, then $(q_0, \underline{\#}w\#) \vdash_T^* (h, \#N\underline{\#})$.

EXAMPLE 28.11.:
Let $\Sigma_0 = \{a\}$ and let $L = \{w \in \Sigma_0^* \mid$ length of w is even$\}$. Then the following Turing machine $T = \langle K,\ \Sigma,\ q_0,\ \delta\rangle$ is a decision procedure for L: $K = \{q_0, q_1, q_2\}$, $\Sigma = \{a, Y, N, \#\}$, and δ is defined as follows:

$\delta(q_0, \#) = (q_1, \#, R)$, $\delta(q_1, a) = (q_2, \#, R)$,
$\delta(q_2, a) = (q_1, \#, R)$, $\delta(q_1, \#) = (h, Y, R)$, $\delta(q_2, \#) = (h, N, R)$.

Then, for instance,

$$\begin{aligned}(q_0, \underline{\#}a\ a\ a\#) &\vdash_T^* (q_1, \#\underline{a}\ a\ a\#)\\ &\vdash_T^* (q_2, \#\#\underline{a}\ a\#)\\ &\vdash_T^* (q_1, \#\#\#\underline{a}\#)\\ &\vdash_T^* (q_2, \#\#\#\#\underline{\#})\\ &\vdash_T^* (h, \#\#\#\#N\underline{\#})\end{aligned}$$

and

$$\begin{aligned}(q_0, \underline{\#}a\ a\#) &\vdash_T^* (q_1, \#\underline{a}\ a\#)\\ &\vdash_T^* (q_2, \#\#\underline{a}\#)\\ &\vdash_T^* (q_1, \#\#\#\underline{\#})\\ &\vdash_T^* (h, \#\#\#Y\underline{\#}).\end{aligned}$$

T erases its input string from left to right, keeping track in its finite control of the parity (odd or even) of the number of symbols it has erased. States q_1 and q_2 are used for this purpose.

Thus, if n is even, $(q_0, \underline{\#}a^n\#) \vdash_T^* (h, \#Y\underline{\#})$, and
if n is odd, $(q_0, \underline{\#}a^n\#) \vdash_T^* (h, \#N\underline{\#})$.

Therefore, T decides $\{a^n \mid n$ is even$\}$.

Note that the Turing machine in example 28.9. does not decide $L = \{w \in \{0,1\}^* \mid w$ contains at least one $0\}$ because it does not halt for input strings w not containing 0. However, for this particular language another Turing machine T' can be constructed which does decide L. See exercise 28.8.

If a Turing machine T accepts a language L over Σ_0, then T does not halt for those inputs $w \in \Sigma_0^*$ which are not in L. And if T decides L, then T halts for all inputs $w \in \Sigma_0^*$.

THEOREM 28.12.: Every Turing-decidable language is Turing-acceptable; i.e., if T decides L, then there is another Turing machine T' that accepts L.

PROOF: Let L be a language which is decided by a Turing machine T. Then L is accepted by the Turing machine T', which operates as follows.

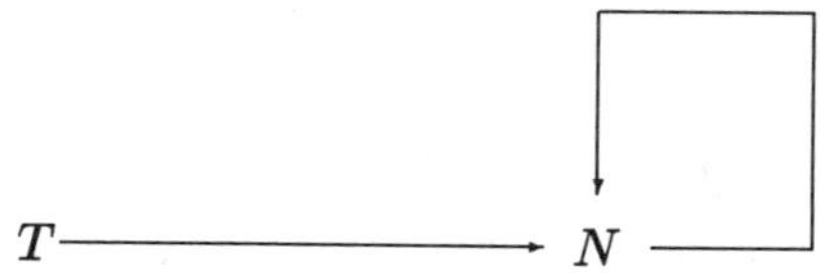

T' simulates T, i.e., T' starts in the initial state of T and operates as T would operate until T halts. When T halts, T' goes into an infinite loop if T leaves the string N on its tape; if T leaves the string Y on its tape, then T' halts. So $w \in L$ if and only if T' halts on input w. Therefore, T' accepts L. (See exercise 28.9.) □

Later we shall show that the converse of Theorem 28.12. does not hold, i.e., that there are Turing-acceptable languages which are not Turing-decidable; see Theorems 29.1. and 29.5.

In particular, we shall see that the set of valid formulas of the *predicate* calculus, although Turing-acceptable, is not Turing-decidable; see Theorem 29.12. and Theorem 29.14. On the other hand, the set of valid formulas of the (classical) *propositional* calculus is Turing-decidable: one can construct a Turing machine T such that

if A is a valid formula, then $(q_0, \#\underline{A}\#) \vdash_T^* (h, \#Y\underline{\#})$, and
if A is a non-valid formula, then $(q_0, \#\underline{A}\#) \vdash_T^* (h, \#N\underline{\#})$.

This Turing machine T simply computes the truth table of A (see § 2) and prints Y(es) if it contains only 1's and N(o) if it contains at least one 0.

In section 22 we presented a procedure of searching for a formal proof of a given formula A of the *predicate* calculus. If there is such a formal proof, our procedure will find one; but if there is no such formal proof, the procedure described in section 22 does not necessarily come to an end. So, we can construct a Turing machine T which accepts the set of all valid formulas of the predicate calculus, but which does not decide this set, since this machine T does not halt on all inputs. In Theorem 29.12. we shall show that no (other) Turing machine can exist which decides the set of valid (or, equivalently, provable) formulas of the classical predicate calculus, in other words, that the classical predicate calculus is undecidable.

THEOREM 28.13.: Let Σ_0 be an alphabet and $L \subseteq \Sigma_0^*$. If L is a Turing-decidable language, then its *complement* $\bar{L}$, i.e., $\{w \in \Sigma_0^* \mid w \notin L\}$, is also Turing-decidable.

PROOF: Suppose Turing machine T decides L. Then $\bar{L}$ is decided by the Turing machine T' illustrated below.

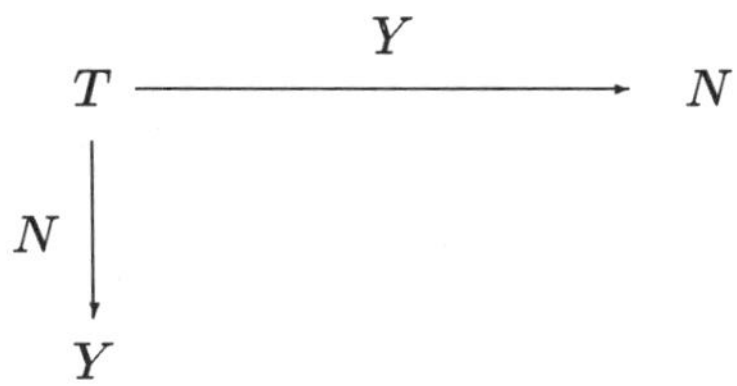

This machine simulates T until it halts, then interchanges the answers Y and N. (See exercise 28.10.) □

Later on we shall show that the analogue of Theorem 28.13. for Turing-acceptable languages does not hold, i.e., there are Turing-acceptable languages, the complement of which is not Turing-acceptable; see Theorem 29.3. and Exercise 29.3.

THEOREM 28.14.: A language is Turing-decidable if and only if both it and its complement are Turing-acceptable.

PROOF: If L is Turing-decidable, then L is Turing-acceptable by Theorem 28.12.; also, $\bar{L}$ is Turing-decidable (Theorem 28.13.), and hence Turing-acceptable.

Conversely, suppose that L is accepted by T_1 and that $\bar{L}$ is accepted by T_2. We can then construct a Turing machine T that decides L as follows. The machine T begins by making two copies of its input string w. Next we let T simulate both T_1 and T_2 in parallel: at each step of the operation of T, one step of T_1's computation is carried out on the first copy of w, and one step of T_2's computation is carried out on the second copy of w. Now, either T_1 or T_2 will accept w, but not both. If $w \in L$ the simulation of T_1 will halt. In this case we let T halt with output Y(es). And if $w \notin L$, the simulation of T_2 will halt. In that case we let T halt with output N(o). □

Instead of "T accepts L", one may also say "T *recognizes* L". So we have defined a Turing machine as a *recognizing device*. However, as mentioned earlier, we can consider a Turing machine as a *procedure*. For example, if we wish to define a procedure for determining whether a number is a prime, we could construct a Turing machine which accepts the set of all primes. (See exercise 28.1.) Note that, in this case, whether the Turing machine is thought of as a recognizer or as a procedure is merely a matter of preference. When we consider a Turing machine as a procedure, if the Turing machine halts for all inputs, then we say that the procedure is an *algorithm*. There are procedures for which there is no corresponding algorithm. An example is the procedure for determining whether there is a perfect number greater than i (see exercise 28.2.); the procedure in section 22 of searching for a formal proof of a given formula A in the predicate calculus is another example. We show in Theorem 29.12. that there is no Turing machine that halts for every input formula A and then tells whether A is valid (or, equivalently, formally provable) or not. Then it follows from Theorem 28.14. that the set of non-valid formulas is not Turing-acceptable.

Working independently, both A. Church (1936) and A. Turing (1936-7) formulated the hypothesis that any process that could naturally be called a procedure can be realized by a Turing machine. This hypothesis is called the *Church-Turing Thesis.* Since this hypothesis identifies a somewhat vague intuitive concept with a mathematically defined concept, it cannot be proved in mathematical terms. Later on in this section some arguments will be given which make the Church-Turing Thesis quite plausible.

TURING-COMPUTABLE FUNCTIONS

In addition to considering a Turing machine as a recognizing device or as a procedure, we can consider a Turing machine as a device to compute a function. Below we define Turing-computability both for (total) functions and for partial functions.

DEFINITION 28.15.

i) Let Σ_0 and Σ_1 be alphabets not containing the blank symbol $\#$. Let τ be a (partial) function from Σ_0^* to Σ_1^*. A Turing machine $T = \langle K,\ \Sigma,\ q_0,\ \delta\rangle$ is said to *compute* τ if $\Sigma_0, \Sigma_1 \subseteq \Sigma$ and for any $w \in \Sigma_0^*$, $\tau(w) = w'$ iff $(q_0, \underline{\#}w\#) \vdash_T^* (h, \#w'\underline{\#})$.

If some such Turing machine T exists, then τ is said to be a *Turing-computable function.*

ii) Let f be a (partial) function from $\mathbb{N}$ to $\mathbb{N}$.
And let $T = \langle K,\ \Sigma,\ q_0,\ \delta\rangle$ be a Turing machine such that Σ contains the symbol 1. T *computes* f := for each $n \in \mathbb{N}$,
$$f(n) = m \text{ iff } (q_0, \underline{\#}1^n\#) \vdash_T^* (h, \#1^m\underline{\#}),$$
where $1^n = 1\ldots1$ (n times). f is *Turing-computable* := there is a Turing machine that computes f.

iii) More generally, let f be a (partial) function from $\mathbb{N}^k$ to $\mathbb{N}$, $k = 1, 2, \ldots$, and let T be a Turing machine.
T *computes* f := for all $n_1, \ldots, n_k \in \mathbb{N}$, $f(n_1, \ldots, n_k) = m$ iff
$$(q_0, \underline{\#}1^{n_1}\#1^{n_2}\#\ldots\#1^{n_k}\#) \vdash_T^* (h, \#1^m\underline{\#}).$$

EXAMPLE 28.16.: The successor function $s : \mathbb{N} \to \mathbb{N}$, defined by $s(n) = n+1$ for each $n \in \mathbb{N}$, is Turing-computable; s is computed by the Turing machine $T = \langle\{q_0, q_1\},\ \{\#, 1\},\ q_0,\ \delta\rangle$ with
$$\delta(q_0, \#) = (q_1, \#, R),$$
$$\delta(q_1, 1) = (q_1, 1, R),$$
$$\delta(q_1, \#) = (h, 1, R).$$
For instance, $(q_0, \underline{\#}11\#) \vdash_T^* (h, \#111\underline{\#})$.

When f is a partial function, say from $\mathbb{N}$ to $\mathbb{N}$, f is Turing-computable if and only if there is some Turing machine T such that

i) if $f(n)$ is defined, then T halts on input $1\ldots1$ (n times) leaving $1\ldots1$ ($f(n)$ times) on the tape as output, and
ii) if $f(n)$ is not defined, then T does not halt on input $1\ldots1$ (n times).

In Definition 28.15. the head in the halted configuration is required to be immediately to the right of the output, being the function value for the given input. Other conventions are possible. If T computes f under one convention, then given any other convention, one can construct a Turing machine T' which computes f under that other convention. For instance, one might agree that in the halted configuration the head is positioned immediately to the left end of the output. This is convenient if one wants to combine Turing machines, the second taking the output of the first as input, since Turing machines start (by convention) with the head positioned immediately to the left end of the input.

DEFINITION 28.17.: *A* function f is called an *initial function* if it is one of the following (for some $k \in \mathbb{N}$).

1. The *zero function* $z : \mathbb{N}^k \to \mathbb{N}$, defined by $z(n_1, \ldots, n_k) = 0$.
2. The *projection function* $p_i : \mathbb{N}^k \to \mathbb{N}$, defined by $p_i(n_1, \ldots, n_k) = n_i$.
3. The *successor function* $s : \mathbb{N} \to \mathbb{N}$, defined by $s(n) = n + 1$.

Next we consider the following operations on functions.

4. $f : \mathbb{N}^k \to \mathbb{N}$ results from $g : \mathbb{N}^m \to \mathbb{N}$ and $h_i : \mathbb{N}^k \to \mathbb{N}$ $(1 \leq i \leq m)$ by *composition* or *substitution* if
$$f(n_1, \ldots, n_k) = g(h_1(n_1, \ldots, n_k), \ldots, h_m(n_1, \ldots, n_k)).$$
5. $f : \mathbb{N}^{k+1} \to \mathbb{N}$ results from $g : \mathbb{N}^k \to \mathbb{N}$ and $h : \mathbb{N}^{k+2} \to \mathbb{N}$ by *primitive recursion* if
$$f(0, n_1, \ldots, n_k) = g(n_1, \ldots, n_k), \text{ and}$$
$$f(n+1,\ n_1, \ldots, n_k) = h(n,\ f(n, n_1, \ldots, n_k),\ n_1, \ldots, n_k).$$
6. $f : \mathbb{N}^k \to \mathbb{N}$ results from $g : \mathbb{N}^{k+1} \to \mathbb{N}$ by *minimization* if $f(n_1, \ldots, n_k)$ is the minimal n such that $g(n,\ n_1, \ldots, n_k) = 0$ if there is such n; otherwise $f(n_1, \ldots, n_k)$ is undefined.

DEFINITION 28.18.: Let $f : \mathbb{N}^k \to \mathbb{N}$.

f is *primitive recursive* := f can be obtained by finitely many applications of composition and primitive recursion to the initial functions.

f is *(general) recursive* := f can be obtained from the initial functions by means of composition, primitive recursion and application of minimization to regular functions, where $g : \mathbb{N}^{k+1} \to \mathbb{N}$ is *regular* iff for all $n_1, \ldots, n_k \in \mathbb{N}$ there is $n \in \mathbb{N}$ such that $g(n,\ n_1, \ldots, n_k) = 0$.

If we also allow minimization to be applied to non-regular functions, then we obtain *partial recursive* functions. (See Exercise 28.22.)

For instance, addition and multiplication of the natural numbers are (primitive) recursive:

$$+(0,m) = m \quad , \qquad *(0,m) = 0 \quad ,$$
$$+(n+1,\ m) = s(+(n,m)) \quad , \qquad *(n+1,\ m) = +(*(n,m),m) \ .$$

It turns out that most commonly occurring computable (total) functions are primitive recursive. However, the following function, called the *Ackermann function*, can be shown to be (general) recursive but not primitive recursive:

$$\begin{aligned} \alpha(0,m) &= m+1, \\ \alpha(n+1,0) &= \alpha(n,1), \\ \alpha(n+1,\ m+1) &= \alpha(n,\ \alpha(n+1,m)). \end{aligned}$$

For more details on the Ackermann function see, for instance, N. Cutland, *Computability*, Example 5.5.

CHURCH-TURING THESIS

At about the same time, 1936, several persons, mostly independently of one another, tried to characterize the intuitive notion of effectively calculable function by means of different mathematical concepts. Here we mention only A. Church's notion of *lambda-definability* (see section 32, in particular Theorems 32.52 and 32.55), the Herbrand-Gödel-Kleene notion of general *recursiveness* (Definition 28.18.) and Turing's notion of *computability* (Definition 28.15.). All these notions, although quite different in formulation, have been proved equivalent. That is,

f is lambda definable iff f is general recursive (Church and Kleene, 1936),

and

f is Turing computable iff f is lambda definable (Turing, 1937).

The fact that quite different mathematical characterizations of the intuitive concept of effective calculability are equivalent strengthens our belief in the *Church-Turing Thesis* which says that we have captured the intuitive notion in precise and adequate mathematical terms.

In our definition of a Turing machine (Definition 28.1.) there was one two-way infinite tape. As becomes clear from the examples, this tape is used in three different ways: 1) to give the machine its input, 2) to serve as working memory and 3) to present the output. So, it seems attractive to consider Turing machines with one read-only input tape, one or more memory tapes to work on and one write-only output tape. Such a machine would look as follows.

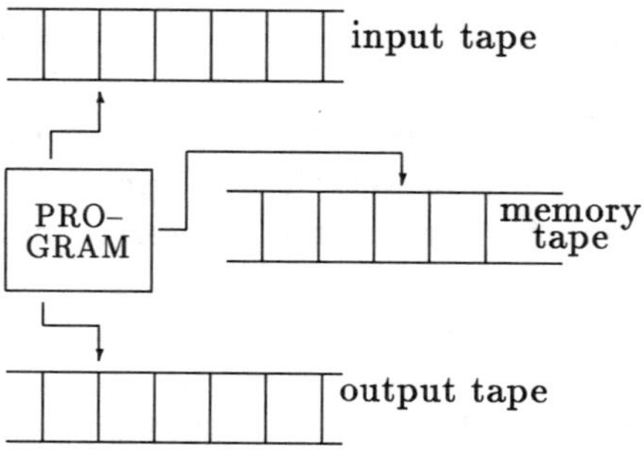

One can think of many other varieties of Turing machines. One might consider machines
- which can move any number of squares to the left or right, or
- which operate with several tapes instead of just one, or
- which operate on a two-dimensional tape, or
- which have two or more heads.

Although at first sight such machines may seem more powerful than our standard machine (Definition 28.1.), it turns out that they yield the same class of acceptable languages and/or the same class of computable functions. Any computation that can be performed by such a machine T can also be carried out by a standard Turing machine T' (as defined in Definition 28.1.), allowing one step in the computation of T to be simulated by one or more steps in the computation of T'. (See, for instance, H.R. Lewis & C.H. Papadimitriou, *Elements of the Theory of Computation*, section 4.5.)

The fact that all kinds of theoretical computing devices – which at first sight seem to possess greater computational power than the standard Turing machines (as defined in Definition 28.1.) – turn out to compute the same functions as the standard Turing machines is another argument in favour of the *Church-Turing Thesis*. It strengthens our belief that we have captured the intuitive notion of effective calculability in precise and adequate mathematical terms.

It can be shown that the machine language for any standard computer can be translated into Turing machine language. Since programming languages like PROLOG, PASCAL, LISP, etc., can be compiled into machine language, it follows that every Pascal program can be translated into an equivalent Turing machine.

It is easy to show that the initial functions (Definition 28.17.) are Turing-computable (1); see exercise 28.5., 28.6. and 28.7. It is more difficult to show that if f is obtained by composition, primitive recursion or minimalisation from Turing-computable functions, then f is also Turing-computable (2). From (1) and (2) it follows that any recursive function is Turing computable. Since the converse also holds, it follows that

f is (general) recursive iff f is Turing-computable.

This explains why Turing-computable functions are also called *recursive*.

The following theorem explains

i) why Turing-decidable languages are also called *recursive*, and

ii) why Turing-acceptable languages are also called *recursively enumerable.*

THEOREM 28.19.: Let Σ_0 be an alphabet and L be a language over Σ_0.

i) L is Turing-decidable iff its characteristic function K_L is Turing-computable.

ii) L is Turing-acceptable iff there is a Turing-computable function f such that f enumerates L, i.e., $L = \{f(0), f(1), f(2), \ldots\}$.

PROOF:

i) Remember that $K_L(w) = 1$ if $w \in L$ and $K_L(w) = 0$ if $w \notin L$. Suppose Turing machine T decides L. Let T' simulate T except that it replaces Y by 1 and N by 0. Then T' computes K_L.

Conversely, suppose T_1 computes K_L. Let T_2 simulate T_1 except that it replaces 1 by Y and 0 by N. Then T_2 decides L.

ii) Suppose f enumerates $L = \{f(0), f(1), \ldots\}$ and T computes f. Let T' determine for any $w \in \Sigma_0^*$ the natural number n, if there is any, such that $w = f(n)$ and next let T' simulate T on input n. If there is no natural number n such that $w = f(n)$, then we let T' go into an infinite loop. Clearly, T' accepts L.

Conversely, suppose T_1 accepts L. Note that Σ_0^* is enumerable, say $\Sigma_0^* = \{w_0, w_1, w_2, \ldots\}$. Design Turing machine T_2 as follows. Given input $n \in \mathbb{N}$, T_2 carries out

- one step of the computation of T_1 on w_0,

- two steps of the computation of T_1 on w_0 and w_1,
- three steps of the computation of T_1 on w_0, w_1 and w_2,

and so on. If T_2, given input n, discovers the nth w that is accepted by T_1, then it leaves that w as output and halts. Clearly, T_2 computes a function f that enumerates L. □

EXERCISES

28.1. i) Give an algorithm for determining if an integer greater than one is prime. ii) Give a procedure which determines, for a given prime number p_n, whether there is a prime number greater than p_n. Is the procedure an algorithm or not?

28.2. A *perfect number* is an integer which is equal to the sum of all its divisors except itself. Thus 6 is a perfect number since $1+2+3 = 6$. 12 is not perfect since its divisors are 1, 2, 3, 4, and 6, which sum to 16. The question of whether or not there is an infinity of perfect numbers is an unsolved problem in number theory. Give a procedure which determines, for an integer i, whether there is a perfect number greater than i. Is the procedure an algorithm ?

28.3. Is the following function computable ?

$$g(n) = \begin{cases} 1 & \text{if there is a run of exactly } n \text{ consecutive 7's} \\ & \text{in the decimal expansion of } \pi, \\ 0 & \text{otherwise.} \end{cases}$$

28.4. Design a Turing machine T such that for each natural number n,
i) if n is even, then $(q_0, \#\underbrace{1\ldots1}_{n}\underline{\#}) \vdash_T^* (h, \#\underline{\#})$, and
ii) if n is odd, then $(q_0, \#\underbrace{1\ldots1}_{n}\underline{\#}) \vdash_T^* (h, \#1\underline{\#})$.

28.5. Design a Turing machine that computes the zero function $z : \mathbb{N} \rightarrow \mathbb{N}$, defined by $z(n) = 0$ for all $n \in \mathbb{N}$. More precisely, design a Turing machine T such that for all $n \in \mathbb{N}$,

$$(q_0, \#\underbrace{1\ldots1}_{n}\underline{\#}) \vdash_T^* (h, \#\underline{\#}).$$

28.6. Design a Turing machine that computes the projection function $p_2 : \mathbb{N}^3 \to \mathbb{N}$, defined by $p_2(n_1, n_2, n_3) = n_2$. More precisely, design a Turing machine T such that

$$(q_0, \underline{\#}\underbrace{1\ldots1}_{n_1}\#\underbrace{1\ldots1}_{n_2}\#\underbrace{1\ldots1}_{n_3}\#) \vdash_T^* (h, \#\underbrace{1\ldots1}_{n_2}\underline{\#}).$$

28.7. Design a Turing machine that computes the successor function on $\mathbb{N}$, i.e., a Turing machine T such that for all $n \in \mathbb{N}$,

$$(q_0, \underline{\#}\underbrace{1\ldots1}_{n}\#) \vdash_T^* (h, \#\underbrace{1\ldots11}_{n+1}\underline{\#}).$$

28.8. Construct a Turing machine which decides

$$\{w \in \{0,1\}^* \mid w \text{ contains at least one } 0\}.$$

28.9. Following the proof of theorem 28.12., construct a Turing machine T' which accepts $L = \{a^n \mid n \text{ is even}\}$ (Example 28.11.)

28.10. Following the proof of Theorem 28.13., construct a Turing machine T' which decides the complement of the language $L = \{a^n \mid n \text{ is even}\}$ of Example 28.11.

28.11. What function from $\mathbb{N}$ to $\mathbb{N}$ is computed by the Turing machine $T = \langle \{q_0, q_1, q_2, q_3\},\ \{1, \#\},\ q_0,\ \delta \rangle$, where δ is defined as follows:
$\delta(q_0, \#) = (q_1, \#, R)$,
$\delta(q_1, \#) = (h, \#, O)$, $\delta(q_2, \#) = (h, \#, O)$, $\delta(q_3, \#) = (h, \#, O)$,
$\delta(q_1, 1) = (q_2, \#, R)$, $\delta(q_2, 1) = (q_3, \#, R)$, $\delta(q_3, 1) = (q_3, 1, R)$.

28.12. Design a Turing machine which computes the function $f : \mathbb{N}^2 \to \mathbb{N}$, defined by $f(n, m) := n + m$. In other words, show that addition in $\mathbb{N}$ is Turing-computable.

28.13. i) Construct a Turing machine which accepts the one-element language $\{0\ 1\ 0\}$.
ii) Construct a Turing machine which accepts the one-element language $\{1\ 0\ 1\}$.
iii) From (i) and (ii) construct a Turing machine which accepts $\{0\ 1\ 0,\ 1\ 0\ 1\}$, i.e., $\{0\ 1\ 0\} \cup \{1\ 0\ 1\}$.
iv) Make clear that any finite language is Turing acceptable.

28.14. One can represent each natural number n by a finite sequence $a_1, \ldots, a_{k+1}$ of zeros and ones, by writing n as
$$a_1 \cdot 2^k + a_2 \cdot 2^{k-1} + \ldots + a_k \cdot 2^1 + a_{k+1} \cdot 2^0.$$
Construct a Turing machine which decides
$$L = \{w \in \{0,1\}^* \mid w \text{ represents an even natural number}\}.$$

28.15. Construct a Turing machine which computes the proper substraction function $f : \mathbb{N}^2 \to \mathbb{N}$ defined by $f(n_1, n_2) = n_1 - n_2$ if $n_1 > n_2$, and $f(n_1, n_2) = 0$ if $n_1 \leq n_2$.

28.16. Construct a Turing machine which computes the function $f : \mathbb{N} \to \mathbb{N}$, defined by $f(n) = 2n$ for all $n \in \mathbb{N}$.

28.17. Construct a Turing machine which accepts $L = \{a^n b^n \mid n \geq 0\}$.

28.18. Construct a Turing machine which accepts
$$L = \{a^n b^m \mid n \geq 0,\ m \geq 0\}.$$

28.19. Construct a Turing machine which accepts
$$L = \{w \in \{a, b\}^* \mid \text{ the string } a\ a \text{ does not occur in } w\}.$$

28.20. Construct a Turing machine which accepts $L = \{(a\ b)^n \mid n \geq 0\}$.

28.21. Construct a Turing machine which accepts
$$L = \{w \in \{a, b\}^* \mid \text{ the string } a \text{ occurs in } w\}.$$

28.22. (R. Sommerhalder) If we also consider partial recursive functions, computability of a function f, obtained from g by minimization, is only guaranteed if g is total (i.e., always defined); regularity of g does not suffice. In order to see this, let $f_0, f_1, \ldots$ be an enumeration of the partial recursive functions. And let $g(0, y) = 0$ if $f_y(y)$ has been defined, $g(x, y) = 0$ for all $x > 0$. Thus, $g(x, y)$ is undefined iff $x = 0$ and $f_y(y)$ is undefined. Then g is (partially) computable and regular, but not total. For f, obtained from g by minimization, we have $f(y) = 0$ if $f_y(y)$ has been defined; otherwise, $f(y) = 1$. So, f is not computable.

§29 UNDECIDABILITY:

A UNIVERSAL TURING MACHINE; THE HALTING PROBLEM; PREDICATE LOGIC IS UNDECIDABLE; NONDETERMINISTIC TURING MACHINES

While any particular Turing machine corresponds to a particular computer program, the universal Turing machine is such that it can run any program. Although the universal Turing machine accepts $H_0 := \{\rho(T)\rho(w) \mid \text{Turing machine } T \text{ halts on input } w\}$, there can be no Turing machine that decides this set. Accepting the Church-Turing Thesis, this result says that the Halting problem – i.e., the problem to determine whether any given Turing machine, given any input w, will halt – is unsolvable. By reducing this Halting problem to the validity problem in classical predicate logic, we conclude that classical predicate logic is undecidable too. We also introduce nondeterministic Turing machines. Their usefulness in the context of complexity problems will become clear in the next section.

A UNIVERSAL TURING MACHINE

Let $K_{00} := \{q_0, q_1, q_2, \ldots\}$ and $\Sigma_{00} := \{s_0, s_1, s_2, \ldots\}$. Without loss of generality, we can assume for every Turing machine $T = \langle K,\ \Sigma,\ q_0,\ \delta\rangle$ that the state set K and the tape alphabet Σ of T are finite subsets of K_{00} and Σ_{00} respectively. As a consequence of this convention, all component symbols of any Turing machine are symbols from the list $L,\ R,\ O,\ h,\ s_0,\ q_0,\ s_1,\ q_1, \ldots$. In order to encode Turing machines as strings over the alphabet $\{c, 1\}$, we define the function $\rho : \{L,\ R,\ O,\ h,\ s_0,\ q_0,\ s_1,\ q_1, \ldots\} \to \{1\}^*$ as follows.

$$\begin{array}{ll} \rho(L) = 1 & ,\ \rho(h) = 1 \quad , \\ \rho(R) = 1\ 1 & ,\ \rho(q_i) = 1^{i+2} \quad . \\ \rho(O) = 1\ 1\ 1 & , \\ \rho(s_i) = 1^{i+4} & , \end{array}$$

Below we will indicate how one can extend ρ to an encoding of Turing machines over the two-symbol alphabet $\{c, 1\}$ such that

1. if $T_1 \neq T_2$, then $\rho(T_1) \neq \rho(T_2)$; and
2. for every string over $\{c, 1\}$ one can determine whether it is the encoding of a Turing machine T and if so, one can easily reconstruct T from it.

Encoding of instruction $\delta(p,a) = (q,b,m)$:

$$c\ \rho(p)\ c\ \rho(a)\ c\ \rho(q)\ c\ \rho(b)\ c\ \rho(m)\ c.$$

Now let T be a Turing machine and let $S_1, \ldots, S_k$ be the encodings of the instructions $\delta(p,a) = (q,b,m)$ in T; and let S_0 be the encoding of the initial state of T. Then we take

$$c\ S_0\ c\ S_1\ \ldots\ S_k\ c$$

as the encoding $\rho(T)$ of Turing machine T.

EXAMPLE: Let $T = \langle\{q_0\}, \{s_0, s_2, s_5\},\ q_0,\ \delta\rangle$ be a Turing machine with
$\delta(q_0, s_0) = (h, s_2, L)$,
$\delta(q_0, s_2) = \delta(q_0, s_5) = (q_0, s_2, R)$.
Encoding of $\delta(q_0, s_0) = (h, s_2, L)\ :\ c\ 1^2\ c\ 1^4\ c\ 1\ c\ 1^6\ c\ 1\ c.$
Encoding of $\delta(q_0, s_2) = (q_0, s_2, R) :\ c\ 1^2\ c\ 1^6\ c\ 1^2\ c\ 1^6\ c\ 1^2\ c.$
Encoding of $\delta(q_0, s_5) = (q_0, s_2, R) :\ c\ 1^2\ c\ 1^9\ c\ 1^2\ c\ 1^6\ c\ 1^2\ c.$
Consequently, the encoding $\rho(T)$ of T is
$c\,1^2\,c\,c\,1^2\,c\,1^4\,c\,1\,c\,1^6\,c\,1\,c\,c\,1^2\,c\,1^6\,c\,1^2\,c\,1^6\,c\,1^2\,c\,c\,1^2\,c\,1^9\,c\,1^2\,c\,1^6\,c\,1^2\,c\,c.$

Once we have encoded Turing machines as strings over the alphabet $\{c, 1\}$, it is not difficult to see that Turing machines can be encoded by natural numbers too: any string over $\{c, 1\}$ can be encoded by a natural number; for instance, the string $1\ c\ 1\ 1\ c\ 1\ c$ can be encoded by $2^1 \cdot 3^0 \cdot 5^1 \cdot 7^1 \cdot 11^0 \cdot 13^1 \cdot 17^0$. Note that 2, 3, 5, 7, 11, 13 and 17 are prime and that there are infinitely many prime numbers. Consequently, there cannot be more Turing machines than there are natural numbers.

Again, our encoding ρ' of Turing machines as natural numbers is such that

(i) if $T_1 \neq T_2$, then $\rho'(T_1) \neq \rho'(T_2)$; and
(ii) for every natural number one can determine whether it is the encoding of a Turing machine T, and if so, one can easily reconstruct T from it.

In order to discuss a *universal Turing machine*, we need not only an encoding of Turing machines but also an encoding of input strings w. So, let $w = b_1 \ldots b_n$ be a string over the alphabet $\Sigma_{00} = \{s_0, s_1, s_2 \ldots\}$. Then $\rho(w) := c\ \rho(b_1)\ c\ \rho(b_2)\ c\ \ldots\ c\ \rho(b_n)\ c$ is the encoding of w as a string over the alphabet $\{c, 1\}$. Note that $\rho(w)$ contains no blanks, even if w does.

Since any Turing machine is describable by a finite string of letters of the infinite alphabet

$$L,\ R,\ O,\ h,\ s_0,\ q_0,\ s_1,\ q_1,\ \ldots$$

and the set of all such strings is enumerable (see exercise 16.7.), the set of all Turing machines is enumerable.

Let $T_0, T_1, T_2, \ldots$ be an enumeration of all Turing machines. This list of Turing machines yields a list of all Turing-computable functions, say

$$f_{n_0}, f_{n_1}, f_{n_2}, \ldots.$$

On the other hand, it is known from set theory that the set of all functions from positive integers to positive integers is not enumerable (see Theorem 16.7.). So there must be many functions $f : \mathbb{N} \to \mathbb{N}$ which are not Turing-computable. Can we exhibit any particular function $f : \mathbb{N} \to \mathbb{N}$ which is not Turing-computable?

Let $g_0, g_1, g_2, \ldots$ be an enumeration of all Turing-computable functions and let $g : \mathbb{N} \to \mathbb{N}$ be defined by $g(n) := g_n(n) + 1$. Suppose g were Turing-computable. Then $g = g_k$ for some natural number k and hence, for every natural number n, $g(n) = g_n(n) + 1 = g_k(n)$. So, in particular for $n = k$, $g_k(k) + 1 = g_k(k)$. Contradiction. Therefore, the function $g : \mathbb{N} \to \mathbb{N}$, defined by $g(n) := g_n(n) + 1$, is *not Turing-computable.*

Now it may seem that g is intuitively computable, i.e., that we can actually compute $g(n)$ for any positive integer n: determine g_n and add 1 to $g_n(n)$. However, the question is whether one can determine the enumeration $f_{n_0}, f_{n_1}, f_{n_2}, \ldots$ of Turing-computable functions from the enumeration $T_0, T_1, T_2, \ldots$ of all Turing machines. In other words, the crucial problem is to specify a uniform routine which, given any Turing machine and given any input of an unbroken string of n 1's on an otherwise blank tape, determines after some finite number of steps whether the machine would eventually halt (with an unbroken string of 1's on an otherwise blank tape).

If there were such a routine, then there would be computable functions which are not Turing computable, and in particular, g would be such a function. In that case, Turing's notion of computability would not capture the intuitive notion. On the other hand, if we accept the Church-Turing Thesis, it follows that the problem mentioned above, called the *halting problem*, is unsolvable.

Further on in this section we shall show that

$H_0 := \{\rho(T)\rho(w) \mid \text{Turing machine } T \text{ halts on (accepts) input string } w\}$ is not Turing-decidable, i.e., there is no Turing machine that decides, for an arbitrary Turing machine T and input string w whether or not T halts on (accepts) input w.

Now the *halting problem* for Turing machines is the problem to determine, for arbitrary Turing machine T and input w whether T will eventually halt on input w. If we accept the Church-Turing Thesis, the result mentioned above says that *the halting problem is unsolvable* i.e., there is no algorithm that decides for an arbitrary given Turing machine T and input string w whether or not T accepts w.

Since there are only countably many Turing machines, there are accordingly only countably many Turing-decidable and Turing-acceptable languages over any given alphabet. On the other hand, since a language over an alphabet Σ is any subset of Σ^*, there are undenumerably many languages over a given alphabet Σ (see exercise 16.8.). So Turing machines accept or decide only an infinitesimal fraction of all the possible languages. Nevertheless, we have yet to exhibit any particular language that is not Turing-decidable or not Turing-acceptable. Further on in this section we shall show that
1. H_0 (defined above) is Turing acceptable, but not Turing-decidable; and
2. $H_1 := \{\rho(T) \mid \text{Turing machine } T \text{ accepts input string } \rho(T)\}$ is Turing-acceptable, while $\bar{H}_1$ (the complement of H_1) is not Turing-acceptable.

Up till now we have designed a particular Turing machine for each problem. This is not in conformity with daily practice, where we have one computer for many purposes. However, even Turing machines are in essence programmable. One can construct a Turing machine T_U which, when presented with an encoding of an arbitrary Turing machine T and an encoding of an input string w, will simulate the behaviour of T when T is presented with w. We call such a Turing machine a *universal Turing machine.* One can think of a universal Turing machine as a general-purpose computer which is powerful enough to simulate any computer, including itself.

By a *universal Turing machine* T_U we mean a Turing machine such that for any Turing machine $T = \langle K, \Sigma, q_0, \delta\rangle$ and for any strings $w, w' \in \Sigma^*$

1. T_U halts on input $\rho(T)\rho(w)$ if and only if T halts on input w, and
2. $(q_0, \underline{\#}\rho(T)\rho(w)\#) \vdash_{T_U}^* (h, \#\rho(w')\underline{\#})$ if and only if $(q_0, \underline{\#}w\#) \vdash_T^* (h, \#w'\underline{\#})$, i.e., the output of T_U given input $\rho(T)\rho(w)$ is the encoding of the output w' of T given input w.

THEOREM 29.0.: One can construct a universal Turing machine T_U.

PROOF: We only give a description of a universal Turing machine without giving the cumbersome technical details.

Given any input string over the alphabet $\{c, 1\}$, T_U starts to analyse whether this string is of the form $\rho(T)\rho(w)$ for some Turing machine T and input string w. If there are such T and w, T_U can determine where $\rho(T)$ ends and $\rho(w)$ begins because this is the only place in $\rho(T)\rho(w)$ where three consecutive c's occur. If the input of T_U is not of the form $\rho(T)\rho(w)$, the behaviour of T_U is unimportant. In the event that the input of T_U is of the form $\rho(T)\rho(w)$ for some Turing machine T and input string w, T_U determines (computes) the initial state of T, the instructions of T and input w of T. Next, T_U simulates the steps of the computation of T in terms of the encodings of the states, instructions and symbols T is working with. □

THE HALTING PROBLEM

THEOREM 29.1.: Let $H_0 := \{\rho(T)\rho(w) \mid \text{Turing machine } T \text{ halts on input } w\}$. H_0 is Turing-acceptable.

PROOF: H_0 is accepted by a universal Turing machine T_U. For T accepts w iff T_U accepts $\rho(T)\rho(w)$, i.e., $\rho(T)\rho(w) \in H_0$ iff T_U accepts $\rho(T)\rho(w)$. □

LEMMA 29.2.: If H_0 were Turing-decidable, then

$$H_1 := \{\rho(T) \mid \text{Turing machine } T \text{ accepts input string } \rho(T)\}$$

would also be Turing-decidable.

PROOF: Suppose T_0 decides H_0, i.e.,
(1) if $\rho(T)\rho(w) \in H_0$, then $(q_0, \underline{\#}\rho(T)\rho(w)\#) \vdash_{T_0}^* (h, \#Y\underline{\#})$, and
(2) if $\rho(T)\rho(w) \notin H_0$, then $(q_0, \underline{\#}\rho(T)\rho(w)\#) \vdash_{T_0}^* (h, \#N\underline{\#})$.
Construct Turing machine T_1 as follows: T_1 transforms its input string

$\underline{\#}w\#$ into $\underline{\#}w\rho(w)\#$ and then yields control to T_0. Then
(3) if $\rho(T) \in H_1$, then $\rho(T)\rho(\rho(T)) \in H_0$ and hence

$$(q_0, \underline{\#}\rho(T)\#) \vdash^*_{T_1} (h, \#Y\underline{\#}), \text{ and}$$

(4) if $\rho(T) \notin H_1$, then $\rho(T)\rho(\rho(T)) \notin H_0$ and hence

$$(q_0, \underline{\#}\rho(T)\#) \vdash^*_{T_1} (h, \#N\underline{\#}).$$

So, T_1 decides H_1. □

THEOREM 29.3.: $\bar{H}_1$, the complement of H_1, is not Turing-acceptable, and hence, by Theorem 28.12., not Turing-decidable.

PROOF: Suppose T^* were a Turing machine that accepted $\bar{H}_1$. Then for any Turing machine T,

$$\rho(T) \in \bar{H}_1 \text{ iff } T^* \text{ accepts } \rho(T), \text{ i.e.,}$$

$$T \text{ does not accept } \rho(T) \text{ iff } T^* \text{ accepts } \rho(T).$$

In particular for $T = T^*$,

$$T^* \text{ does not accept } \rho(T^*) \text{ iff } T^* \text{ accepts } \rho(T^*).$$

Contradiction. □

From Theorem 29.3. and Theorem 28.13. follows
COROLLARY 29.4.: H_1 is not Turing-decidable.

And from this Corollary and Lemma 29.2. follows
THEOREM 29.5.: H_0 is not Turing-decidable, i.e., *there is no Turing machine that decides, for an arbitrary Turing machine T and input string w, whether or not T accepts (halts on) w.*

If one accepts the Church-Turing Thesis, this says that the *Halting Problem is unsolvable*, i.e., there is no algorithm to determine, for arbitrary Turing machine T and input w, whether T will eventually halt on input w.

In Theorem 28.12. we have shown that every Turing-decidable language is also Turing-acceptable, in other words, that every recursive language is also recursively enumerable. We have just seen that H_0 is Turing-acceptable (Theorem 29.1.), but not Turing-decidable (Theorem 29.5.), so the converse of Theorem 28.12. does not hold.

Theorem 28.13. says that if L is Turing-decidable (recursive), then $\bar{L}$, the complement of L, is also Turing-decidable. The analogue of this theorem with "Turing-acceptable (recursively enumerable)" instead of

"Turing-decidable (recursive)" does not hold: H_1 is Turing-acceptable (see exercise 29.3.), while $\bar{H}_1$ is not Turing-acceptable (Theorem 29.3.).

LEMMA 29.6.: If the Turing-acceptable language H_0 were Turing-decidable, then every Turing-acceptable language would be Turing-decidable.

PROOF: Suppose Turing machine T_0 decides H_0, i.e.,
(1) if $\rho(T)\rho(w) \in H_0$, then $(q_0, \underline{\#}\rho(T)\rho(w)\#) \vdash^*_{T_0} (h, \#Y\underline{\#})$, and
(2) if $\rho(T)\rho(w) \notin H_0$, then $(q_0, \underline{\#}\rho(T)\rho(w)\#) \vdash^*_{T_0} (h, \#N\underline{\#})$.
Remember that $\rho(T)\rho(w) \in H_0$ iff T accepts w (3).
Next suppose that T_1 accepts language L, i.e.,
(4) $w \in L$ iff T_1 accepts w. Hence, by (3),
(5) $w \in L$ iff $\rho(T_1)\rho(w) \in H_0$.
Now let T_2 be a Turing-machine, which transforms its input $\underline{\#}w\#$ into $\underline{\#}\rho(T_1)\rho(w)\#$ and then passes control to T_0.
Then by (5) and (1): if $w \in L$, then $(q_0, \underline{\#}w\#) \vdash^*_{T_2} (h, \#Y\underline{\#})$,
and by (5) and (2): if $w \notin L$, then $(q_0, \underline{\#}w\#) \vdash^*_{T_2} (h, \#N\underline{\#})$.
So, T_2 decides L. □

PREDICATE LOGIC IS UNDECIDABLE

Having shown that H_0 is not Turing-decidable, our next goal is to prove that classical predicate logic is not Turing-decidable. We do so by reducing the decision problem for H_0 to the decision problem for classical predicate logic. This leads us to the following definitions.

DEFINITION 29.7.: Let $L \subseteq \Sigma_1^*$ and $L' \subseteq \Sigma_2^*$ be languages. A Turing-computable function $\tau : \Sigma_1^* \to \Sigma_2^*$ is called a *transformation from L to L'* if and only if for each $w \in \Sigma_1^*$ the following holds: $w \in L$ iff $\tau(w) \in L'$.

L is *reducible* to L' := there is a Turing-computable function τ which is a transformation from L to L'.

THEOREM 29.8.: If L is reducible to L', then
i) if L' is Turing-decidable, then L is Turing-decidable too,
or equivalently,
ii) if L is not Turing-decidable, then L' is not Turing-decidable either.

PROOF: Suppose τ is a transformation from L to L'. Then from any decision procedure for L', one can obtain one for L: to see whether any $w \in \Sigma_1^*$ is actually in L, one could compute $\tau(w)$ and see whether that string is in L'. In such a case, we say that we have *reduced* (the decision problem for) L to (the decision problem for) L'. □

Now let **A** be the class of all Turing-acceptable languages and **D** the class of all Turing-decidable languages. Let L' be a language.

DEFINITION 29.9.: L' is **A**-*complete* :=
(i) $L' \in \mathbf{A}$, and
(ii) Any language $L \in \mathbf{A}$ is reducible to L'.

THEOREM 29.10: H_0 is **A**-complete.

PROOF:
(i) By Theorem 29.1., $H_0 \in \mathbf{A}$.
(ii) Suppose $L \in \mathbf{A}$. Then there is a Turing machine T such that for all w, $w \in L$ iff T accepts w, or equivalently,
$w \in L$ iff $\rho(T)\rho(w) \in H_0$.

Now let $\tau(w) := \rho(T)\rho(w)$. Then τ is Turing-computable and a transformation from L to H_0. □

Next we can generalize Lemma 29.6. as follows.
LEMMA 29.11.: Let L' be **A**-complete. Then $L' \in \mathbf{D}$ iff $\mathbf{A} = \mathbf{D}$.

Since H_0 is **A**-complete (Theorem 29.10.) and $H_0 \notin \mathbf{D}$ (Theorem 29.5.) it follows that $\mathbf{A} \neq \mathbf{D}$.

PROOF OF LEMMA 29.11.:
i) Suppose L' is **A**-complete (and hence $L' \in \mathbf{A}$) and $\mathbf{A} = \mathbf{D}$. Then $L' \in \mathbf{D}$.
ii) Suppose L' is **A**-complete (and hence any $L \in \mathbf{A}$ is reducible to L') and $L' \in \mathbf{D}$. Then by Theorem 29.8. any L in **A** also is an element of **D**, i.e., $\mathbf{A} \subseteq \mathbf{D}$. And by Theorem 28.12., $\mathbf{D} \subseteq \mathbf{A}$. So, $\mathbf{A} = \mathbf{D}$. □

In 1936, A. Turing proved that the (first-order, classical) predicate calculus is undecidable by reducing the halting problem for Turing machines to the validity problem for the predicate calculus. In that same year, A. Church also gave, independently, another – less direct – undecidability proof for the predicate logic.

THEOREM 29.12.: *The predicate calculus is undecidable*, i.e., there is no Turing machine to decide for any formula A of the predicate calculus whether A is valid (formally provable) or not.

Since A is satisfiable iff not $\models \neg A$, or equivalently,

$$\models A \text{ iff } \neg A \text{ is not satisfiable,}$$

the following corollary results:

COROLLARY 29.13: There is no Turing machine that decides for any formula A of the predicate calculus whether A is satisfiable or not.

If one accepts the Church-Turing Thesis, Corollary 29.13. says that the *satisfiability problem for the predicate calculus is unsolvable*, i.e., there is no algorithm to determine for an arbitrary formula A of the predicate calculus whether A is satisfiable or not.

Summarizing: The (classical) propositional calculus is (theoretically) decidable (although probably not realistically; see Corollary 30.18.).

In the case of the predicate-calculus, we have given in section 22 a procedure, not being an algorithm, such that

(i) if A is valid, then the procedure stops, giving a formal proof of A and hence saying "yes; A is valid", but

(ii) if A is not valid, the procedure may run forever without ever saying "no; A is not valid".

In fact, we have shown in section 22 the following

THEOREM 29.14.: $\{A \mid A$ is a formula of the (classical) predicate calculus and $\models A\}$ is Turing acceptable.

Theorem 29.12. says that $\{A \mid A$ is a formula of the predicate calculus and $\models A\}$ is not Turing-decidable. Accepting the Church-Turing Thesis this means that in the case of the predicate calculus, no algorithm to decide validity of arbitrary formulas A can exist.

It was in order to prove this and related results that Turing had to make the notion of "algorithm" mathematically precise, and introduced his notion of Turing machine, accepting what we now call the Church-Turing Thesis that the latter mathematical notion is equivalent to the intuitive notion of algorithm.

PROOF OF THEOREM 29.12.: Since H_0 is not Turing-decidable (Theorem 29.5.), in view of Theorem 29.8., in order to show that $\{A \mid A$ is valid$\}$ is not Turing-decidable, it suffices to show the following
LEMMA: H_0 is reducible to $\{A \mid A$ is a valid formula of the predicate calculus$\}$.

So, we have to construct a transformation τ from H_0 to $\{A \mid A$ is valid$\}$, i.e., for any Turing machine T and any input w we have to compute a formula $\tau(\rho(T)\rho(w)) = A_{T,w}$ such that

$$\rho(T)\rho(w) \in H_0 \text{ iff } \tau(\rho(T)\rho(w)) \in \{A \mid A \text{ is valid}\},$$

in other words, such that

(*) T accepts w iff $A_{T,w}$ is valid.

The construction is essentially that of Turing; see A. Turing, *On computable numbers with an application to the Entscheidungsproblem*, section 11. The underlying idea is to construct a formal language and a finite set Δ of formulas in that language, which in the intended interpretation give a description of the operation of the given Turing machine T and a description of the initial configuration $(q_0, \underline{\#}w\#)$; H is to be a formula in the language expressing – in the intended interpretation – that the given Turing machine halts. Then one can show that T accepts w iff $\Delta \vdash H$. Since Δ is a finite set of formulas, $\Delta \vdash H$ is equivalent to $\models \& \Delta \supset H$, where $\& \Delta :=$ the conjunction of all formulas in Δ.

In order to prove the lemma, we suppose that for any Turing machine $T = \langle K, \Sigma, q_0, \delta \rangle$ the state set K is a finite subset of $\{q_0, q_1, q_2, \ldots\}$ and the alphabet Σ is a finite subset of $\{s_0, s_1, s_2, \ldots\}$, where s_0 is the blank. We construct a formal language consisting of the following symbols.

(bound) individual variables	$t_1, x_1, t_2, x_2, \ldots$
one individual constant	0
two binary predicate symbols	$P, <$
binary predicate symbols	$H, Q_0, Q_1, Q_2, \ldots$
binary predicate symbols	$S_0, S_1, S_2, \ldots$

Given any Turing machine T and input $w = s_{n_1} \dots s_{n_k}$, the intended interpretation of the symbols above is the following.

$t_1,\ x_1,\ t_2,\ x_2,\ \dots$	:	integers
0	:	zero
P	:	is the (immediate) predecessor of
$<$	:	is less than
$Q_i(t, x)$	:	at time t machine T is in state q_i scanning square number x
$H(t, x)$	:	at time t machine T is in the halt state h scanning square number x
$S_j(t, x)$	:	at time t symbol s_j is in square number x.

Given any Turing machine $T = \langle K,\ \Sigma,\ q_0,\ \delta\rangle$ and any input $w = s_{n_1} \dots s_{n_k}$, the set $\Delta_{T,w}$, to be defined below, gives – under the intended interpretation – a description of the operation of the given Turing machine on the given input w. $\Delta_{T,w}$ consists of the following formulas.

(1) For every instruction of the form $\delta(q_i, s_j) = (q_m, s_n, L)$ $\Delta_{T,w}$ contains the formula $\forall t_1 \forall x_1 [Q_i(t_1, x_1) \mathbin{\&} S_j(t_1, x_1) \supset$

$$\exists t_2 \exists x_2 [P(t_1, t_2) \mathbin{\&} P(x_2, x_1) \mathbin{\&} Q_m(t_2, x_2) \mathbin{\&} S_n(t_2, x_1) \mathbin{\&}$$
$$\forall y [y \neq x_1 \supset (S_0(t_1, y) \supset S_0(t_2, y)) \mathbin{\&}$$
$$\vdots$$
$$S_k(t_1, y) \supset S_k(t_2, y))\ ]\ ]\ ]\ ,$$

supposing that the alphabet Σ of T is a subset of $\{s_0, \dots, s_k\}$.

(2) For every instruction of the form $\delta(q_i, s_j) = (q_m, s_n, R)$, $\Delta_{T,w}$ contains a similar formula.

(3) Similarly, for instructions of the form $\delta(q_i, s_j) = (q_m, s_n, O)$.

(4) The following formula gives – under the intended interpretation – a description of time zero, i.e., a description of the initial configuration of T given input $w = s_{n_1} \dots s_{n_k}$. (Remember that s_0 is the blank symbol.)

$$Q_0(0,0) \mathbin{\&} S_0(0,0) \mathbin{\&} \exists x_1 \dots \exists x_k [P(0, x_1) \mathbin{\&} P(x_1, x_2) \mathbin{\&} \dots \mathbin{\&}$$
$$P(x_{k-1}, x_k) \mathbin{\&} S_{n_1}(0, x_1) \mathbin{\&} S_{n_2}(0, x_2) \mathbin{\&} \dots \mathbin{\&} S_{n_k}(0, x_k) \mathbin{\&}$$
$$\forall y [y \neq x_1 \mathbin{\&} \dots \mathbin{\&} y \neq x_k \supset S_0(0, y)]].$$

(5) Appropriate axioms for P, $<$ and $=$:

$\forall x \forall y \forall z [x\ P\ y \mathbin{\&} x\ P\ z \supset y = z]$,
$\forall x \forall y \forall z [y\ P\ x \mathbin{\&} z\ P\ x \supset y = z]$,
$\forall x \forall y \forall z [x < y \mathbin{\&} y < z \supset x < z]$,

$\forall x \forall y[x \; P \; y \supset x < y]$,
$\forall x \forall y[x < y \supset x \neq y]$.

This completes the definition of the finitely many formulas in $\Delta_{T,w}$. Let $H := \exists t \exists x[H(t,x)]$. Under the intended interpretation, the formula H expresses that the given Turing machine will halt on the given input.

Given Turing machine T and input w, the following holds:
(*) T halts on input w iff $\Delta_{T,w} \vdash H$, or equivalently,
(*1) if $\Delta_{T,w} \vdash H$, then T accepts w, and
(*2) if T accepts w, then $\models \; \& \; \Delta_{T,w} \supset H$.

Proof of (*1): Suppose $\Delta_{T,w} \vdash H$. Because the intended interpretation is a model of $\Delta_{T,w}$, it also makes H true. That is, T will halt on input w.
Proof of (*2): Remember that a Turing machine T will *stop* only when it enters the halt state h or when it arrives into a state q scanning a symbol s, while $\delta(q, s)$ has not been defined. In order to prove (*2), it suffices to prove (**) for each $n \in \mathbb{N}$, if T has not stopped before time n given input w, then $\Delta_{T,w}$ implies a description of time n.

By a description of time n, we mean any formula of the form

$$\exists t_1 \ldots \exists t_n[P(0,t_1) \;\&\; \ldots \;\&\; P(t_{n-1},t_n) \;\&\; Q_i(t_n,x) \;\&\; S_{j_1}(t_n,x_1) \;\&\; \ldots \; S_j(t_n,x) \;\&\; \ldots S_{j_l}(t_n,x_l) \;\&\; \forall y[y \neq x_1 \;\&\; \ldots \; y \neq x \;\&\; \ldots \; y \neq x_l \supset S_0(t_n,y)]\,].$$

The proof of (**) is by induction on n.
$n = 0$: Formula (4) is a description of time 0 and is an element of $\Delta_{T,w}$.
Induction step: Suppose T has not stopped before time $k+1$. Then T has not stopped before time k. Hence, by the induction hypothesis, $\Delta_{T,w}$ implies a description of time k. Since the intended interpretation is a model of $\Delta_{T,w}$, it also makes the description of time k true. So, at time k the machine is in some state q_i scanning some square x and reading symbol s_j in it. Since T does not stop at time k, T must contain an instruction of the form $\delta(q_i, s_j) = (q_m, s_n, z)$, where $z \in \{L, R, O\}$. Because of the formulas (1), (2) and (3) in $\Delta_{T,w}$, it follows that $\Delta_{T,w}$ implies a description of time $k+1$.

This finishes the proof of Theorem 19.12. □

By adding to $\Delta_{T,w}$ the formula

$$\forall x \forall y \forall t[H(t,x) \supset \neg Q_{i_1}(t,y) \;\&\; \ldots \;\&\; \neg Q_{i_n}(t,y)],$$

where $q_{i_1}, \ldots, q_{i_n}$ are supposed to be the internal states of T, one obtains a finite set $\Delta'_{T,w}$ of formulas such that

THEOREM 29.15.: T accepts w iff $\Delta'_{T,w} \cup \{H\}$ is finitely satisfiable.

Making use of (**) (in the proof of Theorem 29.12.), the proof of Theorem 29.15. is straightforward (see exercise 29.9.). From this result and Theorem 29.8. it follows that $\{A \mid A \text{ is finitely satisfiable}\}$ is not Turing decidable.

The proof of Theorem 29.15. is similar to the proof of Theorem 4.9. in which we showed that a given Traveling Salesman Problem has a solution if and only if a certain formula E is satisfiable. The difference is that the formula E in Theorem 4.9. is a propositional formula, while the formula $\& \Delta'_{T,w} \,\&\, H$ in the proof of Theorem 29.15. is a formula of the predicate calculus. The proof of Theorem 29.15. is also similar to the proof of Theorem 30.17., in which it is shown that the satisfiability problem for the propositional calculus is **NP**-complete. So, there is a general parallelism between the undecidability of the predicate calculus and the **NP**-completeness of the propositional calculus: essentially the same proof can establish both.

NONDETERMINISTIC TURING MACHINES

A nondeterministic Turing machine is like a (deterministic) Turing machine with the important difference that for a given state and tape symbol scanned by the tape head, the machine may have several choices for the next move. For a deterministic Turing machine $T = \langle K, \Sigma, q_0, \delta\rangle$, δ is a partial function, i.e., with each pair (p, a), $p \in K$ and $a \in \Sigma$, δ associates at most one triple (q, b, m) with $q \in K \cup \{h\}$, $b \in \Sigma$ and $m \in \{L, R, O\}$. But for a nondeterministic Turing machine $\langle K, \Sigma, q_0, \Delta\rangle$, Δ is a relation associating with each pair (p, a), $p \in K$, $a \in \Sigma$, none, one or more triples (q, b, m), $q \in K \cup \{h\}$, $b \in \Sigma$ and $m \in \{L, R, O\}$.

DEFINITION 29.16.: A *nondeterministic Turing machine* is a quadruple $\langle K, \Sigma, q_0, \Delta\rangle$ where K, Σ and q_0 are as for deterministic Turing machines and Δ is a *relation* between $K \times \Sigma$ and $(K \cup \{h\}) \times \Sigma \times \{L, R, O\}$.

Let T be a nondeterministic Turing machine. Then given input w, there may be two or more possible computations. The relations $\vdash_T$ and $\vdash_T^*$ are defined as before, but for nondeterministic T there may be several moves or one-step computations. By definition, T accepts w if and only if at least one of the possible computations starting with $(q_0, \#w\underline{\#})$ results in a halted configuration.

DEFINITION 29.17.: Let Σ_0 be some alphabet not containing $\#$, $w \in \Sigma_0^*$, and L a language over Σ_0. Let $T = \langle K,\ \Sigma,\ q_0,\ \Delta\rangle$ be a nondeterministic Turing machine with $\Sigma_0 \subseteq \Sigma$.

T accepts w := there is at least one computation starting from $(q_0, \#w\underline{\#})$ which leads to some halted configuration. In other words, T accepts w iff for some $\alpha \in \Sigma^*$ and $i \in \mathbb{N}$, $(q_0, \#w\underline{\#}) \vdash_T^* (h, \alpha, i)$.

T accepts L := for all $w \in \Sigma_0^*$, $w \in L$ iff T accepts w.

EXAMPLE 29.18.: Using an obvious extension of the flow graphs we have already used for deterministic Turing machines, the following nondeterministic Turing machine T accepts the language

$$L = \{w \in \{a,b\}^* \mid \text{ the string } a\ b\ a\ a \text{ does occur in } w\}.$$

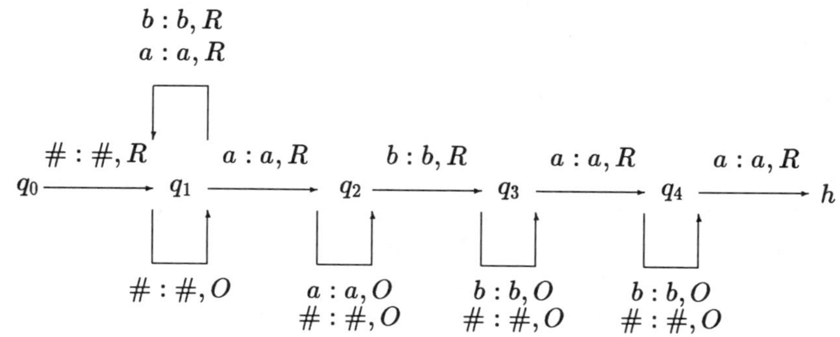

Starting from the initial configuration $(q_0, \#w\underline{\#})$ with $w \in \{a,b\}^*$, this Turing machine scans from left to right and nondeterministically guesses a point in w to start checking for the pattern $a\ b\ a\ a$. This Turing machine is nondeterministic because, when in state q_1 and scanning symbol a, it has the following options:

1. enter state q_2, leaving a in the scanned square and move one cell to the right; i.e., T starts to check for the string $a\ b\ a\ a$; or

2. remain in state q_1, leaving a in the scanned square and move one cell to the right.

We leave it to the reader to give a precise definition of this nondeterministic Turing machine and to verify that T accepts w if and only if w contains a substring $a\ b\ a\ a$ (see exercise 29.5.).

REMARK: From the definition it is clear that any (deterministic) Turing machine (Definition 28.1.) is a special case of a nondeterministic Turing machine (Definition 29.16.).

EXAMPLE 29.19.: The set of all *satisfiable* formulas of the propositional calculus is accepted by a nondeterministic Turing machine T. This machine operates as follows, when given input $\underline{\#}A\#$.

1. Determine the atomic formulas occurring in A and nondeterministically choose a line in the truth table for A, i.e., a valuation for the atomic formulas in A.
2. Compute the truth value of A in the line chosen.
3. If A is 1 in the line chosen, then halt; otherwise continue computing forever in some trivial fashion.

Any language accepted by a nondeterministic Turing machine is accepted by a deterministic Turing machine. More precisely, we prove

THEOREM 29.20.: For every nondeterministic Turing machine T_1 one can construct a standard (deterministic) Turing machine T_2 such that for any string w not containing a blank,

$$T_1 \text{ halts on input } w \text{ iff } T_2 \text{ halts on input } w.$$

PROOF: Let T_1 be a nondeterministic Turing machine and let w be any input. Then one can make a tree of all possible computations of T_1 starting from the initial configuration.

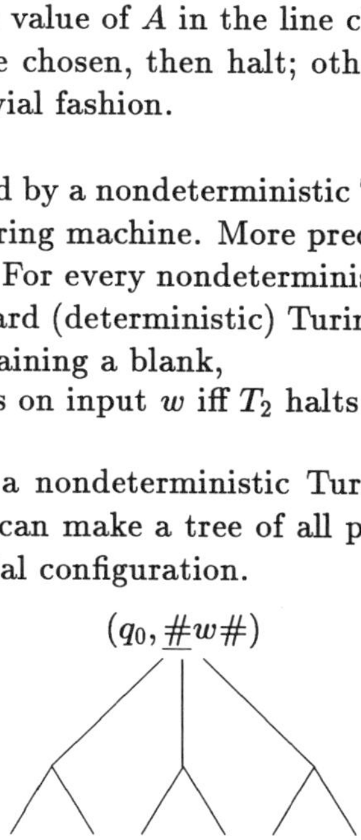

Each node in this tree has only finitely many successors. Now let T_2

be a deterministic Turing machine that searches systematically through this tree of computations by T_1. We let T_2 halt as soon as it discovers a halting computation of T_1. If there is no halting computation of T_1, T_2 will run forever. (Do exercise 29.6.!) □

EXERCISES

29.1. Let T be the Turing machine $\langle\{q_0\},\ \{s_0, s_2, s_5\},\ q_0,\ \delta\rangle$, where δ is defined by $\delta(q_0, s_0) = (h, s_2, O)$ and $\delta(q_0, s_2) = \delta(q_0, s_5) = (q_0, \#, R)$. Following the conventions of § 29 and supposing that the blank # is s_0, determine the encoding $\rho(T)$ of Turing machine T.

29.2. Following the conventions of § 29, determine whether the following strings over $\{c, 1\}$ are in fact the encodings of a Turing machine; and if so, reconstruct T from the encoding $\rho(T)$.
i) $c\ 1^2\ c\ c\ 1^2\ c\ 1^2\ c\ 1^2\ c\ 1^4\ c\ 1^2\ c\ c\ 1\ c\ 1^3\ c\ 1\ c\ 1^3\ c\ c\ 1^2\ c\ 1^2\ c\ c$
ii) $c\ 1^2\ c\ c\ 1^2\ c\ 1^4\ c\ 1^3\ c\ 1^5\ c\ 1^2\ c\ c\ 1^3\ c\ 1^5\ c\ 1\ c\ 1^5\ c\ 1^3\ c\ c$

29.3. Let $H_1 := \{\rho(T) \mid T \text{ accepts } \rho(T)\}$.
Show that H_1 is Turing-acceptable. (In Theorem 29.3. we have shown that $\bar{H}_1$ is not Turing-acceptable.)

29.4. Prove that $\{A \mid A$ is a formula of the predicate calculus and A is satisfiable$\}$ is not Turing-acceptable.

29.5. Give a precise definition of the nondeterministic Turing machine illustrated in Example 29.18.

29.6. Given the nondeterministic Turing machine T in Example 29.18., accepting the language $L = \{w \in \{a, b\}^* \mid$ the string $a\ b\ a\ a$ does occur in $w\}$, construct a deterministic Turing machine T_2 accepting the same language L (see Theorem 29.20).

29.7. Let L be the set of all words over the alphabet $\{a, b\}$ which contain at least two consecutive occurrences of a or at least two consecutive occurrences of b. Construct a nondeterministic Turing machine which accepts L.

29.8. Let L be the set of all words over the alphabet $\{a, b\}$ which contain at least one occurrence of the substring $a\ b$ or at least one occurrence of the substring $b\ a$. Construct a nondeterministic Turing machine which accepts L.

29.9. Prove Theorem 29.15.

§30 COMPLEXITY:

TIME BOUNDED TURING MACHINES; THE CLASSES **P** AND **NP**; **NP**-COMPLETENESS; THE SATISFIABILITY PROBLEM FOR PROPOSITIONAL LOGIC IS **NP**-COMPLETE

In § 3 and § 4 we have seen that the satisfiability problem for the propositional calculus and the Traveling Salesman Problem (TSP) can be solved theoretically, but that so far no *realistic* algorithm is known to decide satisfiability for formulas of the propositional calculus or – in the case of the TSP – to produce, given a map with thirty cities and distances in miles, the itinerary that minimizes the total distance traversed.

In this section we shall try to capture the notion of "realistic algorithm" in precise mathematical terms. The *time* used by a computation is the number of instructions executed during the computation. The *space* used by a computation equals the total number of tape squares scanned during the computation.

P is the class of all languages that are decided by a Turing machine in some polynomial time bound. A quantitative analogon of the Church-Turing Thesis contends that every realistic (practical or efficient) algorithm can be rendered as a polynomially time bounded Turing machine. **NP** is the class of all languages that are accepted by a nondeterministic Turing machine in some polynomial time bound. **NP** contains many natural, practical problems for which membership in **P** is in serious doubt. We prove that the satisfiability problem for the (classical) propositional calculus is **NP**-complete, i.e., i) this problem is in **NP** and ii) any other problem in **NP** is reducible to this problem in polynomial time.

TIME BOUNDED TURING MACHINES

DEFINITION 30.1.:

i) Let $T = \langle K,\ \Sigma,\ q_0,\ \delta\rangle$ be a Turing machine and let C, C' be configurations of T. For $n \in \mathbb{N}$, we define $C \vdash_T^n C' := C \vdash_T^* C'$ by a computation of exactly n steps.
ii) Let t be a function from $\mathbb{N}$ to $\mathbb{N}$. Let $\Sigma_0 \subseteq \Sigma$ and L a language over Σ_0. *T decides L in time t* := for any $w \in \Sigma_0^*$,
if $w \in L$, then for some $n \leq t(|w|)$ $(q_0, \underline{\#}w\#) \vdash_T^n (h, \#Y\underline{\#})$, and
if $w \notin L$, then for some $n \leq t(|w|)$ $(q_0, \underline{\#}w\#) \vdash_T^n (h, \#N\underline{\#})$.

iii) We say that L is *decidable in time t* if there is some Turing machine that decides L in time t. The class of all languages decidable in time t is denoted by TIME(t).

So, in Definition 30.1. ii) the number of steps of a Turing machine is bounded in terms of a function t of the length of the input and not by a constant.

EXAMPLE 30.2.: Let $L = \{w \in \{a,b\}^* \mid w$ has no occurrence of the string $a\ a\}$. Let $T = \langle\{q_0, q_1, q_2, q_3\}, \{\#, a, b\},\ q_0,\ \delta\rangle$ be a Turing machine with δ defined as follows: $\delta(q_0, \#) = (q_1, \#, R)$,

$$\delta(q_1, a) = (q_2, \#, R),\ \delta(q_1, b) = (q_1, \#, R),\ \delta(q_1, \#) = (h, Y, R),$$
$$\delta(q_2, a) = (q_3, \#, R),\ \delta(q_2, b) = (q_1, \#, R),\ \delta(q_2, \#) = (h, Y, R),$$
$$\delta(q_3, a) = (q_3, \#, R),\ \delta(q_3, b) = (q_3, \#, R),\ \delta(q_3, \#) = (h, N, R).$$

Then T decides L in time t, where $t(n) = n + 2$.

For instance, $(q_0, \underline{\#}\ a\ b\ a\ \#) \vdash_T^* (q_1, \#\ \underline{a}\ b\ a\ \#)$
$(q_2, \#\#\ \underline{b}\ a\ \#)$
$(q_1, \#\#\#\ \underline{a}\ \#)$
$(q_2, \#\#\#\#\underline{\#})$
$(h, \#\#\#\#Y\underline{\#})$.

THE CLASSES P AND NP

DEFINITION 30.3.: i) Let p be a function from $\mathbb{N}$ to $\mathbb{N}$. p is called a *polynomial* function if for all $n \in \mathbb{N}$, $p(n) = a_0 + a_1 n + a_2 n^2 + \ldots + a_k n^k$ for some $k \in \mathbb{N}$ and integers $a_0, a_1, \ldots, a_k$ with $a_k > 0$; the number k is called the *degree* of p, and the a_j are called its *coefficients*.
ii) *T decides L in polynomial time* := there is a polynomial function p such that T decides L in time p. In that case, we call L *decidable in polynomial time.*

In order for Definition 30.3. ii) to make sense, this definition should not depend on the particular conventions we made for presenting the input/output and for the number of tapes. And indeed, one can show that if a language is decided by one type of Turing machine in polynomial time, then it is also decided by any other type of Turing machine in polynomial time, where – of course – the degree of the latter polynomial may be higher than that of the former. For details the reader is referred to C.I. Lewis and C.H. Papadimitriou, *Elements of the Theory of Computation*, § 7.3. As a matter of fact, all known practical algorithms run in polynomial time with low degree polynomials; and all known impractical algorithms take exponential time, where $t : \mathbb{N} \rightarrow \mathbb{N}$ is called an *exponential* function if $t(n)$ is of the form r^n for some $r > 1$.

The considerations above lead us to the following definition.

DEFINITION 30.4.: $\mathbf{P} := \cup\{\text{TIME}(p) \mid p \text{ is a polynomial}\}$.

In words: **P** is the class of all languages that are decided by Turing machines in polynomial time.

EXAMPLES 30.5.:

i) Let u assign the value 1 (true) or 0 (false) to each atomic formula P_j. Then $\{A \mid A$ is a formula of the propositional calculus and u assigns to A the value $1\} \in \mathbf{P}$. In other words, given u one can construct a Turing machine T which decides $u(A) = 1$ in polynomial time.

ii) Given n cities to be traversed, given the distances between the cities and given a natural number k, $\{t \mid t$ is tour with length $\leq k\} \in \mathbf{P}$. In other words, given the cities with their mutual distances and given k, $k \in \mathbb{N}$, one can construct a Turing machine T which decides in polynomial time whether or not the length of a tour t is equal to or less than k.

iii) Let A be the matrix $\left[\begin{smallmatrix}1&0&2\\0&3&2\end{smallmatrix}\right]$ and let b be the vector $\left[\begin{smallmatrix}3\\2\end{smallmatrix}\right]$.
Then $Ax = b$ is short for $1x_1 + 0x_2 + 2x_3 = 3$, and

$$0x_1 + 3x_2 + 2x_3 = 2.$$

More generally, m simultaneous equations of the form

$$a_{11}x_1 + a_{12}x_2 + \ldots + a_{1n}x_n = b_1$$
$$\vdots$$
$$a_{m1}x_1 + a_{m2}x_2 + \ldots + a_{mn}x_n = b_m,$$

where all a_{ij}'s and b_i's are natural numbers, are usually written in the form $Ax = b$, where A is a matrix capturing all a_{ij}'s in an $m \times n$ table, x is a vector $(x_1, \ldots, x_n)$ and b is a vector $(b_1, \ldots, b_m)$.
Given a matrix A and a vector b, $\{x \mid x$ is binary and $Ax = b\} \in \mathbf{P}$, where x is binary if and only if each x_j equals 0 or 1. So, given A and b, one can construct a Turing machine T which decides in polynomial time whether $Ax = b$ or not.

According to the *Church-Turing Thesis*, every algorithm can be rendered as a Turing machine. A quantitative analogon of this Thesis contends that every *realistic* (*practical* or *efficient*) *algorithm* can be rendered as a polynomially time bounded Turing machine.

Below we study nondeterministic Turing machines restricted by some polynomial time bound.

DEFINITION 30.6.:

i) Let $t : \mathbb{N} \to \mathbb{N}$. Let $L \subseteq \Sigma_0^*$ be a language and $T = \langle K,\ \Sigma,\ q_0,\ \Delta \rangle$ be a nondeterministic Turing machine. We say that T *accepts* L *nondeterministically in time* t if for all $w \in \Sigma_0^*$,
$w \in L$ iff for some $n \leq t(|w|)$, T halts on input w in n steps,
or equivalently,
$w \in L$ iff for some $n \leq t(|w|)$, $v, u \in \Sigma^*$, $\sigma \in \Sigma$,
$(q_0, \underline{\#}w\#) \vdash_T^n (h, v\, \underline{\sigma}\, u)$.

ii) We say that L is *acceptable nondeterministically in time* t if there is a nondeterministic Turing machine T that accepts L (nondeterministically) in time t.

iii) The class of languages acceptable nondeterministically in time t is denoted by NTIME(t).

DEFINITION 30.7.: We define **NP** (for *Nondeterministically in Polynomial time*) to be the class of all languages that are acceptable nondeterministically in some polynomial time bound. In other words,

$$\mathbf{NP} = \cup\{\text{NTIME}(p) \mid p \text{ is polynomial}\}.$$

EXAMPLES 30.8.:

i) $Sat := \{A \mid A$ is a *satisfiable* formula of the propositional calculus$\}$. $Sat \in$ **NP**. Consider the nondeterministic Turing machine T described in Example 29.19. and note that the computation of the truth value of a given formula A *in a given line of the truth table* can be carried out in polynomial time. (See also Example 30.5. i).)

In other words, given an assignment u for A (see Section 2), $u(A)$ can be computed in polynomial time (see Exercise 4.11.). So, once a line in the truth table of A has been *guessed*, the truth value of A in that line can be computed in polynomial time. The problem with satisfiability is that the number of lines in the truth table for A

is not polynomial, but exponential: if A contains m atomic formulas, the truth table of A has 2^m lines.

ii) In order to formalize the *Traveling Salesman Problem* (TSP), we assume that the cities are named $1, 2, \ldots, n$ and that the distances d_{ij} between any two cities i and j are given in an $n \times n$ matrix D. Given a tour t, let $D(t)$ denote the length of t. Given n cities with distances d_{ij} and given an integer b, called the *budget*, the Traveling Salesman Problem is the question whether there is a tour t such that $D(t) \leq b$. The corresponding language is

$$TSP = \{1^n @\, \alpha(D) @\, 1^b \mid \text{ there is a tour } t \text{ such that } D(t) \leq b\},$$

where $\alpha(D)$ is an appropriate encoding of D.

It is easy to see that $TSP \in \mathbf{NP}$: one can design a nondeterministic Turing machine T which "guesses" some tour and then computes in polynomial time whether the length of the tour chosen is equal to or less than b. (See Example 30.5. ii).)

iii) Let $B := \{\alpha(A) @ \alpha(b) \mid \text{ there is a binary } x \text{ such that } Ax = b\}$, where $\alpha(A)$ and $\alpha(b)$ are encodings of the matrix A and vector b respectively. (See Example 30.5. iii).)

Again, it is easy to see that $B \in \mathbf{NP}$: one can construct a nondeterministic Turing machine that "guesses" some binary x and then computes in polynomial time whether $Ax = b$. The trouble with this problem is again that the number of binary x's – if $x = (x_1, \ldots, x_n)$ – is 2^n. So, it seems that any deterministic Turing machine would in the worst case require an exponential amount of time in order to decide this problem by trying all possibilities for x.

Summarizing: In examples 30.5. and 30.8. we have seen:

i) $\{<A, u> \mid u(A) = 1\} \in \mathbf{P}$ and $\{A \mid \exists u[u(A) = 1]\} \in \mathbf{NP}$,

ii) $\{<n, D, b, t> \mid D(t) \leq b\} \in \mathbf{P}$ and

$$\{<n, D, b> \mid \exists t[D(t) \leq b]\} \in \mathbf{NP},$$

iii) $\{<A, b, x> \mid Ax = b\} \in \mathbf{P}$ and $\{<A, b> \mid \exists x[Ax = b]\} \in \mathbf{NP}$.

In general, one can show the following.

THEOREM 30.9.: Let L be a language over an alphabet Σ.

$L \in \mathbf{NP}$ iff there is a language L' and a polynomial p such that

i) $L' \in \mathbf{P}$, and

ii) $L = \{x \mid \exists y[<x, y> \in L' \;\&\; \text{length}(y) \leq p(\text{length}(x))]\}$.

PROOF: Let $L' \in \mathbf{P}$, p a polynomial and

$$x \in L \text{ iff } \exists y[\ |y| \leq p(|x|) \ \&\ <x,y> \in L'].$$

Let T' decide L' in polynomial time p'. Let T be a nondeterministic Turing machine which given input x first "guesses" a string y with $|y| \leq p(|x|)$ and next simulates the behaviour of T' on $<x,y>$. If T' halts on input $<x,y>$ and answers Y(es), then we let T halt; if T' halts on input $<x,y>$ and answers N(o), then we let T go into an infinite loop. Then there is a polynomial t such that

$$x \in L \text{ iff for some } n \leq t(|x|),\ T \text{ halts on input } x \text{ in } n \text{ steps.}$$

I.e., T nondeterministically accepts L in polynomial time t. In other words, $L \in \mathbf{NP}$.

Conversely, suppose $L \in \mathbf{NP}$, i.e., L is accepted by a nondeterministic Turing machine T in polynomial time t. Without loss of generality, we can assume that for any configuration C there are at most two configurations that may follow C.

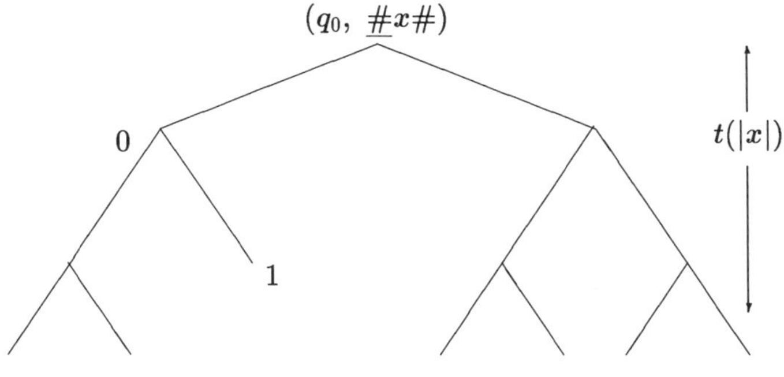

Then, given input x, the sequence of choices of configurations made by T in a computation on x can be encoded as a string y of 0's and 1's. Let $<x,y> \in L'$ iff the computation of T on input x encoded by y halts in less than $t(|x|)$ steps. Let T' be the deterministic Turing machine that given input $<x,y>$ simulates the computation of T on input x with the sequence of choices y. If this computation halts in less than $t(|x|)$ steps, we let T' halt and say Y(es); if not, then we let T' halt and say N(o). Then, clearly, there is some polynomial p such that T' decides L' in time p. Also $x \in L$ iff $\exists y$[the computation of T on input x encoded by y halts in less than $t(|x|)$ steps]. Therefore

$$x \in L \text{ iff } \exists y[<x,y> \in L' \ \&\ |y| \leq p(|x|)].$$

□

The following Theorem is immediate.

THEOREM 30.10.: $\mathbf{P} \subseteq \mathbf{NP}$.

The obvious question now is whether $\mathbf{P} = \mathbf{NP}$. Intuitively, one would guess that $\mathbf{P} \neq \mathbf{NP}$, arguing as follows. Consider the satisfiability problem for the propositional calculus or the corresponding language $Sat := \{A \mid A$ is a satisfiable formula of propositional logic$\}$. In Example 30.8. i) we have seen that $Sat \in \mathbf{NP}$, i.e., there is some nondeterministic Turing machine T that recognizes Sat in polynomial time. T achieves this by choosing nondeterministically a line in the truth table for the given formula A and next computing the truth value of A in the chosen line in polynomial time. The problem for a deterministic Turing machine is that if A is built from n atomic formulas, then there are 2^n lines in the truth table for A. And the obvious way for a deterministic Turing machine to decide Sat is to try all possible lines in the truth table for a given formula A, which in the worst case would require an exponential amount of time (2^n). So far no one has discovered an algorithm to decide satisfiability in polynomial time. This suggests, but does not prove, that $Sat \notin \mathbf{P}$. So far nobody has been able to prove this.

The problem whether $Sat \in \mathbf{P}$ and, more generally, whether $\mathbf{P} = \mathbf{NP}$, is still open.

Another, but similar, way to see the $\mathbf{P} = \mathbf{NP}$ problem is by looking at the tree of possible computations of a nondeterministic Turing machine; the nodes in the tree correspond to reachable configurations, the top node being the initial configuration.

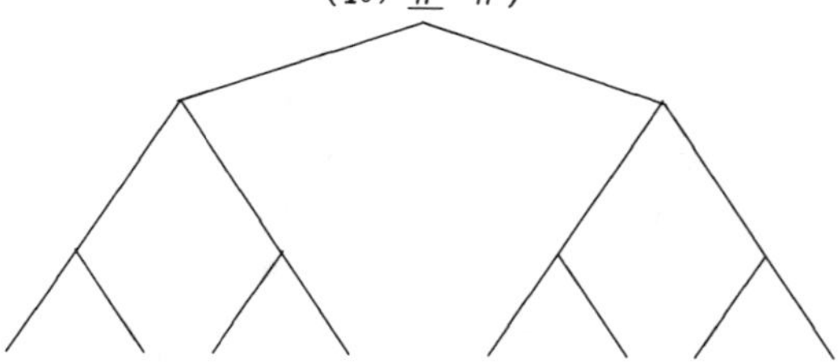

A binary tree of depth n has 2^n branches (i.e., computations). And the only obvious way for a deterministic Turing machine to simulate a nondeterministic one seems to be by an exhaustive examination of all possible computations.

The class **NP** is important because it contains many natural practical problems, such as *Sat*, *TSP*, *B* (see Example 30.8.) and many others, which are generally believed not to be in **P**.

It was shown in Theorem 29.1. and 29.5. that the class **D** of Turing decidable languages is not the same as the class **A** of Turing acceptable languages (see also Lemma 29.11.). The quantitative analogue, $\mathbf{P} \neq \mathbf{NP}$, is still unsettled.

NP-COMPLETENESS

In the preceding section we have introduced the notion of **A**-completeness and have shown:

Theorem 29.10.: $H_0 := \{\rho(T)\rho(w) \mid T \text{ accepts } w\}$ is **A**-complete;

Lemma 29.11.: if L' is **A**-complete, then $\mathbf{D} = \mathbf{A}$ iff $L' \in \mathbf{D}$.

In Theorem 29.5. we proved that $H_0 \notin \mathbf{D}$, and hence by Theorem 29.10. and Lemma 29.11., that $\mathbf{D} \neq \mathbf{A}$.

Since the question whether $\mathbf{P} = \mathbf{NP}$ is the quantitative analogue of the question whether $\mathbf{D} = \mathbf{A}$, answered in the negative in Section 29, we try to settle the question $\mathbf{P} = \mathbf{NP}$ in an analogous way. Therefore, we introduce the notion of **NP**-completeness and show:

Theorem 30.15.: $N_0 := \{\rho(T)\rho(w)@1^n \mid$ the nondeterministic Turing machine T accepts the input w in n or fewer steps$\}$ is **NP**-complete; and

Theorem 30.16.: if L' is **NP**-complete, then $\mathbf{P} = \mathbf{NP}$ iff $L' \in \mathbf{P}$.

However, we shall see that we cannot prove that $N_0 \notin \mathbf{P}$ and hence we cannot conclude that $\mathbf{P} \neq \mathbf{NP}$.

DEFINITION 30.11.: Let Σ_1 and Σ_2 be alphabets. Let $f : \Sigma_1^* \to \Sigma_2^*$ and let $T = \langle K,\ \Sigma,\ q_0,\ \delta\rangle$ be a deterministic Turing machine.
Let $t : \mathbb{N} \to \mathbb{N}$.

i) *f is computed in time t by T* := for all $x \in \Sigma_1^*$ there is some $n \leq t(|x|)$ such that $(q_0, \#x\underline{\#}) \vdash_T^n (h, \#f(x)\underline{\#})$.

ii) f is *computable in time t* if there is a Turing machine T that computes f in time t.

iii) f is *polynomial-time computable* := there is a polynomial t such that f is computable in time t.

If f is computable in time t, then for all $x \in \Sigma_1^*$, $|f(x)| \leq t(|x|) + |x|$, because the machine computing f cannot in time $t(|x|)$ produce an output $f(x)$ of length greater than $t(|x|) + |x|$.

Notice that the characteristic function f_L of any language L in $\mathbf{P}$ (defined by $f_L(x) = 1$ if $x \in L$ and $f_L(x) = 0$ if $x \notin L$) is polynomial-time computable.

The following definition is the analogue of Definition 29.7.

DEFINITION 30.12.: Let $L \subseteq \Sigma_1^*$ and $L' \subseteq \Sigma_2^*$ and $\tau : \Sigma_1^* \rightarrow \Sigma_2^*$.
τ is a *polynomial-time transformation from L to L'* iff
1. τ is a polynomial-time computable function, and
2. for each $x \in \Sigma_1^*$, $x \in L$ iff $\tau(x) \in L'$.

L is reducible to L' in polynomial time := there exists a polynomial-time transformation τ from L to L'.

EXAMPLE: Actually, we have seen in Section 4 that the Traveling Salesman Problem is reducible to the Satisfiability Problem for the propositional calculus in polynomial time.

THEOREM 30.13. (cf. Theorem 29.8.): If L is reducible to L' in polynomial time, then

i) if L' is Turing decidable in polynomial time, then so is L; in other words, if $L' \in \mathbf{P}$, then $L \in \mathbf{P}$; or equivalently,
ii) if L is not Turing-decidable in polynomial time, then neither is L'.

PROOF: Suppose there is a polynomial-time transformation τ from L to L' and suppose L' is Turing-decidable in polynomial time. Then to see whether any $x \in \Sigma_1^*$ is a member of L one can compute $\tau(x)$ in polynomial time and next decide in polynomial time whether $\tau(x) \in L'$. If so, then $x \in L$; if not, then $x \notin L$. In this way one can decide L in polynomial time. □

DEFINITION 30.14. (cf. Definition 29.9.): L' is **NP**-*complete* :=

i) $L' \in \mathbf{NP}$, and
ii) Any language $L \in \mathbf{NP}$ is reducible to L' in polynomial time.

THEOREM 30.15. (cf. Theorem 29.10.):

Let $N_0 := \{\rho(T)\rho(w)@1^n \mid$ the nondeterministic Turing machine T accepts the input w in n or fewer steps$\}$. N_0 is **NP**-complete.

PROOF: In order to show that $N_0 \in \mathbf{NP}$, we have to construct a nondeterministic Turing machine T_u and a polynomial p such that for all nondeterministic Turing machines T, for all inputs w and for all $n \in \mathbb{N}$, $\rho(T)\rho(w)@1^n \in N_0$ iff T_u accepts $\rho(T)\rho(w)@1^n$ in polynomial time p

or, equivalently,

T accepts w in n or fewer steps iff T_u accepts $\rho(T)\rho(w)@1^n$ in polynomial time p.

We can take T_u to be a nondeterministic universal Turing machine which, when presented with an encoding of an arbitrary nondeterministic Turing machine T, an encoding of an input string w and an encoding of a natural number n will simulate the behaviour of T when T is presented with w and will go into an infinite loop if T does not accept w in n or fewer steps. The polynomial p can be constructed from the specification of the universal Turing machine.

Next we have to show that any language L in **NP** is reducible to N_0 in polynomial time. So suppose $L \in \mathbf{NP}$, i.e., there is a nondeterministic Turing machine T_L and a polynomial p_L such that for all w,

$w \in L$ iff T_L accepts w in $p_L(|w|)$ or fewer steps,

or equivalently,

$w \in L$ iff $\rho(T_L)\rho(w)@1^{p_L(|w|)} \in N_0$.

Now let $\tau(w) := \rho(T_L)\rho(w)@1^{p_L(|w|)}$.

Then τ is a polynomial-time transformation from L to N_0, in other words, τ is a polynomial time computable function such that $w \in L$ iff $\tau(w) \in N_0$. So, L is reducible to N_0 in polynomial time. □

THEOREM 30.16. (cf. Lemma 29.11.): Let L' be an **NP**-complete language. Then $\mathbf{P} = \mathbf{NP}$ iff $L' \in \mathbf{P}$.

PROOF: Let L' be **NP**-complete and hence $L' \in \mathbf{NP}$. Then if $\mathbf{P} = \mathbf{NP}$, it follows that $L' \in \mathbf{P}$. Conversely, suppose $L' \in \mathbf{P}$ and L' is an **NP**-complete language. Then by definition, any language L in **NP** is in polynomial time reducible to L'. Hence, by Theorem 30.13., any language L in **NP** is in **P**, i.e., $\mathbf{NP} \subseteq \mathbf{P}$. Since also $\mathbf{P} \subseteq \mathbf{NP}$ (Theorem 30.10.), it follows that $\mathbf{P} = \mathbf{NP}$. □

Theorem 30.16. states that if any **NP**-complete language L' would turn out to be decidable in polynomial time ($L' \in \mathbf{P}$), then any problem in **NP** would be decidable in polynomial time ($\mathbf{NP} = \mathbf{P}$). For instance, in Corollary 30.18. we will prove that $Sat := \{A \mid A$ is a satisfiable formula of the propositional calculus$\}$ is **NP**-complete. Then Theorem 30.16. states that if someone ever succeeds in finding an algorithm that decides Sat in polynomial time, then he or she will have discovered at the same time algorithms that decide TSP, B and all other practical problems in **NP** in polynomial time. Since, in fact, nobody expects this to be the case, one might try to prove that $N_0 \notin \mathbf{P}$ in a similar way as we have shown that $H_0 \notin \mathbf{D}$ (Theorem 29.5.); then it would follow from Theorem 30.16. that $\mathbf{P} \neq \mathbf{NP}$. Below we will try to do so. However, this attempt will fail.

In order to prove that H_0 is not Turing-decidable ($H_0 \notin \mathbf{D}$; Theorem 29.5.), we defined $H_1 := \{\rho(T) \mid T \text{ accepts } \rho(T)\}$ and showed that $\bar{H}_1$, the complement of H_1, is not Turing-acceptable ($\bar{H}_1 \notin \mathbf{A}$ and hence $\bar{H}_1 \notin \mathbf{D}$; Theorem 29.3.). Then it followed that H_1 is not Turing-decidable ($H_1 \notin \mathbf{D}$) and, consequently, by Lemma 29.2., that H_0 is not Turing-decidable ($H_0 \notin \mathbf{D}$).

Following the analogy between H_0 and N_0 we define
$N_1 := \{\rho(T) \mid$ nondeterministic Turing machine T accepts input $\rho(T)$ within $|\rho(T)|$ steps$\}$ and try, unsuccessfully, to show that $\bar{N}_1 \notin \mathbf{NP}$. So, suppose that $\bar{N}_1 \in \mathbf{NP}$, i.e., that $\bar{N}_1$ is accepted by a nondeterministic Turing machine T^* in polynomial time p, i.e., for all nondeterministic Turing machines T,

$$\rho(T) \in \bar{N}_1 \text{ iff } T^* \text{ accepts } \rho(T) \text{ within } p(|\rho(T)|) \text{ steps}$$

or, equivalently,

T does not accept $\rho(T)$ within $|\rho(T)|$ steps iff T^* accepts $\rho(T)$ within $p(|\rho(T)|)$ steps. We would hope to derive a contradiction by taking
$T = T^*$: T^* does not accept $\rho(T^*)$ within $|\rho(T^*)|$ steps iff T^* does accept $\rho(T^*)$ within $p(|\rho(T^*)|)$ steps. But no contradiction results; for it could very well be that T^* accepts $\rho(T^*)$ in exactly $|\rho(T^*)| + 1$ steps and no less!

So, this approach fails to prove that $\bar{N}_1 \notin \mathbf{NP}$ and, consequently, it fails to prove $\mathbf{P} \neq \mathbf{NP}$. So far no one has been able to settle the $\mathbf{P} = \mathbf{NP}$ question either way. It is generally believed that $\mathbf{P} \neq \mathbf{NP}$ because **NP** contains many practical problems such as TSP, Sat and B for which no polynomial decision procedure is known. But no one has been able to prove that $\mathbf{P} \neq \mathbf{NP}$.

THE SATISFIABILITY PROBLEM IS NP-COMPLETE

In the preceding section we showed that the Halting problem is reducible to the validity problem for the predicate calculus (Theorem 29.12.) and also that the Halting problem is reducible to the finitely satisfiability problem for the predicate calculus (Theorem 29.15). Consequently, the predicate calculus is undecidable. The proof we gave goes back to Alan Turing's original paper of 1936, in which he described the operation of a Turing machine in terms of a formula of the predicate calculus.

In Section 4 we have seen that the Traveling Salesman Problem can be reduced in polynomial time to the satisfiability problem for the propositional calculus. In 1971 Stephen Cook showed that N_0, and, hence, any problem in **NP** can be reduced in polynomial time to the satisfiability problem in the propositional calculus; in other words, that N_0 is reducible to $Sat := \{A \mid A$ is a satisfiable formula of the *propositional* calculus$\}$ in polynomial time. Consequently, Sat is **NP**-complete too. In fact, satisfiability in the propositional calculus is the granddaddy of all the **NP**-complete problems; historically, it was the first to be observed. Cook's proof is similar to Turing's: he describes the operation of a nondeterministic Turing machine in polynomial time by means of an expression of the propositional calculus.

So, there is a general parallelism between unsolvability and **NP**-completeness: in some cases essentially the same proof can establish both.

THEOREM 30.17.: N_0 is reducible to $Sat := \{A \mid A$ is a satisfiable formula of the *propositional* calculus$\}$ in polynomial time.

Since $Sat \in$ **NP** (Example 30.8. i)) and N_0 is **NP**-complete (Theorem 30.15.), we immediately have
COROLLARY 30.18.: Sat is **NP**-complete.

PROOF OF THEOREM 30.17.: (S. Cook, 1971):
We have to construct a polynomial-time transformation τ from N_0 to Sat, i.e., τ should be a polynomial-time computable function such that

$$\rho(T)\rho(w)@1^n \in N_0 \text{ iff } \tau(\rho(T)\rho(w)@1^n) \in Sat$$

or equivalently, using $A_{T,w,n}$ for $\tau(\rho(T)\rho(w)@1^n)$,

T nondeterministically accepts w in n or fewer steps iff $A_{T\,w,n}$ is satisfiable.

For $T = \langle K,\ \Sigma,\ q_0,\ \Delta\rangle$ a nondeterministic Turing machine, w an input and $n \in \mathbb{N}$, we construct the formula $A_{T,w,n}$ as follows. Suppose $K = \{q_0, q_1, \ldots, q_k\}$, $\Sigma = \{s_0, s_1, \ldots, s_l\}$ and $w = s_{n_1} \ldots s_{n_j}$. We suppose that the tape-squares are numbered by the integers and that in the initial configuration (time $t = 0$) square number 0 is scanned. Then it follows that in a computation of n or fewer steps no tape square beyond number n is scanned.

We introduce a propositional language with the following atomic formulas.

i) Q^i_t for $0 \leq i \leq k$ and $0 \leq t \leq n$. Intended meaning of Q^i_t: at time t machine T is in state q_i,

ii) H_t for $0 \leq t \leq n$. Intended meaning of H_t: at time t Turing machine T is in the halt state h.

iii) $S^j_{t,m}$ for $0 \leq j \leq l$ and $0 \leq t, m \leq n$. Intended meaning of $S^j_{t,m}$: at time t tape square number m contains the symbol s_j.

iv) $P_{t,m}$ for $1 \leq t, m \leq n$. Intended meaning of $P_{t,m}$: at time t square number m is scanned by the tape head.

The formula $A_{T,w,n}$ is a conjunction B & C & D & E & F & H defined as follows.

B expresses that at each time t one and only one square is scanned. B is a conjunction B_0 & B_1 & $\ldots$ & B_n, where B_t asserts that at time t one and only one square is scanned:

$$B_t = (P_{t,0} \vee P_{t,1} \vee \ldots \vee P_{t,n})\ \&\ \forall^n_{i=1}\forall^n_{j=1}[j \neq i\ \&\ P_{t,i} \supset \neg P_{t,j}].$$

For $1 \leq t, m \leq n$ let $C_{t,m}$ express that at time t square number m contains one and only one symbol. C is the conjunction of all the $C_{t,m}$ for $1 \leq t, m \leq n$.

D expresses that for each time t, $1 \leq t \leq n$, there is one and only one state.

E is a description of the initial configuration, given input w. $E :=$ Q^0_0 & $P_{0,0}$ & $S^0_{0,0}$ & $S^{n_1}_{0,1}$ & $\ldots$ & $S^{n_j}_{0,j}$ & $S^0_{0,j+1}$ & $\ldots$ & $S^0_{0,n}$, supposing that $w = s_{n_1} \ldots s_{n_j}$ and s_0 is the blank symbol #.

F gives a description in terms of the Q's, H's, S's and P's of the quintuples in Δ. More precisely, suppose $(q_i,\ s_j,\ q_{m_1},\ s_{n_1},\ R) \in \Delta$, and

$$\vdots$$

$(q_i,\ s_j,\ q_{m'},\ s_{n'},\ L) \in \Delta$; then

$$F^t_{i,j} := \forall^n_{m=1}[(Q^i_t \ \& \ P_{t,m} \ \& \ S^j_{t,m} \supset Q^{m_1}_{t+1} \ \& \ P_{t+1,m+1} \ \& \ S^{n_1}_{t+1,m}) \vee$$
$$\vdots$$
$$(Q^i_t \ \& \ P_{t,m} \ \& \ S^j_{t,m} \supset Q^{m'}_{t+1} \ \& \ P_{t+1,m-1} \ \& \ S^{n'}_{t+1,m})].$$

And F is the conjunction of all formulas $F^t_{i,j}$.

Finally, the formula H expresses that T reaches the halting state at some time, i.e., $H := H_0 \vee H_1 \vee \ldots \vee H_n$.

This completes the definition of $A_{T,w,n}$.

It is now straightforward to verify that $A_{T,w,n}$ has the property that T nondeterministically accepts w in n or fewer steps iff $A_{T,w,n}$ is satisfiable. Now let $\tau(\rho(T)\rho(w)@1^n) := A_{T,w,n}$. Then τ is a polynomial-time computable function and τ is a polynomial-time transformation from N_0 to Sat. □

Once again, notice the analogy between the proof that the finite satisfiability problem for the predicate logic is unsolvable (Theorem 29.15.) and the proof of the **NP**-completeness of the satisfiability problem for the propositional logic (Theorem 30.17.). In both proofs a description of the operation of a Turing machine is given in terms of a logical formula. But only the operation of a *polynomially bounded* (nondeterministic) Turing machine can be expressed by a formula in the *propositional* calculus.

EXERCISES

30.1. A *composite* number is one that is the product of two natural numbers, each greater than one; for example, 4, 6, 8, 9, 10 and 12 are composite, but 1, 2, 3, 5, 7 and 11 are not. In other words, a composite number is a non-prime other than one or zero.
Let $L = \{1^n \mid n \text{ is a composite number}\}$. Show that $L \in \mathbf{NP}$.

30.2. For f, g functions from $\mathbb{N}$ to $\mathbb{N}$, let $f \leq g :=$ there is a constant $c > 0$ and an integer $n_0 \in \mathbb{N}$ such that for all $n \geq n_0$, $f(n) \leq c \cdot g(n)$; $f \approx g := f \leq g$ and $g \leq f$. Prove that $\approx$ is an equivalence relation. The equivalence class of a function $f : \mathbb{N} \to \mathbb{N}$ under $\approx$ is called the *rate of growth* of f.

30.3. Prove that

(a) $a_0 + a_1 n + \ldots + a_k n^k \approx n^k$,
(b) $a_0 + a_1 n + \ldots + a_k n^k \leq n^{k+1}$, but not $a_0 + a_1 n + \ldots + a_k n^k \approx n^{k+1}$.
(For the definition of $\leq$ and $\approx$ see Exercise 30.2.) This shows that the rate of growth of a polynomial can be identified with its degree.

30.4. In Definition 30.1. we defined the time complexity of a Turing machine in terms of a function t of the length of the input. In practice, *space complexity* is often just as important. The space used is defined to be the number of distinct tape squares visited by the read-x write head. For instance, the Turing machine T of Example 30.2. decides L in space s, where $s(n) = n + 3$.
Prove: any problem solvable in polynomial time is also solvable in polynomial space. (It is still an unresolved question whether problems exist solvable in polynomial space that cannot be solved in polynomial time.)

§31 GRAMMARS AND AUTOMATA:
TYPE 0, 1, 2 and 3 GRAMMARS; AUTOMATA

Chomsky's notion of (type 0) *grammar* is introduced and we prove that the languages generated by these grammars are precisely the languages recognized by Turing machines. In order to explain that some sentences which superficially have the same structure may have quite different meanings, Chomsky distinguishes between the *surface structure* and the *deep structure* of a sentence. By imposing certain restrictions on the type 0 grammars, Chomsky obtained type 1, 2 and 3 *grammars*. Correspondingly, one may impose certain restrictions on Turing machines, resulting in *linear bounded automata* , *pushdown automata* and *finite automata*. Without proof, we mention that the languages generated by type 1, 2 or 3 grammars turn out to be precisely the languages recognized by linear bounded automata, pushdown automata or finite automata, respectively.

GRAMMARS

In the sixties Noam Chomsky developed the notion of grammar which turned out to be important not only for linguistics but also for computer science, for instance, in building parsers and compilers.

One of the main reasons for Chomsky to introduce the notion of grammar was to explain the linguistic competence of people; that they are able to produce new sentences they have never read or heard before. In order to do so, Chomsky assumed that everybody is equipped with certain grammatical rules which can be applied again and again to produce more and more linguistic expressions.

It is well-known that a sentence (S) can be built from a noun phrase (NP) and a verb phrase (VP). Chomsky's basic idea was to represent this fact as a *production rule* $S \rightarrow NP + VP$. This rule should be read as follows: whenever symbol S occurs, it is allowed to rewrite S as the string consisting of the symbols NP and VP. Similar rewrite or production rules, also called *phrase structure rules*, exist for NP and VP. For instance, $NP \rightarrow Art + N$, expressing that a noun phrase (NP) can be built from an article (Art) and a noun (N); and $VP \rightarrow Aux + V + NP$, expressing that a verb phrase (VP) may consist of an auxiliary verb (Aux), a main verb (V) and a noun phrase (NP).

In order to produce English sentences, we also need rewrite or production rules of the form $Aux \to$ can, $Aux \to$ may, $Aux \to$ will, $Aux \to$ must, usually represented by $Aux \to$ (can, may, will, must) for short. And we also need rules such as $V \to$ (read, hit, eat), $Art \to$ (a, the) and $N \to$ (boy, man, frog).

The expressions "can", "read", "a", "boy", etc., are called *terminals* because there are no production rules starting with these expressions. The symbols S, NP, VP, Art, N, Aux, V, on the other hand, are called *non-terminals* for obvious reasons. Starting with the symbol S, we can, by repeated application of the production rules above, generate terminal strings to which no production rules can be applied. For instance, starting with S, the production rules mentioned above can generate the English sentence "the boy must eat the frog". The *derivation tree* or *phrase marker* for this sentence looks as follows.

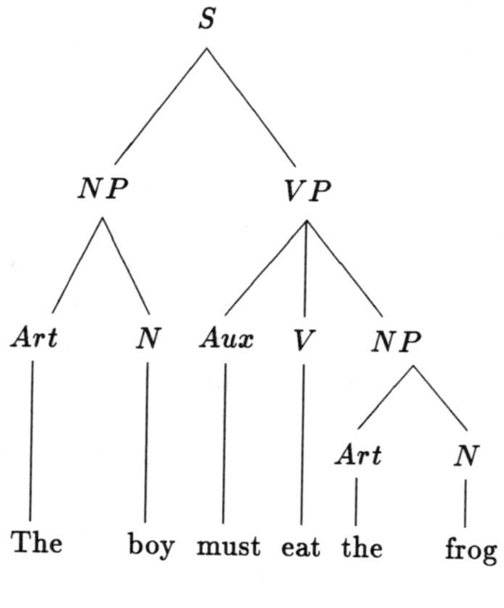

In this way the production rules above can generate a small fragment of English. Together they form what Chomsky called a *grammar*. In his view, the production rules in a grammar represent the linguistic competence of a speaker and are part of everyone's innate mental equipment.

DEFINITION 31.1.:
A (type 0) *grammar* G is a quadruple $\langle V_N,\ V_T,\ P,\ S\rangle$, where

i) V_N is a finite set; the elements of V_N are called *non-terminals.*
ii) V_T is a finite set such that V_N and V_T have no elements in common; the elements of V_T are called *terminals.*
iii) P is a finite set of expressions of the form $\alpha \to \beta$, where α and β are strings of finite length composed of symbols of V_N and/or V_T (i.e., $\alpha, \beta \in (V_N \cup V_T)^*$) and the length of α is at least 1 (i.e., $|\alpha| \geq 1$); the elements of P are called *productions.*
iv) $S \in V_N$; S is called the sentence- or *start-symbol.*

EXAMPLES 31.2.:

1. Let $G_1 = \langle V_N,\ V_T,\ P,\ S\rangle$ where
   $V_N = \{S, NP, VP, Art, N, Aux, V\}$,
   $V_T = \{$can, may, will, must, read, hit, eat, a, the, boy, man, frog$\}$, and
   P consists of the following productions:
   $S \to NP$ - VP , $Aux \to$ (can, may, will, must) ,
   $NP \to Art$ - N , $V \to$ (read, hit, eat) ,
   $VP \to Aux$ - V - NP , $Art \to$ (a, the) ,
   $N \to$ (boy, man, frog) .
2. Let $G_2 = \langle\{S\},\ \{0,1\},\ \{S \to 0S1, S \to 01\},\ S\rangle$.
   Here, S is the only non-terminal, 0 and 1 are terminals and there are two productions, $S \to 0S1$ and $S \to 01$.

For many more examples, we refer the reader to the exercises.

Further on in this section we shall put some restrictions on the productions in P and hence obtain grammers of type 1, 2 and 3, depending on the kind of restrictions.

DEFINITION 31.3.: Let $G = \langle V_N,\ V_T,\ P,\ S\rangle$ be a grammar and let α and β be finite strings composed of elements of V_N and/or V_T ($\alpha, \beta \in (V_N \cup V_T)^*$). $\alpha \overset{*}{\underset{G}{\Rightarrow}} \beta$ (β is *derivable from* α in grammar G) := β can be obtained from α by application of one or more productions in P. By convention, $\alpha \overset{*}{\underset{G}{\Rightarrow}} \alpha$ for each string α.

EXAMPLES 31.2. (continued): $S \overset{*}{\underset{G_1}{\Rightarrow}} Art$ - N - Aux - eat - NP and $VP \overset{*}{\underset{G_1}{\Rightarrow}} Aux$ - eat - the - frog. $S \overset{*}{\underset{G_2}{\Rightarrow}} 00S11$.

DEFINITION 31.4.: Let $G = \langle V_N,\ V_T,\ P,\ S\rangle$ be a grammar.

$$L(G) := \{w \in V_T^* \mid S \overset{*}{\underset{G}{\Rightarrow}} w\}, \text{ i.e.,}$$

$L(G)$, called *the language generated by G*, is by definition the set of all strings (or words) w of terminals such that $S \overset{*}{\underset{G}{\Rightarrow}} w$.

So, a string w is in $L(G)$ iff 1. w consists solely of terminals and
2. w is derivable from S in G.

EXAMPLES 31.2. (continued):

1) The reader should verify that the sentences "the boy must eat the frog" and "a man may hit the boy" both are in $L(G_1)$.
2) $L(G_2) = \{0^n1^n \mid n \geq 1\}$, i.e., the language generated by G_2 consists of all finite sequences composed of n 0's followed by the same number n of 1's ($n \geq 1$).

Consider the following three sentences.

1. Locusta is an alleged poisoner.
2. Locusta is a Roman poisoner.
3. Locusta is a skillful poisoner.

These three sentences have similar surface structures, but still intuitively we feel they are quite different in meaning. (An alleged poisoner is not another kind of poisoner along with Roman and Carthaginian or skillful and clumsy; a Roman poisoner is one who is both Roman and a poisoner, but a skillful strangler may be a clumsy poisoner.) In order to explain this, Chomsky distinguishes the *deep structure* of a sentence, determined by the production rules of a grammar and the *surface structure* of a sentence which results by applying to the derivation tree of the sentence certain transformation rules. Corresponding with their quite different meanings, the sentences 1, 2 and 3 above have radically different deep structures. Transformation rules transform these different deep structures into similar surface structures. (Do Exercise 31.1.) Postulating that the deep structure determines the meaning of a sentence, Chomsky explains in this way that the sentences 1, 2 and 3 above have quite different meanings although they have a similar surface structure.

The sentences "a man may hit the boy" and "the boy may be hit by a man", on the other hand, have the same meaning, although they are syntactically different. This can also be explained in terms of deep and surface structures. These sentences have the same deep structure and hence the same meaning. And the surface structure of one of these sentences is obtained by applying certain transformation rules to its deep structure. (Do Exercise 31.2.)

So, in Chomsky's view, the syntax of a language consists of two components:

i) a *base component*, containing the production (or phrase structure) rules. These rules generate the *deep structure* of each sentence.

ii) a *transformational component*, containing the transformation rules which map derivation trees into (other) derivation trees. The transformation rules take as input a deep structure and generate as output a *surface structure.*

The deep structure determines the meaning of a sentence; the surface structure determines its sound.

In the case of the sentences "a man will hit the boy", in the active mood, and "the boy will be hit by a man", in the passive mood, two surface structures are derived from one deep structure. And in the case of "Locusta is an alleged/Roman/skillful poisoner", similar surface structures are derived from several different deep structures.

In the following two theorems we shall show that the languages recognized by a Turing machine are precisely the languages generated by a (type 0) grammar.

THEOREM 31.5.: Let L be a language over the alphabet Σ_0. If L is Turing-acceptable, then there is a grammar G such that G generates L.

PROOF: Let $L \subseteq \Sigma_0^*$ and suppose L is Turing-acceptable. Then one can construct a Turing machine $T = \langle K,\ \Sigma,\ q_0,\ \delta\rangle$ such that $\Sigma_0 \subseteq \Sigma$ and for all $w \in \Sigma_0^*$,

$$w \in L \text{ iff } (q_0, \underline{\#}w\#) \vdash_T^* (h, \#w\underline{\#}).$$

Let $G := \langle V_N,\ V_T,\ P,\ S\rangle$ be the grammar with
$V_N := K \cup \{h,\ X,\ Y,\ S,\ \#\}$,
$V_T := \Sigma_0$, and P consisting of the following productions.

(1) $S \rightarrow \#X\#$,
$X \rightarrow Xa$ for each $a \in \Sigma_0$,
$\#X \rightarrow q_0\#$.

The productions in group (1) guarantee that $S \overset{*}{\underset{G}{\Rightarrow}} q_0\#w\#$ for any word $w \in \Sigma_0^*$. Remember that a configuration $(q, a_1 \ldots \underline{a_i} \ldots a_n)$ can be written as $a_1 \ldots a_{i-1}\; q\; a_i \ldots a_n$. So, the expression $q_0\#w\#$ corresponds with the configuration $(q_0, \underline{\#}w\#)$.

(2) For any instruction $\delta(p, a) = (q, b, O)$ in T, P contains the production $p\ a \rightarrow q\ b$.
For any instruction $\delta(p, a) = (q, b, R)$ in T, P contains the production $p\ a \rightarrow b\ q$.
For any instruction $\delta(p, a) = (q, b, L)$ in T, P contains the production $c\ p\ a \rightarrow q\ c\ b$ for every $c \in \Sigma_0 \cup \{\#\}$.

The productions in group (2) guarantee that
$(p, a_1 \ldots \underline{a_i} \ldots a_n) \vdash_T^* (q, b_1 \ldots \underline{b_j} \ldots b_m)$ iff

$$a_1 \ldots\ p\ a_i \ldots a_n \overset{*}{\underset{G}{\Rightarrow}} b_1 \ldots\ q\ b_j \ldots b_m,$$

and, in particular, that $(q_0, \underline{\#}w\#) \vdash_T^* (h, \#w\underline{\#})$ iff $q_0\#w\# \overset{*}{\underset{G}{\Rightarrow}} \#wh\#$.

(3) $h\# \rightarrow Y$,
$aY \rightarrow Ya$ for every $a \in \Sigma_0$,
$\#Y \rightarrow e$ (the empty string).

The productions in group (3) guarantee that $\#wh\# \overset{*}{\underset{G}{\Rightarrow}} w$. Now it is straightforward to check that $L(G) = L$, i.e., the language generated by G equals L. (Do Exercise 31.3.) □

Conversely, we also show that any language generated by a (type 0) grammar is recognized by some Turing machine.

THEOREM 31.6.: Let $G = \langle V_N,\ V_T,\ P,\ S\rangle$ be a grammar. Then there is a Turing machine T such that T recognizes $L(G)$.

PROOF: Let $G = \langle V_N,\ V_T,\ P,\ S\rangle$ be a grammar. We informally describe a Turing machine $T = \langle K,\ \Sigma,\ q_0,\ \delta\rangle$ that recognizes $L(G)$.
$\Sigma := V_N \cup V_T \cup \{\#\}$. We let δ be such that

i) $(q_0, \underline{\#}w\#) \vdash_T^* (q, \#w\#\underline{S}\#)$ for some $q \in K$, and
ii) T simulates the derivations in G in a systematic way, starting with S. So, $S \overset{*}{\underset{G}{\Rightarrow}} \alpha$ iff $(q_0, \underline{\#}w\#) \vdash_T^* (q', \#w\#\alpha\underline{\#})$ for some $q' \in K$.
iii) If α is a string of terminals (i.e., $\alpha \in V_T^*$), then T compares α to w and halts if they are equal; otherwise T continues to simulate the next derivation in G.

From the informal description of T it follows that T recognizes $L(G)$, i.e., for all $w \in V_T^*$, $w \in L(G)$ iff T halts on input w, or equivalently, $S \overset{*}{\underset{G}{\Rightarrow}} w$ iff T halts on input w. □

From the preceding two theorems the following corollary results.
COROLLARY 31.7.: Let L be a language. L is recognized by some Turing machine iff L is generated by some (type 0) grammar.

DEFINITION 31.8.: Let $G = \langle V_N,\ V_T,\ P,\ S\rangle$ be a (type 0) grammar.

i) G is *type* 1 or *context-sensitive* := all productions in P are of the form $\alpha_1 X \alpha_2 \to \alpha_1 \beta \alpha_2$ with $\alpha_1, \alpha_2, \beta \in (V_N \cup V_T)^*$, $X \in V_N$ and $|\beta| \geq 1$.
The production $\alpha_1 X \alpha_2 \to \alpha_1 \beta\ \alpha_2$ expresses that in the *context* of α_1 and α_2, X may be replaced by β.

ii) G is *type* 2 or *context-free* := all productions in P are of the form $X \to \beta$ with $X \in V_N$, $\beta \in (V_N \cup V_T)^*$ and $|\beta| \geq 1$.
The production $X \to \beta$ expresses that X may be replaced by β independent of the context in which X appears.

iii) G is *type* 3 or *regular* := all productions in P are of the form $X \to aY$ or $X \to a$ with $X, Y \in V_N$ and $a \in V_T$.

Notice that every regular (or type 3) grammar is context-free (or type 2), that every context-free (or type 2) grammar is context-sensitive (or type 1) and that every context-sensitive (or type 1) grammar is type 0. For examples of type 1, 2 and 3 grammars the reader is referred to the exercises, in particular Exercise 31.14.

DEFINITION 31.9.: Let L be a language over an alphabet Σ_0.
L is *recursively enumerable, context-sensitive, context-free* or *regular* := L is generated by a type 0, 1, 2 or 3 grammar respectively. (See Theorem 28.19 and Corollary 31.7. for an explanation of the expression "recursively enumerable".)

AUTOMATA

In Corollary 31.7. we have seen that the languages generated by a (type 0) grammar are precisely the languages recognized by a Turing machine.

Corresponding to the restrictions on the nature of the productions in type 1, 2 and 3 grammars, one can also put certain restrictions on Turing machines.

A *linear bounded automaton* is a nondeterministic Turing machine the read/write head of which never leaves those cells that contain the input. It turns out that

1. a language is accepted by a linear bounded automaton if and only if it is generated by a type 1 grammar, and
2. the context-sensitive (type 1) languages are a proper subclass of the resursively enumerable (type 0) languages.

A *pushdown automaton* can be conceived of as a nondeterministic Turing machine with a read-only input on which the input head can only move in one direction, plus a storage tape such that if the storage tape head moves left, it must print a blank. Consequently, the storage tape to the right of the head contains only blanks. It turns out that

1. a language is accepted by a pushdown automaton if and only if it is generated by a type 2 grammar, and
2. the context-free (type 2) languages are a proper subset of the context-sensitive (type 1) languages. In particular, $\{a^n b^n c^n \mid n \geq 1\}$ is context-sensitive, but not context-free. (See Exercise 31.12.)

A *finite automaton* is a quintuple $\langle K,\ \Sigma,\ q_0,\ \delta,\ F\rangle$, where $\delta : K \times \Sigma \rightarrow K$ and $F \subseteq K$ is a set of final states. $\delta(p, a) = q$, with $p, q \in K$ and $a \in \Sigma$, is to be understood as follows: when in state p and scanning the input symbol a, the finite automaton moves its read head one cell to the right and enters state q. Initially, the finite automaton is in state q_0 and the finite control is scanning the left-most symbol of the input. The input string is accepted when the read head reaches the end of the input string, the finite automaton being in a final state (i.e., an element of F).

If $\delta(p, a) = q$, we say that $(p, aw) \vdash (q, w)$. And the finite automaton $T = \langle K,\ \Sigma,\ q_0,\ \delta,\ F\rangle$ accepts w if and only if there is a state $q \in F$ such that $(q_0, w) \vdash_T^* (q, e)$.

For examples of finite automata, see Exercises 31.18. and 31.19.

It turns out that

1. a language is accepted by a finite automaton if and only if it is generated by a type 3 grammar, and
2. the regular (type 3) languages are a proper subset of the context-free (type 2) languages. In particular, $\{w\ c\ w^R \mid w \in \{0, 1\}^*\}$ and

$\{a^n b^n \mid n \geq 1\}$ are context-free, but not regular. (See Exercises 31.10. and 31.11.)

For more details, we refer the reader to J.E. Hopcroft and J.D. Ullman, *Introduction to Automata Theory, Languages and Computation* , or to H.R. Lewis and C.H. Papadimitriou, *Elements of the Theory of Computation.*

EXERCISES

31.1. Let G be a grammar with the following phrase structure rules (productions):

$S \to S$ - [and] - S
$S \to NP$ - VP
$NP \to (Art)$ - (Adj) - N
$VP \to V$ - [that] - S
$VP \to (Adv)$ - V - N
$VP \to Copula$ - NP
$VP \to Copula$ - Adj

$Art \to$ (a, an, the)
$Adj \to$ (roman, alleged, skilful)
$N \to$ (someone, Locusta, poisoner)
$V \to$ (allege, poison)
$Copula \to$ be
$Adv \to$ skilfully

Generate the phrase markers (derivation trees) in G for the deep structures of the following sentences:

1. Locusta is an alleged poisoner.
2. Locusta is a Roman poisoner.
3. Locusta is a skilful poisoner.

Various transformations give rise to the same surface structure.

31.2. Let G be the same grammar as in exercise 31.1. and consider the following transformation rule:

$$N_1 \text{ - } V \text{ - } N_2 \;\to\; N_2 \text{ - [be] - } V \text{ - [ed] - [by] - } N_1.$$

Check that this transformation rule, applied to the phrase marker (derivation tree) in G for the deep structure of "someone is poisoned by Locusta", which is the same as the one of "Locusta poisons someone", yields the phrase marker for the surface structure of "someone is poisoned by Locusta".

31.3. Let $L := \{w \in \{0,1\}^* \mid w \text{ contains at least one } 0\}$.

i) Construct a Turing machine T such that
$$w \in L \text{ iff } (q_0, \underline{\#}w\#) \vdash_T^* (h, \#w\underline{\#}).$$

ii) Following the proof of Theorem 31.5. construct a grammar G such that G generates L.

31.4. Construct a context-free grammar which generates precisely all formulas of the propositional calculus built from the atomic formulas Q, Q', Q'', ... by means of the connectives $\sim$, $\supset$, & , $\vee$ and $\neg$.

31.5. Let $G = \langle\{S\},\ \{0,1\},\ \{S \to 0S1, S \to 01\},\ S\rangle$. Show that $L(G) = \{0^n1^n \mid n \geq 1\}$, where 0^n stands for 0 repeated n times and similarly for 1^n.

31.6. Give a regular grammar generating the set of all finite strings w of 0's and 1's such that w does not contain two consecutive 1's.

31.7. Give a context-sensitive grammar generating the set of all finite strings w of a's, b's and c's such that w consists of equal numbers of a's, b's and c's.

31.8. Let $\{a,b\}^*$ be the set of all finite strings of a's and b's, including the empty string e of length 0. Let $L = \{w \in \{a,b\}^* \mid w \text{ contains an even number of } b\text{'s}\}$. Check that $L - \{e\}$ is generated by the type 3 (regular) grammar $G = \langle\{S,B\},\ \{a,b\},\ P,\ S\rangle$ with $P = \{S \to aS,\ S \to a,\ S \to bB,\ B \to aB,\ B \to bS,\ B \to b\}$.

31.9. With $\{a,b\}^*$ as in Exercise 31.8., let $L = \{w \in \{a,b\}^* \mid w \text{ does not contain three consecutive } b\text{'s}\}$. Verify that $L - \{e\}$ is generated by the type 3 (regular) grammar $G = \langle\{S,B,C,D\},\ \{a,b\},\ P,\ S\rangle$ with P consisting of the following productions:
$$\begin{array}{llll} S \to aS, & B \to aS, & C \to aS, & D \to aD, \\ S \to a, & B \to a, & C \to a, & D \to bD. \\ S \to bB, & B \to bC, & C \to bD, & \\ S \to b, & B \to b. & & \end{array}$$

31.10. The *reversal* of a string w, denoted by w^R, is the string "spelled backwards". For instance, $dog^R = god$ and $01^R = 10$.

Now let $L = \{w\ c\ w^R \mid w$ is a finite string of a's and b's$\}$. Check that L is generated by the type 2 (context-free) grammar $\langle\{S\},\ \{a,b,c\},\ P,\ S\rangle$ with $P = \{S \to aSa,\ S \to bSb,\ S \to c\}$.

FACT: $\{w\ c\ w^R \mid w \in \{a,b\}^*\}$ cannot be generated by a type 3 grammar. See H.R. Lewis and C.H. Papadimitriou, *Elements of the Theory of Computation*, Theorem 2.6.1., p. 75-76.

31.11. For n a natural number, let a^n be the string consisting of n a's. For instance, $a^5 = aaaaa$. Let $L = \{a^n b^n \mid n \geq 1\}$. So $aabb \in L$, but $abab \notin L$. Check that L is generated by the type 2 (context-free) grammar $\langle\{S\},\ \{a,b\},\ P,\ S\rangle$ with $P = \{S \to aSb, S \to ab\}$.

FACT: $\{a^n b^n \mid n \geq 1\}$ cannot be generated by a type 3 grammar. See H.R. Lewis and C.H. Papadimitriou, *Elements of the Theory of Computation*, Example 2.6.3., p. 75.

31.12. Let $L = \{a^n b^n c^n \mid n \geq 1\}$. (See Exercise 31.11.) Check that L is generated by the type 1 (context-sensitive) grammar $G = \langle\{S,B,C\},\ \{a,b,c\},\ P,\ S\rangle$ with P consisting of the following productions:

$$\begin{array}{llll} S \to aSBC, & CB \to BC, & bB \to bb, & cC \to cc. \\ S \to aBC, & aB \to ab, & bC \to bc, & \end{array}$$

FACT: $\{a^n b^n c^n \mid n \geq 1\}$ cannot be generated by a type 2 grammar. See Exercise 31.17., or H.R. Lewis and C.H. Papadimitriou, *Elements of the Theory of Computation*, Theorem 3.5.4., p. 126.

31.13. Let $L = \{w \in \{a,b\}^* \mid$ the number of a's in w is equal to the number of b's in $w\}$, where $\{a,b\}^*$ is defined as in Exercise 31.8. Check that L is generated by the grammar $G = \langle\{S,A,B\},\ \{a,b\},\ P,\ S\rangle$ with P consisting of the following productions:

$$\begin{array}{lll} S \to aB, & A \to a, & B \to b\ , \\ S \to bA, & A \to aS, & B \to bS\ , \\ & A \to bAA, & B \to aBB. \end{array}$$

What type of grammar is this?

31.14. Which of the following grammars G_i are regular (type 3), context-free (type 2) or context-sensitive (type 1)? Determine the languages $L(G_i)$.

$G_1 = \langle\{S,A\},\ \{a,b\},\ \{S \to aA,\ S \to b,\ A \to Sa\},\ S\rangle$

$G_2 = \langle\{S,A,B\},\ \{a,b\},\ \{S \to a,\ S \to aA,\ A \to aA,\ A \to a,$
$A \to bB,\ S \to b,\ S \to bB,\ B \to bB,\ B \to b\},\ S\rangle$

$G_3 = \langle\{S,B,C\},\ \{a,b,c\},\ \{S \to aSBC,\ S \to abc,\ CB \to BC,$
$bB \to bb,\ C \to c\},\ S\rangle$

$G_4 = \langle\{S,A,B,C,D\},\ \{a,b\},\ \{S \to AC,\ S \to BD,\ S \to AA,$
$S \to BB,\ C \to SA,\ D \to SB,\ A \to a,\ B \to b\},\ S\rangle$

$G_5 = \langle\{S,A,B\},\ \{a,b,c\},\ \{S \to AB,\ B \to bBc,\ B \to bc,$
$A \to aA,\ A \to a\},\ S\rangle$

$G_6 = \langle\{S\},\ \{a,b\},\ \{S \to aS,\ S \to Sb,\ S \to ab\},\ S\rangle$

$G_7 = \langle\{S,A,B\},\ \{a,b\},\ \{S \to AB,\ A \to aA,\ B \to bB\},\ S\rangle$

31.15. Prove:

i) if L_1 and L_2 are context-free languages, then the union $L_1 \cup L_2$ is context-free.

ii) if L_1 and L_2 are context-free languages, then the concatenation L_1L_2 is context-free.

iii) if L is a context-free language, then the closure (or Kleene star) of that single language, $L^* - \{e\}$, i.e., the set of all strings obtained by concatenating one or more strings from L, is context-free too.

iv) if L is context-free, then L^R, the language which contains the reversals of all strings w in L, is context-free (the reversal w^R of a string w, is the string "spelled backwards").

v) for some context-free grammars G_1 and G_2, $L(G_1) \cap L(G_2)$ is not context-free.

vi) not for all context-free languages L, the complement $\overline{L}$ is context-free.

31.16. i) Let $G = \langle\{S,B,C\},\ \{a\},\ \{S \to BC,\ S \to a,\ B \to a\},\ S\rangle$ be a context-free grammar. Convert G to an equivalent grammar G_1 with no *useless* symbols. A symbol $\alpha \in V_N \cup V_T$ is *useful* if there are $\beta_1, \beta_2 \in (V_N \cup V_T)^*$ and $w \in V_T^*$ such that $S \overset{*}{\underset{G}{\Rightarrow}} \beta_1\alpha\beta_2 \overset{*}{\underset{G}{\Rightarrow}} w$; otherwise α is *useless*.

ii) Let $G = \langle\{S,A,B\},\ \{b\},\ \{S \to BA,\ A \to B,\ B \to BB,$ $B \to b\},\ S\rangle$ be a context-free grammar. Convert G to an equivalent grammar G_2 with no *unit* productions and no *useless* symbols. A *unit* production is of the form $A \to B$ with $B \in V_N$.

iii) The *Chomsky Normal Form Theorem* states that every context-free language can be generated by a grammar for which all productions are of the form $A \to BC$ or $A \to a$; here A, B and C are non-terminals and a is a terminal.
Now consider the grammar $G = \langle \{S, A, B\},\ \{a, b\},\ P,\ S\rangle$ which has the productions:

$$\begin{array}{llll} S \to bA & , & S \to aB & , \\ A \to a & , & B \to b & , \\ A \to aS & , & B \to bS & , \\ A \to bAA & , & B \to aBB & . \end{array}$$

Find an equivalent grammar G_3 in Chomsky normal form.

31.17. *Pumping Theorem* for context-free languages:
Let $G = \langle V_N,\ V_T,\ P,\ S\rangle$ be a context-free grammar in Chomsky normal form, and let k be the number of elements in V_N. Then for all $w \in L(G)$, if $|w| > 2^{k-1}$, then w may be written as $w = u\ v\ z\ x\ y$, where $|v\ z\ x| \leq 2^k$ and v and x are not both e, such that for each integer $i \geq 0$, $u\ v^i\ z\ x^i\ y \in L(G)$. For a proof see J. E. Hopcroft and J. D. Ullman, *Formal Languages and their relation to Automata*, Theorem 4.7. Using the Pumping Theorem prove that $\{a^n b^n c^n \mid n \geq 1\}$ is not context-free.

31.18. Let $L = \{w \in \{a, b\}^* \mid w$ has an even number of b's$\}$. Check that L is accepted by the finite automaton
$T = \langle \{q_0, q_1\},\ \{a, b\},\ q_0,\ \delta,\ \{q_0\}\rangle$ with δ defined by
$\delta(q_0, a) = q_0,\ \delta(q_1, a) = q_1,\ \delta(q_0, b) = q_1,\ \delta(q_1, b) = q_0$.
(Confer Exercise 31.8.)

31.19. Let $L = \{w \in \{a, b\}^* \mid w$ does not contain three consecutive b's$\}$. Check that L is accepted by the finite automaton $T = \langle \{q_0, q_1, q_2, q_3\},\ \{a, b\},\ q_0,\ \delta,\ \{q_0, q_1, q_2\}\rangle$ with δ defined by
$\delta(q_0, a) = q_0,\quad \delta(q_1, a) = q_0,\quad \delta(q_2, a) = q_0,\quad \delta(q_3, a) = q_3,$
$\delta(q_0, b) = q_1,\quad \delta(q_1, b) = q_2,\quad \delta(q_2, b) = q_3,\quad \delta(q_3, b) = q_3.$
(Confer Exercise 31.9.)

§32 UNTYPED λ-CALCULUS:

INTRODUCTION; SYNTAX; THE POWER OF THE λ-CALCULUS; COMBINATORY LOGIC; REDUCTION AND CHURCH-ROSSER; REPRESENTABILITY OF RECURSIVE FUNCTIONS; REDUCTION STRATEGIES; MODELS OF THE λ-CALCULUS

J.H. GEUVERS and R.P. NEDERPELT

INTRODUCTION

The λ-calculus is a formal system that deals with two basic operations on functions: application and abstraction. It was developed by A. Church in the fourth decade of this century as a system for representing computable functions. There is, in fact, a large collection of λ-calculi, differing in the kind of abstractions and applications that are allowed and in the specific additional constants that are part of the system. In this section we are mainly concerned with the *pure* λ-calculus. In the next section we shall deal with systems of *typed λ-calculus*, which can be seen as subsystems of the pure one in the sense that application and abstraction are restricted by the types.

One of the main reasons for introducing the λ-calculus is to have a flexible formal system for denoting functions. For example, the function f that takes a number to the power two and then adds three is usually denoted by

$$f : x \mapsto x^2 + 3 \text{ or } f(x) := x^2 + 3.$$

In the λ-calculus this function is denoted by

$$\lambda x \,.\, (x^2 + 3),$$

which eliminates the need to introduce a new specific symbol for any newly defined function and also makes it easier to speak about functions of a higher order such as

$$\lambda f.\lambda x.f(f(f(x))),$$

which takes a function f as input and gives as output the function f iterated three times. This is the operation of *abstraction*: if t is an expression representing, say, a natural number, the expression $\lambda x.t$ denotes the function that maps an input value c to the output value $t[x := c]$, where $t[x := c]$ denotes t with c substituted for x. If in the first example we take $x^2 + 3$ for t, then we obtain by abstraction (over x) the function $\lambda x \,.\, (x^2 + 3)$. This function maps c to $c^2 + 3$, and also x to $x^2 + 3$ and $y + 3$ to $(y + 3)^2 + 3$.

The character λ was chosen by Church as an auxiliary symbol to denote the abstraction operation. Moreover, Church used a pair of extra parentheses, writing $(\lambda x.t)$ instead of $\lambda x.t$.

The other basic operation on functions is *application*, which takes a function f and an input t to form (ft). In the first example we can apply $\lambda x \,.\, (x^2 + 3)$ to the value 5 to obtain $((\lambda x \,.\, (x^2 + 3))5)$. Normally, this is written as $(\lambda x \,.\, (x^2 + 3))\,(5)$, but one of the conventions of the λ-calculus is to place the brackets as above. We discuss this convention in more detail later when we give the precise definitions.

The operations of abstraction and application allow the formation of new *λ-terms* from old ones. Following the intuitive interpretations of the operations of abstraction and application as discussed above, there is an equivalence relation on the set of λ-terms which is generated by the base equality

$$(\lambda x.t)s = t[x := s].$$

This is the so-called *β-rule*, which states that the function $(\lambda x.t)$ applied to s is equal to $t[x := s]$, the result of substituting s for x in t. If we think of the λ-calculus as a system for representing computations, the β-rule is not a symmetric equality but should be directed as a *rewrite rule* as follows

$$(\lambda x.t)s \rightarrow t[x := s].$$

Terms that contain subterms of the form $(\lambda x.t)s$ are seen as 'intermediate results', leading to a value by consecutive rewritings.

In this section we are mainly concerned with the pure (untyped) λ-calculus, whose terms are built up from variables by using application and abstraction only. This system, which can be seen as a small programming language, has a remarkably strong expressive power in terms of representability of (recursive) functions. It will be shown that all partial recursive functions can be represented in it. Owing to the exploitation of this fact, the λ-calculus has become the theoretical basis of functional programming languages. Looking at the system from this point of view, the β rewrite rule discussed above corresponds in an obvious way to evaluation of a program. Therefore, the λ-calculus serves as a good theoretical framework for studying different orders of evaluation, of which we shall discuss one at the end of this section. Finally, we devote one subsection to the mathematical structures that can serve as models for the untyped lambda calculus.

SYNTAX

We define the untyped lambda calculus and the relevant notions of equality related to it. It is built up from a base set of variables by using application and abstraction. Therefore, in the following, let Var be a countable set of variables.

DEFINITION 32.1.: The set of *λ-terms* of the untyped λ-calculus, Λ, is inductively defined by the following clauses:

$$\begin{array}{rl} \text{if } x \in \mathsf{Var}, \text{ then} & x \in \Lambda, \\ \text{if } t \in \Lambda,\ x \in \mathsf{Var}, \text{ then} & (\lambda x.t) \in \Lambda\ (\textit{abstraction}), \\ \text{if } s,t \in \Lambda, \text{ then} & (st) \in \Lambda\ (\textit{application}). \end{array}$$

In abstract syntax this reads as follows:

$$\Lambda ::= \mathsf{Var} \mid (\lambda \mathsf{Var}.\Lambda) \mid (\Lambda\Lambda).$$

In the following, x, y, z, p, q denote arbitrary variables and s, t, u, v denote arbitrary λ-terms. For convenience, we omit some of the brackets in a term. Here we follow the usual conventions for denoting λ-terms.

NOTATION 32.2.: Brackets are to be inferred by associating to the left, so stu denotes $((st)u)$. The outermost brackets are not written, so, for example, instead of $(\lambda x.(\lambda z.(\lambda y.zx)(\lambda q.xq)))$, we write $\lambda x.(\lambda z.(\lambda y.zx)(\lambda q.xq))$. Furthermore, a sequence of λ's is compressed to one, so $\lambda xyz.t$ denotes $\lambda x.(\lambda y.(\lambda z.t))$. Application has priority over abstraction; so, $\lambda x.yz$ means $\lambda x.(yz)$ and not $(\lambda x.y)z$.

The λ is a *variable-binder*, just like the ∀ in predicate logic. The distinction between *free* and *bound* variables is very important in the λ-calculus; first, because terms that only differ in their bound variables are identified: the term $\lambda x.yx$ is the same as $\lambda z.yz$, and second, because substitution should be done with the usual care: a term may not be substituted for a bound variable, nor are free variables allowed to become bound by a substitution. An example of the first feature is

$$(\lambda x.xy)[x := \lambda z.z],$$

where no substitution is performed because there is no free x in $\lambda x.xy$. So we see that $(\lambda x.xy)[x := \lambda z.z]$ is $\lambda x.xy$. An example of the second feature is

$$(\lambda x.xy)[y := x],$$

where the free variable x in $[y := x]$ would become bound after the substitution. The solution is to rename the bound x, for instance, by p, and then perform the substitution. Thus, $(\lambda x.xy)[y := x]$ is the same as $(\lambda p.py)[y := x]$, which in its turn is the same as $\lambda p.px$. Of course, the use of λ as a variable binder is not different from the use of $\forall$ in predicate logic, but we give some explicit definitions because the notion of substitution is so prominent in the λ-calculus.

DEFINITION 32.3.: For t a λ-term, the set of free variables $FV(t)$, respectively the set of bound variables $BV(t)$, is defined inductively as follows:

$$\begin{array}{rclrcl} FV(x) & = & \{x\}, & BV(x) & = & \emptyset, \\ FV(\lambda x.t) & = & FV(t) - \{x\}, & BV(\lambda x.t) & = & BV(t) \cup \{x\}, \\ FV(st) & = & FV(s) \cup FV(t), & BV(st) & = & BV(s) \cup BV(t). \end{array}$$

Two λ-terms s and t are equal by *α-conversion* or *syntactically equal*, notation $s \equiv t$, if one is obtained from the other by renaming bound variables.

Similarly, we can speak about the *free occurrence* of a variable in a specific term, meaning that at that place the variable does not appear 'under a λ'. For example, the first occurrence of x in the term $\lambda y.x(\lambda x.yx)$ is free, the second and third are bound. A *closed term* or *combinator* is a λ-term in which all variables are bound.

We do not define substitution explicitly but just note that it should be done with the necessary care, preventing free variables from becoming bound after substitution. In practice this means that we adopt the so-called *variable convention*, which says that if at some place we discuss the terms $s_1, s_2, \ldots, s_n$, then all the bound variables in these terms are different from the free ones. This implies that we can always perform the substitution of t for x in s without first having to rename bound variables. We also freely adopt the usual notation for simultaneous substitution, writing $s[\vec{x} := \vec{t}]$ for the result of substituting t_i for x_i in s for each $i \in \{1, \ldots, n\}$ (assuming $\vec{x}$ to stand for $x_1, \ldots, x_n$).

EXAMPLES 32.4.:

1. $(\lambda y.x(\lambda x.yx))[x := \lambda z.xz] \equiv \lambda y.(\lambda z.xz)(\lambda x.yx)$,
2. $(\lambda y.yx)[y := t] \equiv \lambda y.yx$,
3. $\lambda xy.yx \equiv \lambda yx.xy$,
4. $(\lambda x.yxy)[y := zx] \equiv (\lambda p.ypy)[y := zx] \equiv \lambda p.zxp(zx)$.

The really interesting equality on Λ is the one generated by the β-rule.

DEFINITION 32.5.: On Λ, *β-equality* or *β-conversion*, denoted by $=_\beta$, is the equivalence relation axiomatised as follows.

$$\begin{array}{lrcl}
\beta & (\lambda x.s)t & =_\beta & s[x := t], \\
\xi & s =_\beta t & \Rightarrow & \lambda x.s =_\beta \lambda x.t, \\
 & s =_\beta t,\ u =_\beta v & \Rightarrow & su =_\beta tv, \\
 & s & =_\beta & s, \\
 & s =_\beta t,\ t =_\beta u & \Rightarrow & s =_\beta u, \\
 & s =_\beta t & \Rightarrow & t =_\beta s.
\end{array}$$

The theory of β-equality on Λ is referred to as $\boldsymbol{\lambda}_\beta$.

The β-rule is the basic one, representing the intuitive meaning of the operations of abstraction and application. The last three rules turn the relation $=_\beta$ into an equivalence relation. The ξ-rule says that β-equality is compatible with abstraction and the third rule says that β-equality is compatible with application.

Besides the β-rule, there is another basic equality, which is given by the *η-rule*. This rule equals any two λ-terms that have the same behaviour as functions. The idea is simple: for any term s and fresh variable x (i.e., x is not free in s), the term s has the same *applicative behaviour* as $\lambda x.sx$. That is, for any term t

$$(\lambda x.sx)t =_\beta st.$$

So, $\lambda x.sx$ and s are the same functions. (In general, $\lambda x.sx$ and s are not β-equal.) The η-rule equates the terms $\lambda x.sx$ and s.

DEFINITION 32.6.: The theory $\boldsymbol{\lambda}_{\beta\eta}$ is the extension of $\boldsymbol{\lambda}_\beta$ with the η-rule. The equality in this theory is denoted by $=_{\beta\eta}$.

$$\eta \qquad\qquad \lambda x.sx =_{\beta\eta} s \text{ if } x \notin FV(s).$$

The η-rule represents a form of *extensionality*. It is not difficult to show that we have: if $su =_{\beta\eta} tu$ for all λ-terms u, then $s =_{\beta\eta} t$. This is left as an exercise (see Exercise 32.3).

NOTATION 32.7.: In the following we shall mainly be concerned with the theory $\boldsymbol{\lambda}_\beta$. Often we shall write $s =_\beta t$ instead of $\boldsymbol{\lambda}_\beta \vdash s =_\beta t$.

THE POWER OF THE λ-CALCULUS

Although the language of the λ-calculus is simple, it is still remarkably powerful. In this subsection we discuss representations of data such as numerals and booleans in the λ-calculus, and how to operate (i.e., write programs) with them. In the next subsection we look at *combinatory logic*, a system whose development is closely connected to that of the λ-calculus and which is equally powerful (and simple). But first we treat the crucial feature of the λ-calculus that establishes its power: the fact that every term has a *fixed point* and that there is a *fixed point combinator*. Because of this feature many functions can be defined and many equations can be solved. On the other hand, this general aspect of the λ-calculus is counter-intuitive because we certainly do not expect all functions to have a fixed point.

DEFINITION 32.8.: Let s and t be λ-terms. t is a *fixed point* of s if

$$st =_\beta t.$$

The λ-term F is a *fixed point combinator* if for each λ-term s, Fs is a fixed point of s. That is, for all s, $Fs =_\beta s(Fs)$.

The presence of a fixed point combinator makes it possible to solve equations like

$$(1)\ sx =_\beta xs.$$

(That is, find a λ-term s such that $sx =_\beta xs$.) Suppose we have a fixed point combinator F. Then this equation is solved by noting that if s is a solution to

$$(2)\quad s =_\beta (\lambda px.xp)s,$$

then it is also a solution to (1). Now, a solution to (2) is easily found by taking a fixed point of $\lambda px.xp$. So, we can take

$$s \equiv F(\lambda px.xp).$$

We could now proceed by simply giving a fixed point combinator: the most well-known one, Y, is rather short and it is easy to check that it is indeed a fixed point combinator. However, we proceed a little differently, following P. Odifreddi [1989] and show how this fixed point combinator arises quite naturally from the operations of abstraction and application.

First note that the set Λ is countable. Say

$$t_1, t_2, t_3, \ldots$$

is an enumeration of Λ. Let p be a variable. (It is instructive to think of p as the term whose fixed point we are going to construct.) Now consider the following two-dimensional diagram, consisting of all the possible applications of a t_i to a t_j.

$$\begin{array}{cccccc}
 & 1 & 2 & 3 & j & \\
1 & t_1t_1 & t_1t_2 & t_1t_3 & t_1t_j & \ldots \\
2 & t_2t_1 & t_2t_2 & t_2t_3 & t_2t_j & \ldots \\
3 & t_3t_1 & t_3t_2 & t_3t_3 & t_3t_j & \ldots \\
 & & & & & \\
i & t_it_1 & t_it_2 & t_it_3 & t_it_j & \ldots \\
\vdots & \vdots & \vdots & \vdots & \vdots & \ddots
\end{array}$$

The diagonal of this diagram consists of the terms

$$t_1t_1,\ t_2t_2,\ t_3t_3, \ldots\ .$$

Among the t_i we also find the term $\lambda x.p(xx)$, say it is t_n. At place n in the diagonal we now find t_nt_n as a fixed point of p because

$$t_nt_n \equiv (\lambda x.p(xx))t_n =_\beta p(t_nt_n).$$

So, every variable p has $(\lambda x.p(xx))(\lambda x.p(xx))$ as a fixed point and if we abstract from p, we find that

$$\lambda p.(\lambda x.p(xx))(\lambda x.p(xx))$$

is a fixed point combinator.

The fixed point combinator constructed above is called Y or the *paradoxical combinator*. The construction can be seen as an application of Cantor's diagonal method which is used here to derive a positive result.

THEOREM 32.9.: The term $\mathsf{Y} := \lambda p.(\lambda x.p(xx))(\lambda x.p(xx))$ is a fixed point combinator.

PROOF: Indeed, for any $s \in \Lambda$, $\mathsf{Y}s =_\beta (\lambda x.s(xx))(\lambda x.s(xx)) =_\beta s((\lambda x.s(xx))(\lambda x.s(xx))) =_\beta s(\mathsf{Y}s)$. □

We have seen the usefulness of a fixed point combinator for solving the equation $sx =_\beta xs$. This was done by first finding an equation of the form $ts =_\beta s$ (t not containing s) with the property that, if s is a solution of $ts =_\beta s$, then it is also a solution of $sx =_\beta xs$. Then one can take s to be $\mathsf{Y}t$.

To be able to program in the λ-calculus, it is necessary to have at one's disposal, for example, the booleans with an *if - then - else* - construction and the numerals. They are given by the following definition.

DEFINITION 32.10.: The booleans T (true) and F (false) and the *Church numerals* c_n for $n \in \mathbb{N}$ are given by the following combinators. Let $n \in \mathbb{N}$.

$$\begin{array}{rcll} \text{(i)} & \mathsf{T} & := & \lambda xy.x, \\ \text{(ii)} & \mathsf{F} & := & \lambda xy.y, \\ \text{(iii)} & c_n & := & \lambda fx.f^n(x), \end{array}$$

where $f^0(x) := x$, $f^1(x) := fx$ and $f^{n+1}(x) := f(f^n(x))$ for $n > 0$. The combinators T and F are also known as K and K_* respectively.

Other standard combinators are S and I, defined by

$$\begin{array}{rcll} \text{(iv)} & \mathsf{S} & := & \lambda xyz.xz(yz), \\ \text{(v)} & \mathsf{I} & := & \lambda x.x. \end{array}$$

Thus, the combinator T is a first projection and F is a second projection ($\mathsf{T}st =_\beta s$ and $\mathsf{F}st =_\beta t$). The number n is represented by the combinator that iterates a function n times. To see why these definitions make sense, some examples are treated below.

EXAMPLES 32.11.:

1. If s, t and u are terms, the term 'If s then t else u' is represented by stu. This makes sense because $\mathsf{T}tu =_\beta t$ and $\mathsf{F}tu =_\beta u$ in $\boldsymbol{\lambda}_\beta$.

2. The successor is represented by the combinator $\mathsf{S}^+ := \lambda cfx.f(cfx)$. Make sure that $\mathsf{S}^+ c_n =_\beta c_{n+1}$. Note that this is not the only possibility. The combinator $\mathsf{S'}^+ := \lambda cfx.cf(fx)$ also represents the successor.

3. Addition can be defined on the Church numerals by taking

   $\mathsf{A} := \lambda cdfx.cf(dfx)$. Check that $\mathsf{A}c_n c_m =_\beta c_{n+m}$ in $\boldsymbol{\lambda}_\beta$.

4. The booleans T and F can also be used as projections in combination with a pairing combinator. For s and t λ-terms, the pair consisting of s and t is given by $[s,t] := \lambda z.zst$. Then $[s,t]\mathsf{T} =_\beta s$ and

   $[s,t]\mathsf{F} =_\beta t$ in $\boldsymbol{\lambda}_\beta$.

COMBINATORY LOGIC

In the same period in which Church developed the λ-calculus, Schönfinkel and Curry developed the system of *combinatory logic* . It was meant to be a formalism for studying manipulations of functions, but also to serve as a framework for the foundation of mathematics. Because of its strong connections with the λ-calculus, we treat it here.

DEFINITION 32.12.: The system of *combinatory logic*, **CL**, has as terms the set $\mathcal{C}$, inductively defined by the following clauses:

$$\begin{array}{l} \mathbf{K},\ \mathbf{S} \in \mathcal{C}, \\ \text{if } x \in \mathsf{Var}, \text{ then } x \in \mathcal{C}, \\ \text{if } s,t \in \mathcal{C}, \text{ then } (st) \in \mathcal{C}. \end{array}$$

For **CL** we have the same notions and conventions as we have for the λ-calculus. So, for example, a *closed term* (or *combinator*) is a term that has no free variables.

DEFINITION 32.13.: The equality on $\mathcal{C}$, $=_w$, is the equivalence relation axiomatised as follows:

$$
\begin{array}{rcl}
\mathbf{K}st & =_w & s, \\
\mathbf{S}stu & =_w & su(tu), \\
s =_w t,\ u =_w v & \Rightarrow & su =_w tv, \\
s & =_w & s, \\
s =_w t,\ t =_w u & \Rightarrow & s =_w u, \\
s =_w t & \Rightarrow & t =_w s.
\end{array}
$$

If the terms s and t are equal via this equality relation, this is denoted by

$$\mathbf{CL} \vdash s =_w t.$$

Note the correspondence of **K** and **S** with K and S of Definition 32.10. This correspondence can be made precise by defining equality-preserving mappings between Λ and $\mathcal{C}$. These mappings will show that the λ-calculus and combinatory logic are equally powerful (with respect to the representation of recursive functions) and also that all closed terms of the λ-calculus can be generated (up to β-conversion) from the two combinators K and S.

REMARK 32.14.: It is convenient to have some extra combinators defined in terms of **S** and **K**. The most well-known one is the identical combinator **I**.

$$\mathbf{I} := \mathbf{SKK}$$

EXAMPLES 32.15.:

1. $\mathbf{CL} \vdash \mathbf{SKK}x =_w x.$
2. $\mathbf{CL} \vdash \mathbf{S}(\mathbf{KS})\mathbf{K}xyz =_w x(yz).$
3. $\mathbf{CL} \vdash \mathbf{SK}xy =_w \mathbf{KI}xy.$

See also exercise 32.7. It is obvious that the functions definable in **CL** can also be defined in the λ-calculus: just replace **K** by $\lambda xy.x$ and **S** by $\lambda xyz.xz(yz)$. In the opposite direction, the first step is taken by showing how to do abstraction in combinatory logic.

DEFINITION 32.16.: For s and t, terms of **CL**, and x, a variable, we define the **CL** term $\lambda^* x.t$ as follows:

$$\begin{array}{lcll} \lambda^* x.x & := & \mathbf{SKK}, & \\ \lambda^* x.s & := & \mathbf{K}s & \text{if } x \notin FV(s), \\ \lambda^* x.st & := & \mathbf{S}(\lambda^* x.s)(\lambda^* x.t) & \text{if } x \in FV(st). \end{array}$$

That this definition makes sense is shown by the following lemma.

LEMMA 32.17.:

1. $FV(\lambda^* x.s) = FV(s) - \{x\}$,
2. $\mathbf{CL} \vdash (\lambda^* x.u)v =_w u[x := v]$.

PROOF: See exercise 32.8. □

The above definition of abstraction does not always produce what one might expect. For example, $\lambda^* y.x \equiv \mathbf{K}x$ and hence $\lambda^* x.(\lambda^* y.x) \equiv \mathbf{S}(\mathbf{KK})\mathbf{I}$, whereas one would expect $\lambda^* x.\lambda^* y.x \equiv \mathbf{K}$.

The abstraction operation in **CL** is used in defining the mapping from the λ-calculus to combinatory logic.

DEFINITION 32.18.: The mappings $(-)_\lambda$ from $\mathcal{C}$ to Λ and $(-)_{\mathbf{CL}}$ from Λ to $\mathcal{C}$ are defined as follows:

$$\begin{array}{lcl} (x)_\lambda & := & x, \\ (\mathbf{K})_\lambda & := & \lambda xy.x, \\ (\mathbf{S})_\lambda & := & \lambda xyz.xz(yz), \\ (st)_\lambda & := & (s)_\lambda (t)_\lambda. \end{array} \qquad \begin{array}{lcl} (x)_{\mathbf{CL}} & := & x, \\ (st)_{\mathbf{CL}} & := & (s)_{\mathbf{CL}}(t)_{\mathbf{CL}}, \\ (\lambda x.s)_{\mathbf{CL}} & := & \lambda^* x.(s)_{\mathbf{CL}}. \end{array}$$

THEOREM 32.19.: For s and t in $\mathcal{C}$,

$$\text{If } \mathbf{CL} \vdash s =_w t, \text{ then } \boldsymbol{\lambda}_\beta \vdash (s)_\lambda =_\beta (t)_\lambda.$$

PROOF: By an easy induction on the proof of $\mathbf{CL} \vdash s =_w t$. □

The reverse is not valid.

FACT 32.20.: It is not true that for every s and t in Λ,

$$\text{If } \boldsymbol{\lambda}_\beta \vdash s =_\beta t, \text{ then } \mathbf{CL} \vdash (s)_{\mathbf{CL}} =_w (t)_{\mathbf{CL}}.$$

A possible proof breaks down because the ξ-rule is not valid in **CL**. The following is a counterexample to the reverse of Theorem 32.19:

$\boldsymbol{\lambda}_\beta \vdash \lambda x.y =_\beta \lambda x.(\lambda z.y)x$, but $\mathbf{CL} \not\vdash \mathbf{K}y =_w \mathbf{S}(\mathbf{K}(\mathbf{K}y))\mathbf{I}$. (See exercise 32.9). It is also true that in general

$$\mathbf{CL} \not\vdash ((s)_\lambda)_{\mathbf{CL}} =_w s.$$

(See exercise 32.9 for the construction of a counterexample.)

It is not difficult to extend **CL** such that this extended theory is equivalent to $\boldsymbol{\lambda}_{\beta\eta}$.

DEFINITION 32.21.: The *extensionality rule* in **CL** is

$$(ext) \quad sx =_w tx \ (x \notin FV(st)) \Rightarrow s =_w t.$$

The theory **CL** extended with this rule is denoted by $\mathbf{CL}^{ext}$.

THEOREM 32.22.:

1. For s in $\mathcal{C}$, $\mathbf{CL}^{ext} \vdash ((s)_\lambda)_{\mathbf{CL}} =_w s$.
2. For s in Λ, $\boldsymbol{\lambda}_\beta \vdash ((s)_{\mathbf{CL}})_\lambda =_\beta s$.

PROOF: Both by induction on the structure of terms. We treat some of the details for the case of **CL**.

- $s \equiv \mathbf{K}$. Then $((\mathbf{K})_\lambda)_{\mathbf{CL}} \equiv \mathbf{S}(\mathbf{KK})\mathbf{I}$. Now $\mathbf{S}(\mathbf{KK})\mathbf{I}x =_w \mathbf{K}x$, and so by (ext), $\mathbf{S}(\mathbf{KK})\mathbf{I} =_w \mathbf{K}$ in $\mathbf{CL}^{ext}$.
- $s \equiv \mathbf{S}$. A long calculation shows that $((\mathbf{S})_\lambda)_{\mathbf{CL}}x =_w \mathbf{S}x$; use (ext) as above.
- $s \equiv ut$. Then $((ut)_\lambda)_{\mathbf{CL}} =_w ((u)_\lambda)_{\mathbf{CL}}((t)_\lambda)_{\mathbf{CL}}$ and we are done by induction hypothesis. □

The equivalence of $\mathbf{CL}^{ext}$ and $\boldsymbol{\lambda}_{\beta\eta}$ is now shown by the following theorem.

THEOREM 32.23.:

For $s, t \in \Lambda$, $\boldsymbol{\lambda}_{\beta\eta} \vdash s =_{\beta\eta} t$ iff $\mathbf{CL}^{ext} \vdash (s)_{\mathbf{CL}} =_w (t)_{\mathbf{CL}}$.

For s and t in $\mathcal{C}$, $\mathbf{CL}^{ext} \vdash s =_w t$ iff $\boldsymbol{\lambda}_{\beta\eta} \vdash (s)_\lambda =_{\beta\eta} (t)_\lambda$.

PROOF: The implication from left to right is in both cases by induction on the derivation that proves the equality in $\boldsymbol{\lambda}_{\beta\eta}$, respectively $\mathbf{CL}^{ext}$. The implication from right to left for the first case uses the implication left to right for the second case and Theorem 32.22. The implication from right to left for the second case is similar. □

REDUCTION AND CHURCH-ROSSER

As has already been pointed out, the intuition behind the equality in the λ-calculus is reduction. This conforms with the idea that a term that contains a subterm of the form $(\lambda x.s)t$ is a kind of 'intermediate value'. The subterm $(\lambda x.s)t$ evaluates to $s[x := t]$. The reduction approach to the equality also gives a method for deciding whether two terms are equal.

DEFINITION 32.24.: The notion of *one-step β-reduction* on λ-terms, $\rightarrow_\beta$, is defined as follows:

$$\begin{array}{lrcl} (\beta) & (\lambda x.s)t & \rightarrow_\beta & s[x := t], \\ (\xi) & s \rightarrow_\beta t & \Rightarrow & \lambda x.s \rightarrow_\beta \lambda x.t, \\ & s \rightarrow_\beta t & \Rightarrow & su \rightarrow_\beta tu, \\ & s \rightarrow_\beta t & \Rightarrow & us \rightarrow_\beta ut. \end{array}$$

The last three rules are also called 'compatibility rules' because they ensure that the reduction is *compatible* with the operations on λ-terms. A term of the form $(\lambda x.s)t$ is called a *β-redex* or just *redex* (*reducible expression*).

DEFINITION 32.25.: Similar to the one-step β-reduction is the *one-step η-reduction*, $\rightarrow_\eta$, and *one-step βη-reduction*, $\rightarrow_{\beta\eta}$. For one-step η-reduction replace in Definition 32.24 the case (β) by

$$(\eta) \quad \lambda x.sx \quad \rightarrow_\eta \quad s \quad \text{if } x \notin FV(s).$$

The notion of one-step βη-reduction, $\rightarrow_{\beta\eta}$, is defined as the union of $\rightarrow_\beta$ and $\rightarrow_\eta$. A term of the form $\lambda x.sx$ with $x \notin FV(s)$ is called an *η-redex.*

Owing to the compatibility rules, the reduction of a term s to a term t is determined by rewriting a redex in s, say $(\lambda x.u)v$, to its *reduct* $u[x := v]$. The redex $(\lambda x.u)v$ is said to be *contracted* in this process. Below are some examples of reductions.

EXAMPLES 32.26.:

1. $\mathsf{K}st \rightarrow_\beta (\lambda y.s)t \rightarrow_\beta s.$
2. $(\lambda xy.y(yx))\mathsf{KI} \rightarrow_\beta (\lambda y.y(y\mathsf{K}))\mathsf{I} \rightarrow_\beta \mathsf{I}(\mathsf{IK}).$

3. $c_1 \equiv \lambda fx.fx \rightarrow_\eta \lambda f.f \equiv \mathsf{I}$.

4. $\lambda x.\mathsf{K}zxx \rightarrow_\beta \lambda x.(\lambda y.z)xx \rightarrow_\beta \lambda x.zx \rightarrow_\eta z$.

5. For $x \notin FV(s)$, $\lambda x.(\lambda y.s)x \rightarrow_\beta \lambda x.s[y := x]$ and
$\lambda x.(\lambda y.s)x \rightarrow_\eta \lambda y.s$.

Like the conversion (the equality of $\boldsymbol{\lambda}_\beta$ or $\boldsymbol{\lambda}_{\beta\eta}$), reduction is also done 'modulo renaming of bound variables'. So in fact we see $\rightarrow_\beta$ and $\rightarrow_{\beta\eta}$ as relations between equivalence classes modulo α-conversion. Note that in the last example the term $\lambda x.(\lambda y.s)x$ reduces to the same term, regardless of whether we contract the β-redex or the η-redex.

The relation $\rightarrow_\beta$ (one-step β-reduction) is extended to the relation $\twoheadrightarrow_\beta$, called *β-reduction*, by taking the transitive reflexive closure. This is again extended to the equivalence relation $\simeq_\beta$ by taking the transitive symmetric closure of $\twoheadrightarrow_\beta$. A more perspicuous definition is the following.

DEFINITION 32.27.: The relations $\twoheadrightarrow_\beta$ and $\simeq_\beta$ on Λ are defined as follows:

$$\begin{array}{rcl} & & s \twoheadrightarrow_\beta s, \\ s \rightarrow_\beta t & \Rightarrow & s \twoheadrightarrow_\beta t, \\ s \twoheadrightarrow_\beta t,\ t \twoheadrightarrow_\beta u & \Rightarrow & s \twoheadrightarrow_\beta u. \\ s \twoheadrightarrow_\beta t & \Rightarrow & s \simeq_\beta t, \\ s \simeq_\beta t & \Rightarrow & t \simeq_\beta s, \\ s \simeq_\beta t,\ t \simeq_\beta u & \Rightarrow & s \simeq_\beta u. \end{array}$$

The notions $\twoheadrightarrow_\eta$, $\twoheadrightarrow_{\beta\eta}$, $\simeq_\eta$ and $\simeq_{\beta\eta}$ are defined analogously, following Definition 32.25. In the following, when we write $\rightarrow$ ($\twoheadrightarrow$ respectively), we arbitrarily refer to either $\rightarrow_\beta$ or $\rightarrow_{\beta\eta}$ ($\twoheadrightarrow_\beta$ or $\twoheadrightarrow_{\beta\eta}$ respectively).
Note that the relation $\twoheadrightarrow_\beta$ is transitive and reflexive (being the transitive reflexive closure of $\rightarrow_\beta$) and that the relation $\simeq_\beta$ is symmetric, transitive and reflexive (being the transitive symmetric closure of $\twoheadrightarrow_\beta$). Hence, $\simeq_\beta$ is an equivalence relation.

The equality $=_{\beta(\eta)}$ of the theory $\boldsymbol{\lambda}_{\beta(\eta)}$ and the equality $\simeq_{\beta(\eta)}$ are the same. Before giving the proof of this statement, we will first give some examples of reductions.

EXAMPLES 32.28.: Recall the definition of the fixed point combinator Y of Definition 32.9.

1. $\lambda x.z(\lambda y.xy) \twoheadrightarrow_\eta z$.
2. $\mathsf{Y}f \simeq_\beta f(\mathsf{Y}f)$. It is in general not true that $\mathsf{Y}f \twoheadrightarrow_\beta f(\mathsf{Y}f)$.
3. For $\Theta := (\lambda xy.y(xxy))(\lambda xy.y(xxy))$ we have $\Theta f \twoheadrightarrow_\beta f(\Theta f)$.

THEOREM 32.29.: For s and t λ-terms,

$$\boldsymbol{\lambda}_{\beta(\eta)} \vdash s =_{\beta(\eta)} t \text{ iff } s \simeq_{\beta(\eta)} t.$$

PROOF: The proof of the implication from left to right is by induction on the generation of $s =_{\beta(\eta)} t$. The proof of the reverse implication is by induction on the number of steps from s to t. □

The reduction rules of Definition 32.27 give an operational meaning to the equality $=_{\beta(\eta)}$. Given Theorem 32.29, we can arbitrarily use $=_{\beta(\eta)}$ or $\simeq_{\beta(\eta)}$ to denote the equality of $\boldsymbol{\lambda}_{\beta(\eta)}$. Most of the time we shall be using the notation $=_{\beta(\eta)}$.

The notions of β-reduction and η-reduction can be seen as defining the steps that can be used for a kind of evaluation of expressions to values, the values being the terms that cannot be reduced any further. Terms that cannot be reduced are called *normal forms*.

DEFINITION 32.30.: Let s be a λ-term.

1. s is in *β-normal form* (notation: β-nf) if s does not contain a β-redex. s is in *βη-normal form* (notation: βη-nf) if s does not contain a β-redex or an η-redex.
2. *s has a β-normal form* (*βη-normal form*) if there is a λ-term t in β-normal form (βη-normal form) with $s \simeq_\beta t$ ($s \simeq_{\beta\eta} t$).
3. s is *β-normalizing* (*βη-normalizing*) if there is a λ-term t such that t is in β-normal form (t is in βη-normal form) and $s \twoheadrightarrow_\beta t$ ($s \twoheadrightarrow_{\beta\eta} t$).
4. s is *strongly β-normalizing* (*strongly βη-normalizing*) if there are no infinite β-reduction paths (βη-reduction paths) starting from s.

In the following, 'normalizing', 'normal form', etc. without a prefix always refers to β-normalizing, β-normal form, etc.

EXAMPLES 32.31.: Let Ω be the combinator $(\lambda x.xx)(\lambda x.xx)$.

1. K, S and I are in normal form.
2. Ω is not in normal form and is not normalizing.
3. K I Ω is normalizing but not strongly normalizing.

It is easy to check whether a term is in normal form. To see whether a term is normalizing, one can check all reduction sequences starting from it. It is then easily seen that a term like Ω is not normalizing. To see whether Ω has a normal form is a different matter: as Ω is not normalizing, we know that it cannot reduce to its normal form, but there may still be a term t in normal form with $\Omega \simeq_\beta t$. That this is not the case is an immediate consequence of the so-called 'reduction to normal form' property (Corollary 32.35) saying that if s has a normal form t, then $s \twoheadrightarrow_\beta t$. (The 'reduction to normal form' property is a corollary of the Church-Rosser Theorem.)

One very important question has not been treated so far.

Question: Are there terms s and t such that $\boldsymbol{\lambda}_{\beta(\eta)} \not\vdash s =_{\beta(\eta)} t$?

In other words: Are the equational theories $\boldsymbol{\lambda}_\beta$ and $\boldsymbol{\lambda}_{\beta\eta}$ consistent? The reduction approach to the equality of $\boldsymbol{\lambda}_\beta$ helps solve this problem (in combination with Theorem 32.29).

DEFINITION 32.32.: For V a set with binary relation $\stackrel{R}{\rightarrow}$ on it, $\langle V, \stackrel{R}{\rightarrow}\rangle$ satisfies the *diamond property* if for all $s, t_1, t_2 \in V$, if $s \stackrel{R}{\rightarrow} t_1$ and $s \stackrel{R}{\rightarrow} t_2$, then there is a $t \in V$ such that $t_1 \stackrel{R}{\rightarrow} t$ and $t_2 \stackrel{R}{\rightarrow} t$.

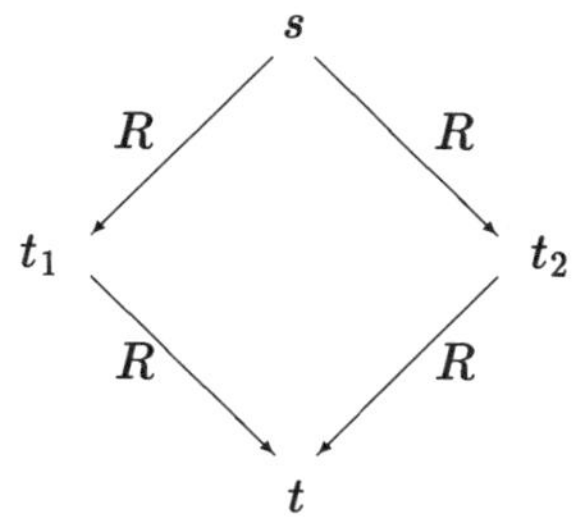

The binary relation $\stackrel{R}{\rightarrow}$ is *Church-Rosser* if it satisfies the diamond property.

One of the most important properties of the λ-calculus is that the relations $\twoheadrightarrow_\beta$ and $\twoheadrightarrow_{\beta\eta}$ on Λ satisfy the diamond property.

THEOREM 32.33. **(Church-Rosser Theorem)**: The relations $\twoheadrightarrow_\beta$ and $\twoheadrightarrow_{\beta\eta}$ on Λ are Church-Rosser.

We do not treat the proof of this theorem because it involves too many new concepts. In fact, there are several proofs of the Church-Rosser theorem. The first proof was given by Curry and Feys [1958]. A relatively short proof of the Church-Rosser theorem for β was given by Tait and Martin-Löf. Recently this proof has been shortened even more and has been made more perspicuous by Takahashi [1989]. Various proofs of the Church-Rosser theorem can be found in Barendregt [1984]. Only some important corollaries are dealt with here.

COROLLARY 32.34. **(Confluence)**: If $s \simeq_\beta t$, then there is a term u such that $s \twoheadrightarrow_\beta u$ and $t \twoheadrightarrow_\beta u$. Such a u is usually called a *common reduct* of s and t. A similar theorem holds for $\beta\eta$ instead of β.

COROLLARY 32.35. **(Reduction to normal form property)**: If s has a normal form t, then $s \twoheadrightarrow_\beta t$.

COROLLARY 32.36. **(Uniqueness of normal forms)**: If s has as normal forms t_1 and t_2, then $t_1 \equiv t_2$.

The proofs of these three corollaries are not difficult. For Corollary 32.34, note that if $s \simeq_{\beta(\eta)} t$, then there is a path through Λ of the following form.

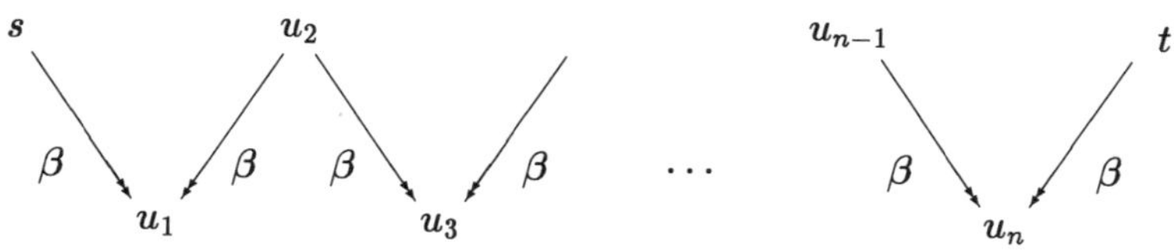

Now one proves the statement of the Corollary by induction on n. For Corollary 32.35 if s has t as normal form, then s and t have a common reduct, say u. But then $t \twoheadrightarrow u$ and, so, $t \equiv u$, because t is in normal form. For Corollary 32.36, if both t_1 and t_2 are normal forms of s, then t_1 and t_2 have a common reduct and, hence, $t_1 \equiv t_2$.

It follows immediately from these corollaries that two terms in normal form that are not syntactically equal (α-convertible) are not $\beta(\eta)$-

equal. So, for example, $\boldsymbol{\lambda}_{\beta(\eta)} \nvdash \mathsf{I} =_{\beta(\eta)} \mathsf{K}$.

THEOREM 32.37.: The λ-calculus is consistent, that is, there are s and t such that $\boldsymbol{\lambda}_{\beta(\eta)} \nvdash s =_{\beta(\eta)} t$.

There is no algorithm for deciding whether two terms are equal, nor for deciding whether a term reduces to another term. This is due to the fact that all partial recursive functions can be represented in the λ-calculus. (The decidability of equality then implies the decidability of the halting problem.)

FACT 32.38.: The equalities $=_\beta$ and $=_{\beta\eta}$ are undecidable.

Actually, equality and (more step) reduction in Λ are recursively enumerable (Σ_1^0). (One-step reduction and syntactic equality are recursive.) The above mentioned fact is an instance of a more general theorem, proved by Scott, which states that for a set X of λ-terms with $X \neq \emptyset$ and $X \neq \Lambda$, if X is closed under $=_\beta$, then X is not recursive. (That is, the problem whether a given λ-term is in X is undecidable.) See Barendregt [1984] for more details.

We have seen that a term that has a normal form reduces to its normal form. If $s \twoheadrightarrow_\beta t$, there are in general many different reduction paths from s to t that establish this fact. It is therefore hard to prove that a term s does *not* reduce to another term t: it has to be shown that there is no reduction path possible from s to t. That it is not necessary to check all possible reduction paths is shown by the Standardization Theorem, which states that if $s \twoheadrightarrow_\beta t$, then there is a 'standard reduction path' from s to t. This standard reduction path is, roughly speaking, a reduction from left to right. To be able to define the notion of standard reduction precisely, some extra notation is required.

NOTATION 32.39.: If $s \rightarrow_\beta t$ by contracting the redex Δ in s, this is denoted by

$$s \xrightarrow{\Delta} t.$$

Similarly, we denote a (possibly infinite) reduction path $s \rightarrow_\beta s_1 \rightarrow_\beta \cdots$ that consecutively contracts $\Delta_0, \Delta_1, \ldots$ by

$$s \xrightarrow{\Delta_0} s_1 \xrightarrow{\Delta_1} s_2 \xrightarrow{\Delta_2} \cdots.$$

DEFINITION 32.40.: For a λ-term s containing the redexes Δ and Δ', *Δ is left (right) of Δ' in s* if the λ of Δ is left (right) of the λ of Δ' in s. Usually the s will not be mentioned explicitly if it is clear from the context.

The redexes $\mathsf{I}x$ and $\mathsf{K}x$ are therefore to the right of the redex $(\lambda y.\mathsf{I}x)(\mathsf{K}x)$ in the term $(\lambda y.\mathsf{I}x)(\mathsf{K}x)$. Note that if the term s is itself a redex, all the other redexes in s are to the right of s.

DEFINITION 32.41.: A reduction path

$$s_0 \xrightarrow{\Delta_0} s_1 \xrightarrow{\Delta_1} s_2 \xrightarrow{\Delta_2} \cdots$$

is *standard* if for any pair (Δ_i, Δ_{i+1}), the λ of the redex Δ_{i+1} comes from a λ in s_i that is to the right of the λ of Δ_i in s_i.

Note that if $s \xrightarrow{\Delta} t$ and Δ' is a redex in t, then there is a unique λ in s that has generated the λ of the redex Δ'.

The intuition behind a standard reduction is that one reduces from left to right in a term, possibly leaving some redexes undone. Another way to formalize this is to say that at every reduction step $s_i \xrightarrow{\Delta_i} s_{i+1}$, all redexes to the left of Δ_i in s_i are 'frozen' and remain undone in the rest of the reduction path. (This 'freezing' of redexes can be made precise by labelling the λ's of the redexes that are to be frozen.) This idea of freezing is a good way of checking whether a reduction path is standard or not.

EXAMPLES 32.42.:

1. $(\lambda x.(\lambda y.xy)z)\mathsf{I} \to_\beta (\lambda y.\mathsf{I}y)z \to_\beta \mathsf{I}z$ is standard.
2. $(\lambda x.(\lambda y.xy)z)\mathsf{I} \to_\beta (\lambda x.xz)\mathsf{I} \to_\beta \mathsf{I}z$ is not standard.
3. $(\lambda x.(\lambda y.xy)z)\mathsf{I} \to_\beta (\lambda y.\mathsf{I}y)z \xrightarrow{\mathsf{I}y}_\beta \mathsf{I}z \to_\beta z$ is not standard.
4. $(\lambda x.(\lambda y.xy)z)\mathsf{I} \to_\beta (\lambda y.\mathsf{I}y)z \xrightarrow{(\lambda y.\mathsf{I}y)z}_\beta \mathsf{I}z \to_\beta z$ is standard.

The main property about standard reductions is given by the following theorem, that we give without proof. (See Barendregt [1984].)

THEOREM 32.43. **(Standardization)**: If $s \twoheadrightarrow_\beta t$, then there is a standard reduction path from s to t.

REPRESENTABILITY OF RECURSIVE FUNCTIONS

To show that the Church numerals introduced in Definition 32.10 really make sense, one has to make sure that the recursive functions are definable on them. This property is usually called the *λ-definability* of the recursive functions. Whether a numeric function is λ-definable obviously depends on the numeral system that one has chosen. Apart from the Church numerals, there are many other numeral systems on which all recursive functions are λ-definable.

DEFINITION 32.44.: A *numeral system* in the λ-calculus is a sequence of closed terms $d_0, d_1, d_2, \ldots$, such that there are closed terms S_d^+ (the successor) and $\mathsf{Z?}_d$ (test for zero) such that

$$\begin{aligned} \mathsf{S}_d^+ d_n &=_\beta d_{n+1}, \\ \mathsf{Z?}_d d_0 &=_\beta \mathsf{T}, \\ \mathsf{Z?}_d d_{n+1} &=_\beta \mathsf{F}. \end{aligned}$$

In exercise 32.11 the reader is asked to show that the Church numerals $c_0, c_1, c_2, \ldots$ indeed form a numeral system.

DEFINITION 32.45.: A function $f : \mathbb{N}^k \to \mathbb{N}$ is *λ-definable* on the numeral system $d_0, d_1, d_2, \ldots$ if there is a λ-term F such that for all $n_1, \ldots, n_k \in \mathbb{N}$

$$F d_{n_1} \cdots d_{n_k} =_\beta d_{f(n_1,\ldots,n_k)}.$$

We say that F *λ-defines* the function f.

As we have seen in 32.11, the addition function is λ-definable on the numeral system c_n. The term $\mathsf{A} := \lambda cdfx.cf(dfx)$ λ-defines addition:

$$\mathsf{A} c_n c_m =_\beta c_{n+m}.$$

A different numeral system is the following, defined in Barendregt [1984].

DEFINITION 32.46.: The numeral system $\ulcorner 0 \urcorner, \ulcorner 1 \urcorner, \ulcorner 2 \urcorner, \ldots$ is defined by taking

$$\begin{aligned} \ulcorner 0 \urcorner &:= \mathsf{I}, \\ \ulcorner n+1 \urcorner &:= [\mathsf{F}, \ulcorner n \urcorner]. \end{aligned}$$

(For the definition of $[-,-]$, see Example 32.11.) The successor Suc and test for zero Zero? are defined as follows.

$$\begin{aligned} \mathsf{Suc} &:= \lambda x.[\mathsf{F}, x], \\ \mathsf{Zero?} &:= \lambda x.x\mathsf{T}. \end{aligned}$$

It is also straightforward to define a predecessor function P^- on these numerals by

$$\mathsf{P}^- := \lambda x.x\mathsf{F}.$$

A definite advantage of the numeral system $\ulcorner - \urcorner$ over the Church numerals is that the proof that all recursive functions are λ-definable is much easier for $\ulcorner - \urcorner$ than for the Church numerals. We therefore treat the case for $\ulcorner - \urcorner$ first and then give two combinators that map c_n to $\ulcorner n \urcorner$ and back to show that all recursive functions are λ-definable on the Church numerals.

We first recall the definition of recursive function. (See Definition 28.18.) To show that all recursive functions are λ-definable on the numerals $\ulcorner - \urcorner$ it suffices to show that the initial functions are λ-definable and that the λ-definable functions are closed under composition, primitive recursion and minimization restricted to regular functions.

LEMMA 32.47.: The initial functions are all λ-definable on $\ulcorner - \urcorner$ by taking

$$\begin{aligned} \mathsf{U}_i^p &:= \lambda x_0 \cdots x_p.x_i, \\ \mathsf{Suc} &:= \lambda x.[\mathsf{F}, x], \\ \mathsf{Z} &:= \lambda x.\mathsf{I}\,. \end{aligned}$$

PROOF: By an easy checking of the required equalities. □

LEMMA 32.48.: The set of λ-definable functions is closed under composition.

PROOF: We have to show that if $g : \mathbb{N}^m \to \mathbb{N}$ and $h_1, \ldots, h_m : \mathbb{N}^k \to \mathbb{N}$ are λ-definable, then $f : \mathbb{N}^k \to \mathbb{N}$ defined by $f(n_1, \ldots, n_k) = g(h_1(n_1, \ldots, n_k), \ldots, h_m(n_1, \ldots, n_k))$ is λ-definable. Now, say that g is λ-defined by G and $h_1, \ldots, h_m$ are λ-defined by $H_1, \ldots, H_m$. Then f is λ-defined by $\lambda x_1 \cdots x_k.G(H_1 x_1 \cdots x_k) \cdots (H_m x_1 \cdots x_k)$. □

LEMMA 32.49.: The set of λ-definable functions is closed under primitive recursion.

PROOF: We have to show that if $g : \mathbb{N}^k \to \mathbb{N}$ and $h : \mathbb{N}^{k+2} \to \mathbb{N}$ are λ-definable, then $f : \mathbb{N}^{k+1} \to \mathbb{N}$ defined by

$$\begin{aligned} f(0, n_1, \ldots, n_k) &= g(n_1, \ldots, n_k), \\ f(m+1, n_1, \ldots, n_k) &= h(m, n_1, \ldots, n_k, f(m, n_1, \ldots, n_k)) \end{aligned}$$

is λ-definable. Now, say that g is λ-defined by G and h is λ-defined by H. Then a term F that satisfies

$$F x \vec{y} =_\beta \text{If Zero?} x \text{ then } G\vec{y} \text{ else } H(\mathsf{P}^- x)\vec{y}(F(\mathsf{P}^- x)\vec{y}) \tag{1}$$

does the job. That is, such a term F satisfies

$$F \ulcorner m \urcorner \ \ulcorner n_1 \urcorner \ \cdots \ \ulcorner n_k \urcorner = \ulcorner f(m, \vec{n}) \urcorner ,$$

which is proved by induction on k. A term F satisfying (1) is found by the Fixed Point Theorem 32.9. To be precise, one can take

$$F := \mathsf{Y}(\lambda f x \vec{y}. \text{ If Zero?} x \text{ then } G\vec{y} \text{ else } H(\mathsf{P}^- x)\vec{y}(f(\mathsf{P}^- x)\vec{y}). \qquad \square$$

LEMMA 32.50.: The set of λ-definable functions is closed under minimization restricted to regular expressions.

PROOF: We have to show that if $g : \mathbb{N}^{k+1} \to \mathbb{N}$ is λ-definable and $\forall n_1, \ldots, n_k \in \mathbb{N}\, \exists m \in \mathbb{N}[g(n_1, \ldots, n_k, m) = 0]$, then $f : \mathbb{N}^k \to \mathbb{N}$ defined by

$$f(n_1, \ldots, n_k) = \mu m[g(n_1, \ldots, n_k, m) = 0]$$

is λ-definable. Now, say that g is λ-defined by G. To λ-define f, we are now looking for a term F that satisfies something like

$$\begin{aligned} F\vec{y} \ =_\beta \ & \text{If Zero?}(G\vec{y} \ulcorner 0 \urcorner) \text{ then } \ulcorner 0 \urcorner \text{ else} \\ & \text{If Zero?}(G\vec{y} \ulcorner 1 \urcorner) \text{ then } \ulcorner 1 \urcorner \text{ else} \\ & \text{If Zero?}(G\vec{y} \ulcorner 2 \urcorner) \text{ then } \ulcorner 2 \urcorner \text{ else} \\ & \vdots \end{aligned}$$

But the length of this expression depends on the point where we first reach the output 0, so we cannot fix this equation. The solution is to define an intermediate term H satisfying

$$H\vec{y}x =_\beta \text{If Zero?}(G\vec{y}x) \text{ then } x \text{ else } H\vec{y}(\mathsf{Suc} x).$$

Using the Fixed Point Theorem 32.9, it is not difficult to find such a H. ($H\vec{y}\ulcorner n \urcorner$ computes the first numeral larger than or equal to n for which G is zero, if it exists.) Now take $F := \lambda\vec{y}.H\vec{y}\ulcorner 0 \urcorner$. □

THEOREM 32.51.: All recursive functions are λ-definable on the numerals $\ulcorner - \urcorner$.

PROOF: Using the Lemmas 32.47, 32.48, 32.49 and 32.50. □

Note that to show that all recursive functions are λ-definable on $\ulcorner - \urcorner$, we have only used that $\ulcorner - \urcorner$ is a numeral system and that one can define a predecessor function on $\ulcorner - \urcorner$. This gives the following general theorem.

THEOREM 32.52.: Let $d_0, d_1, \ldots$ be a numeral system. If the predecessor function is λ-definable on $d_0, d_1, \ldots$, then all recursive functions are λ-definable on $d_0, d_1, \ldots$.

To show that all recursive functions can be λ-defined on the Church numerals, it suffices to define the predecessor function for c_n. However, this is not so easy. We therefore follow Barendregt [1984] and show that there are λ-terms T and T^{-1} that constitute a bijection between the Church numerals and the numerals $\ulcorner - \urcorner$.

LEMMA 32.53.: There are λ-terms T and T^{-1} such that for all $n \in \mathbb{N}$

$$Tc_n =_\beta \ulcorner n \urcorner \text{ and } T^{-1} \ulcorner n \urcorner =_\beta c_n.$$

PROOF: Define

$$\begin{array}{rcl} T & := & \lambda x.x\mathsf{Suc}\ulcorner 0 \urcorner, \\ T^{-1} & := & \mathsf{Y}(\lambda tx.\text{If } \mathsf{Zero?}x \text{ then } c_0 \text{ else } \mathsf{S}^+(t(\mathsf{P}^-x))). \end{array}$$

T^{-1} then satisfies $T^{-1}x =_\beta \text{If } \mathsf{Zero?}x \text{ then } c_0 \text{ else } \mathsf{S}^+(T^{-1}(\mathsf{P}^-x))$. Now the required equalities are easily shown (by induction on n). □

COROLLARY 32.54.: All recursive functions are λ-definable on c_n.

PROOF: Let g be a unary recursive function. Then g is λ-definable on $\ulcorner - \urcorner$ (Theorem 32.51.), say by G. Then $\lambda x.T^{-1}(G(Tx))$ λ-defines g on the Church numerals. Similarly for functions of higher arities. □

To see whether the partial recursive functions can be λ-defined as well, it is necessary to first find a good representation for the notion of 'undefinedness' in the λ-calculus. This falls outside the scope of this section. We just want to note here that it is indeed possible to λ-define the partial recursive recursive functions (on both $\ulcorner - \urcorner$ and the Church numerals). For details we refer to Barendregt [1984].

REDUCTION STRATEGIES

For computing with functions in the λ-calculus, it is important to be able to find a normal form of a term if it exists. This involves a specific *reduction strategy*, a method for telling which redex or redexes should be contracted at a specific point in the evaluation. Besides finding the normal form, there are other reasons for the choice of one reduction strategy or another. For example, we may look for a strategy suited to finding a common reduct of two terms if it exists, or to finding a normal form in the least possible number of reduction steps. In this section only a few reduction strategies are treated, covering the basic ideas. Furthermore, we restrict our attention to β-reduction.

In its most general form, a reduction strategy is just a function F from Λ to Λ such that $t \twoheadrightarrow_\beta F(t)$ for every t. Obviously, one wants F to be effective. Otherwise, F with

$$\begin{aligned} F(t) &= \text{normal form of } t, && \text{if it exists,} \\ &= t, && \text{otherwise} \end{aligned}$$

would be a reduction strategy.

In the literature one also distinguishes between *reduction strategies* and *one-step reduction strategies*, the first giving for every term t a term t' such that $t \twoheadrightarrow_\beta t'$ and the second giving for every term t that is not in normal form a term t' such that $t \rightarrow_\beta t'$. Here we only treat the last type and therefore we use the terminology 'reduction strategy' for one-step strategies only.

DEFINITION 32.55.:

1. A *reduction strategy* is a function F such that $s \rightarrow_\beta F(s)$ for every λ-term s that is not in normal form, and $F(s) \equiv s$ otherwise.

2. A reduction strategy F is *normalizing* if it computes the normal form if it exists. That is, if s has a normal form, then $F^n(s) \equiv \text{nf}(s)$

for some $n \in \mathbb{N}$.

3. A reduction strategy F is *Church-Rosser* if it computes the common reduct, if it exists. That is, if $s \simeq_\beta t$, then $F^m(s) \equiv F^n(t)$ for some $m, n \in \mathbb{N}$.

LEMMA 32.56.: If the reduction strategy F is Church-Rosser, then it is normalizing.

PROOF: If F is Church-Rosser and s has normal form t, then $F^m(s) \equiv t$ for some $m \in \mathbb{N}$. □

Note that a normalizing reduction strategy finds a common reduct for normalizing terms, viz., their common normal form, but may not do so for non-normalizing terms. The normalizing strategy treated here is an example of a reduction strategy that does not always find a common reduct.

DEFINITION 32.57.: The *leftmost reduction strategy* is the reduction strategy that contracts in a term t the redex that is to the left of all other redexes in t. (See Definition 32.40 for the notion of a redex being to the left of another redex.) *Leftmost reduction* (contracting the leftmost redex) is sometimes indicated by $\rightarrow_l$.

THEOREM 32.58.: The leftmost reduction strategy is normalizing.

PROOF: Let s be a term that has a normal form, say t. Then $s \twoheadrightarrow_\beta t$ and by Theorem 32.43 there is a standard reduction path from s to t. At any point in this standard reduction path the leftmost redex is contracted, so the standard reduction is a leftmost reduction. (Suppose that at a specific point in the standard reduction path a redex Δ is contracted that is not leftmost. Then all redexes to the left of Δ are frozen and remain in the term. Hence, the term t still contains redexes, which is not the case.) □

EXAMPLES: 32.59.: $\mathsf{KI}\Omega \rightarrow_l (\lambda y.\mathsf{I})\Omega \rightarrow_l \mathsf{I}$.
$\mathsf{YF} \twoheadrightarrow_l \mathsf{I}$.

If a term is normalizing but not strongly normalizing, the leftmost reduction strategy will always avoid the 'infinite loop'. This provides us with a method for finding a normal form, if it exists. The leftmost reduction can also be used for showing that a term does not have a normal form. There is no normal form if the leftmost reduction is infinite.

This does not mean that leftmost reduction is always the best thing to do. If we already know that a term is strongly normalizing, there may be better strategies (in terms of the number of reduction steps) for reduction. For example, if $s \equiv (\lambda x.u)((\lambda y.v)t)$ with x occurring in u several times, it seems to be a better idea to first reduce $(\lambda y.v)t$ to normal form and then contract s itself to avoid having to reduce $(\lambda y.v)t$ several times. Note that this reasoning only applies if the multiple occurrences of x in u are all 'real'. If, for example, $u \equiv \mathsf{K}\ z(\mathsf{K}\ x(\mathsf{K}\ x(\mathsf{K}\ xx)))$, then none of the x's occurs in the normal form of u and it would be better to first reduce u to normal form and then contract s. The reduction strategy that first reduces the argument $(\lambda y.v)t$ to normal form, say w, and then contracts $(\lambda x.u)w$ is a kind of 'call-by-value' evaluation; an argument is 'called' only if it is a value (a normal form). In contrast, leftmost reduction is a kind of 'call-by-name' evaluation; an argument is called without first computing its value.

As we have seen in the example above, it is important to know whether the variable occurrences in a term are 'real'. For that reason the notion of *strictness* is introduced in programming languages, a program (function) F being strict if for all inputs s, Fs is undefined if s is undefined. The parallel in the λ-calculus should be clear: a term F is *strict* if Fs has no normal form in the event s has no normal form. In general, it is undecidable whether a term is strict, but if we have somehow obtained the knowledge that $\lambda x.u$ is strict, we know that it is safe to first reduce the argument to normal form and then apply $\lambda x.u$ to it. Another important notion is that of a *needed* redex. A redex Δ in s is *needed* if any reduction of s to normal form at some point contracts (a residual of) the redex Δ. So, an optimal strategy of reducing s to normal form is found by contracting each time only a needed redex. Of course, whether a redex is needed in a λ-term is in general undecidable.

MODELS OF THE λ-CALCULUS

It is interesting to see which type of mathematical structures can serve as a model for the theory $\boldsymbol{\lambda}_\beta$. (The same can be said for $\boldsymbol{\lambda}_{\beta\eta}$, but we restrict ourselves here to $\boldsymbol{\lambda}_\beta$.) The way functions are treated in the λ-calculus does not correspond to our set-theoretic intuition according to which functions correspond with their graphs. A denotational semantics for the λ-calculus forces us to find a set D (of interpretations for all λ-terms) such that $D \underset{1}{=} D^D$, because then an interpretation of a λ-term (an element of D) is at the same time a function from D to D. Hence, an application st can be interpreted as the result of applying the function s (in D^D) to the argument t (in D), and an abstraction $\lambda x.u$ lifts a λ-term u (in D) to a function (in D^D). In set theory, the cardinality of D^D (the set of functions from D to D) is always larger than the cardinality of D (for $|D| \geq 2$), so $D \underset{1}{=} D^D$ is impossible. (See section 16, Volume I.)

There are many different ways to solve this paradox, all based on the idea that one does not need the full function space D^D to interpret the λ-calculus. It suffices to find a subset $[D{\rightarrow}D]$ of D^D such that $[D{\rightarrow}D] \underset{1}{=} D$. It even suffices if there is a function G from $[D{\rightarrow}D]$ to D and a function F from D to $[D{\rightarrow}D]$ such that $F \circ G = id_{[D\rightarrow D]}$. Furthermore, it has been observed that it is not at all necessary to take the set-theoretic function space. For example, one can also take a non-extensional viewpoint and not identify all mappings from D to D that are extensionally equal. (The mappings f and g in D^D are extensionally equal if $\forall d \in D[fd = gd]$.)

Here only a very minimal idea of the denotational semantics of the λ-calculus is given. Therefore, we treat a rather non-standard set-theoretic notion of 'model of the λ-calculus', which does not depend on the theory of complete lattices or complete partial orders. Our model notion is not very general and we certainly do not claim to cover any considerable part of the literature on this topic. However, it does give some important intuitions and it is strong enough to include some interesting examples of models. As an instance of the notion of set-theoretic model, we treat the model $P\omega$. For more details on models, we refer to Barendregt [1984] and Hindley and Seldin [1986].

DEFINITION 32.60.: For D a set and $E \subset D^D$, E is called a *retract* of D if there are maps F from D to E and G from E to D such that

$$F \circ G = id_E.$$

Such an E is usually written as $[D{\rightarrow}D]$ and is called a *retract of D by F and G*. The pair (G, F) is called a *retraction*.

The interpretation of λ-terms in D is defined modulo a *valuation* ρ for the term variables in D. So $\rho : \mathsf{Var} \to D$. If $d \in D$, we denote by $\rho(x := d)$ the valuation for which $\rho(x := d)(y) = \rho(y)$ if $y \not\equiv x$ and $\rho(x := d)(x) = d$.

DEFINITION 32.61.: For $[D{\to}D]$ a retract of D by F and G and ρ a valuation, the *interpretation function* $[\![-]\!]^D_\rho$ from Λ to D is defined as follows:

$$\begin{array}{rcl} [\![x]\!]^D_\rho & := & \rho(x), \\ [\![st]\!]^D_\rho & := & F([\![s]\!]^D_\rho)([\![t]\!]^D_\rho), \\ [\![\lambda x.s]\!]^D_\rho & := & G(d \mapsto [\![s]\!]^D_{\rho(x:=d)}). \end{array}$$

Here $d \mapsto [\![s]\!]^D_{\rho(x:=d)}$ denotes the function that maps an element d to $[\![s]\!]^D_{\rho(x:=d)}$. This function is also denoted by $\lambda d \in D.[\![s]\!]^D_{\rho(x:=d)}$, where we have to be careful not to mix up λ-terms and elements of $[D{\to}D]$.

REMARK 32.62.: Note that there is a snag in this definition: In the case that defines the interpretation of $\lambda x.s$, the mapping $\lambda d \in D.[\![s]\!]^D_{\rho(x:=d)}$ may not be in the set $[D{\to}D]$ (which is a real subset of D^D). So, the function $[\![-]\!]^D_\rho$ is in general a *partial* function. We see that to have a model, it is not sufficient to find a subset $[D{\to}D]$ of D^D that is a retract of D, because the interpretation may be undefined for some λ-terms.

Keeping in mind the previous remark, we now define the notion of *set-theoretic λ-model.*

DEFINITION 32.63.: A *set-theoretic λ-model* is a quadruple $\langle D, [D{\to}D], F, G\rangle$ with the properties

1. $[D{\to}D]$ is a retract of D by F and G,
2. the interpretation function $[\![-]\!]^D_\rho$ of Definition 32.61 is total for every valuation ρ.

The superscript D is omitted in $[\![-]\!]^D_\rho$ if it is clear from the context.

In some respects this definition of model of the λ-calculus is not very satisfying. It does not, for example, say how models can be constructed; neither is it obvious that models exist at all. (It is not difficult to give a $[D{\to}D] \subset D^D$ such that $[D{\to}D]$ is a retract of D by F and G, but it is not clear how this can be done in such a way that $\langle D, [D \to D], F, G\rangle$ is also a model.) That there are models is shown later by constructing one. The general problem is that the model notion is very syntactical. In fact, we are saying that D is a model of the λ-calculus if all λ-terms can be interpreted in D. It would be much nicer to give a syntax-independent description of a class of structures for which it can be shown that they are λ-models.

This is not the place to go very deeply into the theory of models of the λ-calculus. The problem that we have just mentioned (how to give a syntax-independent description of the notion of model, in view of Remark 32.62) has been a permanent issue in the development of the denotational semantics of the λ-calculus, giving rise to various notions of model. Some of these notions are quite close to our notion of set-theoretic λ-model, for example, the so-called syntactic λ-algebras. For an overview of the different notions of model, we refer to Barendregt [1984].

Below is a list of some properties of the set-theoretic λ-models.

LEMMA 32.64.: For $\langle D, [D \to D], F, G\rangle$ a set-theoretic λ-model, s a λ-term and ρ and ρ' valuations, if $\rho(x) = \rho'(x)$ for all free variables x of s, then $[\![s]\!]_\rho = [\![s]\!]_{\rho'}$.

PROOF: By induction on the structure of s, using the fact that $[\![t]\!]_\rho$ and $[\![t]\!]_{\rho'}$ are always defined. □

LEMMA 32.65.: For $\langle D, [D \to D], F, G\rangle$ a set-theoretic λ-model, s and t λ-terms and ρ a valuation, $[\![s[x := t]]\!]_\rho = [\![s]\!]_{\rho(x:=[\![t]\!]_\rho)}$.

PROOF: By induction on the structure of s one proves that $[\![s[x := t]]\!]_\rho = [\![s]\!]_{\rho(x:=[\![t]\!]_\rho)}$ for every valuation ρ. The interesting case is when $s \equiv \lambda y.u$. By the variable convention, we may assume that $x \not\equiv y$ and y is not free in t, so we have

$$\begin{aligned}
[\![\lambda y.u]\!]_{\rho(x:=[\![t]\!]_\rho)} &= G(d \mapsto [\![u]\!]_{\rho(x:=[\![t]\!]_\rho)(y:=d)}) \\
&= G(d \mapsto [\![u]\!]_{\rho(y:=d)(x:=[\![t]\!]_\rho)}) \\
&= G(d \mapsto [\![u[x := t]]\!]_{\rho(y:=d)}) \\
&= [\![\lambda y.u[x := t]]\!]_\rho.
\end{aligned}$$

The first and last equality are by definition, the second follows from the variable convention and the previous lemma, and the third equality is the induction hypothesis using the valuation $\rho(y := d)$. (Note that $[\![t]\!]_\rho = [\![t]\!]_{\rho(y:=d)}$.) □

DEFINITION 32.66.: Let $M = \langle D, [D \to D], F, G\rangle$ be a set-theoretic λ-model, s and t λ-terms and ρ a valuation.

1. $s = t$ *is true in* M *under the valuation* ρ, notation $M, \rho \models s = t$, if $[\![s]\!]_\rho = [\![t]\!]_\rho$.
2. $s = t$ *is true in* M, notation $M \models s = t$, if $M, \rho \models s = t$ for every valuation ρ.
3. $s = t$ *is true in all models*, notation $\models s = t$, if $M \models s = t$ for every set-theoretic λ-model M.

THEOREM 32.67. (**soundness**): For s and t λ-terms,

$$\text{If } \boldsymbol{\lambda}_\beta \vdash s =_\beta t, \text{ then } \models s = t.$$

PROOF: Let M be a set-theoretic λ-model. By induction on the derivation of $s =_\beta t$ in the theory $\boldsymbol{\lambda}_\beta$ one shows that $M \models s = t$. The interesting case is when the last rule applied is the ξ-rule. (The β-rule follows immediately from Lemma 32.65.) Suppose $s \equiv \lambda x.u$ and $t \equiv \lambda x.v$ and $\lambda x.u =_\beta \lambda x.v$ has been derived from $u =_\beta v$. Then we know by induction hypothesis that $[\![u]\!]_\rho = [\![v]\!]_\rho$ for every valuation ρ. Let ρ be a valuation. Then $[\![u]\!]_{\rho(x:=d)} = [\![v]\!]_{\rho(x:=d)}$ for any $d \in D$ and, hence, $d \mapsto [\![u]\!]_{\rho(x:=d)}$ and $d \mapsto [\![v]\!]_{\rho(x:=d)}$ are the same (set-theoretic) functions. Hence, $G(d \mapsto [\![u]\!]_{\rho(x:=d)}) = G(d \mapsto [\![v]\!]_{\rho(x:=d)})$; so, $[\![\lambda x.u]\!]_\rho = [\![\lambda x.v]\!]_\rho$. □

The theorem can be strengthened by allowing reasoning from assumptions. This works as follows. A set of equations between closed λ-terms, $\mathcal{T}$, is called a $\boldsymbol{\lambda}_\beta$*-theory* if $\mathcal{T}$ is closed under derivability in $\boldsymbol{\lambda}_\beta$. (That is, if $s =_\beta t$ is an equality between closed λ-terms that can be derived from equalities in $\mathcal{T}$ by making use of the rules of $\boldsymbol{\lambda}_\beta$, then $s =_\beta t \in \mathcal{T}$.) Then one defines, for $\mathcal{T}$ a $\boldsymbol{\lambda}_\beta$-theory and u and v arbitrary λ-terms, that *the equality* $u = v$ *is true in* $\mathcal{T}$ (notation: $\mathcal{T} \models u = v$) if for all models M in which all equalities of $\mathcal{T}$ are true one has $M \models u = v$. The Soundness

Theorem can now be extended to theories as follows: Let $\mathcal{T}$ be a $\boldsymbol{\lambda}_\beta$-theory and let u and v be λ-terms. If $u =_\beta v$ is derivable from $\mathcal{T}$ using the rules of $\boldsymbol{\lambda}_\beta$, then $\mathcal{T} \models u = v$.

The reverse of Theorem 32.67 also holds. This is called 'completeness' and says that if $s = t$ is true in all set-theoretic λ-models, then $\boldsymbol{\lambda}_\beta \vdash s =_\beta t$. The proof is accomplished by describing a set-theoretic λ-model in which only the equations of the theory $\boldsymbol{\lambda}_\beta$ itself are true. This model is the so-called *open term model*, consisting of equivalence classes of λ-terms modulo β-equality. It is described in Barendregt [1984], to which we refer for further details.

THEOREM 32.68. (**completeness**): For s and t λ-terms,

$$\text{If } \models s = t, \text{ then } \boldsymbol{\lambda}_\beta \vdash s =_\beta t.$$

There are many different aspects to models. One is that in an actual model there are generally more equalities between λ-terms than just the ones provable in $\boldsymbol{\lambda}_\beta$. This can suggest interesting extensions of the equality $=_\beta$. For example, there are models in which I and Ω have the same value and I and K have distinct values. This implies that by adding the equality $\mathsf{I} =_\beta \Omega$ to the theory $\boldsymbol{\lambda}_\beta$, this theory remains consistent. Another interesting aspect is that with the models one can prove things about the syntax, such as the consistency of the equational theory $\boldsymbol{\lambda}_\beta$. But perhaps most importantly, models provide insight into syntax. They show that syntax is really a formal system about existing mathematical things.

To benefit from models, it is not enough to describe an abstract notion of model. It is also important to construct them. Therefore we now describe the model $P\omega$, developed by Scott and Plotkin independently (See Scott [1974]).

DEFINITION 32.69.: The set $P\omega$ consists of all subsets of $\mathbb{N}$. The set $P\omega$ is always considered together with the topology given by the base sets $O_V = \{X \mid V \subseteq X\}$, for finite subsets V of $\mathbb{N}$.

The reason for considering a topology on $P\omega$ is that the set of functions $[P\omega {\rightarrow} P\omega] \subset P\omega^{P\omega}$ that will be considered is the set of functions that are continuous with respect to this topology.

DEFINITION 32.70.:

1. A subset I of $P\omega$ is *directed* if for all $X, Y \in I$ there exists a $Z \in I$ such that $X \subseteq Z$ and $Y \subseteq Z$.
2. A function f from $P\omega$ to $P\omega$ is *continuous* if for all directed subsets I of $P\omega$,
$$f(\bigcup I) = \bigcup_{X \in I} f(X).$$
Remember that $\bigcup I = \{n \in \mathbb{N} \mid n \in X \text{ for some } X \in I\}$.

So, a function from $P\omega$ to $P\omega$ is continuous if it *preserves directed unions.* In less topological (more technical) terms, this means that a continuous function f is determined by its behaviour on finite sets: for $X \subset \mathbb{N}$, $f(X) = \bigcup\{f(V) \mid V \subseteq X,\ V \text{ finite}\}$, as the following lemma shows.

LEMMA 32.71.: For f a function from $P\omega$ to $P\omega$ we have

1. If f is continuous, then f is monotone (that is, $X \subseteq Y$ implies $f(X) \subseteq f(Y)$).
2. f is continuous if and only if $f(X) = \bigcup\{f(V) \mid V \subseteq X,\ V \text{ finite}\}$ for all $X \subseteq \mathbb{N}$.

PROOF: Both are a consequence of the definition of continuity. □

LEMMA 32.72.: For f a continuous function from $P\omega$ to $P\omega$,
if $n \in f(X)$, then $n \in f(V)$ for some finite $V \subseteq X$.

PROOF: Suppose f is continuous. By Lemma 32.71, if $n \in f(X)$, then $n \in f(V)$ for some finite $V \subseteq X$. □

DEFINITION 32.73.: The subset $[P\omega \rightarrow P\omega]$ of $P\omega^{P\omega}$ is defined as the set of continuous functions from $P\omega$ to $P\omega$.

To make $P\omega$ into a λ-model, the continuous functions on $P\omega$ have to be coded as elements of $P\omega$. This is done by coding the graph of the function ($P\omega$ is also called the *graph model*). It is well known that there are bijective codings from the set $P_{fin}\mathbb{N}$ of finite subsets of $\mathbb{N}$ into

$\mathbb{N}$. Using such an encoding, the continuous functions are represented as elements of $P\omega$. The specific embedding from $P_{fin}\mathbb{N}$ into $\mathbb{N}$ that one chooses has consequences for the properties of the model, so, in fact, this coding is part of the model construction. This dependency is usually not denoted explicitly.

DEFINITION 32.74.: The coding $<-,->$ of pairs of natural numbers into $\mathbb{N}$ and the coding $\langle - \rangle$ of finite subsets of $\mathbb{N}$ into $\mathbb{N}$ are defined as follows: (See the proof of Theorem 16.22.)

1. $<n,m> \;:= \frac{1}{2}(n+m)(n+m+1)+m$,
2. $\langle\{n_1,\ldots,n_k\}\rangle := \Sigma_{1\leq i\leq k} 2^{n_i}$.

We denote $\langle\{n_1,\ldots,n_k\}\rangle$ also by $\langle n_1,\ldots,n_k\rangle$, but note that $n_1,\ldots,n_k$ is not a sequence: a number cannot occur twice.

LEMMA 32.75.: $<-,->\; : \mathbb{N}\times\mathbb{N}\to\mathbb{N}$ and $\langle - \rangle : P_{fin}\mathbb{N}\to\mathbb{N}$ are bijections. The inverse of $\langle - \rangle$ is denoted by e_n; so, for $n\in\mathbb{N}$, e_n is the finite set with $\langle e_n\rangle = n$.

The model that we construct uses the codings of Definition 32.74.

DEFINITION 32.76.: The function $G : [P\omega{\to}P\omega]\to P\omega$ is defined by

$$G(f) := \{<n,m> \mid m\in f(e_n)\}.$$

The function $F : P\omega\to[P\omega{\to}P\omega]$ is defined by

$$F(X)(Y) := \{m \mid \exists e_n\subseteq Y[<n,m>\in X]\}.$$

The G is also called 'graph', the F is also called 'fun'.

It is not immediate that $F(X)$ is indeed a continuous function for every $X\subseteq\mathbb{N}$.

LEMMA 32.77.: For $X\subseteq\mathbb{N}$, $F(X)\in[P\omega{\to}P\omega]$.

PROOF: By Lemma 32.71, it suffices to show that $F(X)(Y) = \bigcup_{e_n\subseteq Y} F(X)(e_n)$ for arbitrary X and Y. Now

$$\begin{array}{rcl} F(X)(Y) & = & \{m \mid \exists e_n\subseteq Y[<n,m>\in X]\} \\ & = & \bigcup_{e_n\subseteq Y}\{m \mid <n,m>\in X\} \\ & = & \bigcup_{e_n\subseteq Y} F(X)(e_n). \end{array}$$

□

The idea behind G should be clear. A continuous function f on $P\omega$ is represented by its graph $\{<X,Y> \mid f(X) = Y\}$, which can again be represented by a relation between $P(\mathbb{N})$ and $\mathbb{N}$, $\{<X,m> \mid m \in f(X)\}$. Because f is continuous, it is determined by its behaviour on $P_{fin}\mathbb{N}$, so it can be represented by the relation $\{<n,m> \mid m \in f(e_n)\}$. The mapping F also uses Lemma 32.71. A set $X \subseteq \mathbb{N}$ consists of codes of pairs $<n,m>$, which can be seen as representing the function $e_n \mapsto \{m \mid <n,m> \in X\}$ from $P_{fin}\mathbb{N}$ to $P(\mathbb{N})$. By Lemma 32.71, this determines a continuous function on $P\omega$.

THEOREM 32.78.: The quadruple $\langle P\omega, [P\omega \to P\omega], F, G\rangle$ is a set-theoretic λ-model.

PROOF: Check that $F \circ G$ is the identity on $[P\omega \to P\omega]$ and that $[\![s]\!]_\rho$ is defined for every valuation ρ and every term s. For the first part, let $f \in [P\omega \to P\omega]$ and $Y \subseteq \mathbb{N}$. Then

$$\begin{aligned} F(G(f))(Y) &= \{m \mid \exists e_n \subseteq Y[<n,m> \in G(f)]\} \\ &= \{m \mid \exists e_n \subseteq Y[m \in f(e_n)]\} \\ &= f(Y) \text{ (by Lemma 32.71).} \end{aligned}$$

That $[\![s]\!]_\rho$ is always defined is shown in the following Lemma 32.79. □

LEMMA 32.79.: For all valuations ρ and all $s \in \Lambda$, $[\![s]\!]_\rho$ is defined and $d \mapsto [\![s]\!]_{\rho(x:=d)}$ is continuous.

PROOF: Simultaneously, by induction on s. □

EXERCISES

32.1. Rewrite the following terms according to the notational conventions of 32.2. (That is, remove brackets as much as possible and compress λ-abstractions.)

1. $s \equiv (((\lambda x.(\lambda y.((yx)y)))(\lambda z.z))(\lambda p.(\lambda q.p)))$.
2. $t \equiv (\lambda x.(\lambda y.(\lambda z.(((xz)z)(y(xz))))))$.

32.2. Give the sets of free and bound variables of the following λ-terms.
1. $s \equiv \lambda z.xy(\lambda xy.z)$. 2. $t \equiv \lambda x.x(\lambda y.yy)z$.

32.3. Show that extensionality holds in $\boldsymbol{\lambda}_{\beta\eta}$:

$$\text{If } tu =_{\beta\eta} su \text{ for all } \lambda\text{-terms } u, \text{ then } t =_{\beta\eta} s.$$

32.4. Show that if the equation $\lambda xy.x =_\beta \lambda x.x$ is added to the theory $\boldsymbol{\lambda}_\beta$, the theory becomes inconsistent. (That is, show that all terms become equal to I if $\lambda xy.x =_\beta \lambda x.x$ is added as an equation in Definition 32.5.)

32.5. Construct a λ-term F for which $Fxy =_\beta xFy$.

32.6. Show that in $\boldsymbol{\lambda}_{\beta\eta}$, for any closed term Z, Z is a fixed point combinator iff Z is a fixed point of SI.

32.7. Find a short term s with $\mathbf{S}(\mathbf{K}(\mathbf{S}(\mathbf{SKK})))(\mathbf{S}(\mathbf{KK})(\mathbf{SKK}))pq =_w s$ in **CL**.

32.8. Prove Lemma 32.17.

32.9. The equality relation $=_w$ in **CL** can be seen as the least equivalence relation generated by the reduction $\rightarrow_w$, which is defined by

$$\begin{aligned} \mathbf{K}st &\rightarrow_w s, \\ \mathbf{S}stu &\rightarrow_w su(tu), \end{aligned}$$

and compatibility rules for the application. The transitive reflexive closure of $\rightarrow_w$ is denoted by $\twoheadrightarrow_w$. It can be shown that $\twoheadrightarrow_w$ is Church-Rosser on **CL**. (i) Show that $\mathbf{CL} \not\vdash \mathbf{K}y =_w \mathbf{S}(\mathbf{K}(\mathbf{K}y))\mathbf{I}$ and (ii) construct a **CL** term s for which $((s)_\lambda)_{\mathbf{CL}} \neq_w s$ in **CL**

32.10. Show that there is no λ-term H for which $Hst =_\beta \mathsf{T}$ if $s =_\beta t$ and $Hst =_\beta \mathsf{F}$ if $s \neq_\beta t$ for all λ-terms s and t.

32.11. Show that the Church numerals form a numeral system. Define a predecessor function P_c^- on the Church numerals (and show that indeed $\mathsf{P}_c^- c_{n+1} =_\beta c_n$).

32.12. Construct terms M and E that λ-define multiplication and exponentiation on the Church numerals.

32.13. Construct a term F for which $Fc_nx =_\beta F(x \cdots x)c_{n+1}$ (n times an x), c_n indicating the Church numerals.

32.14. Let the recursive function $g : \mathbb{N} \to \mathbb{N}$ be λ-defined by G. Construct a term F such that F λ-defines f when $f(n) := \Sigma_{0 \leq i < n} g(i)$.

32.15. Find terms s and t for which there is more than one standard reduction from s to t. (Hint: Consider the term $\mathsf{I}(\mathsf{II})$.)

32.16. Show that $\Theta(\lambda f.f(\mathsf{SSS}))$ has no normal form. (Θ is defined in Example 32.28.)

32.17. Show that, for s and t λ-terms, if st is strongly normalizing, then s and t are both strongly normalizing. Construct λ-terms s and t such that st is normalizing and s and t are both non-normalizing.

32.18. Show that for $M = \langle D, [D \to D], F, G \rangle$ a set-theoretic λ-model, if $G \circ F = id_D$, then the η-rule is valid in M. (The η-rule is valid in M if $M \models s = \lambda x.sx$ for any λ-term s with $x \notin FV(s)$.)

32.19. Is the η-rule valid in the model $P\omega$? (Hint: compute the interpretations of $\lambda x.x$ and $\lambda xy.xy$.)

32.20. Use Theorem 32.67 and the model $P\omega$ to show that there are λ-terms s and t such that $\boldsymbol{\lambda}_\beta \not\vdash s =_\beta t$.

§33 TYPED λ-CALCULUS:

INTRODUCTION; SIMPLY TYPED λ-CALCULUS; POLYMORPHIC λ-CALCULUS; FORMULAS-AS-TYPES; FURTHER DEVELOPMENTS

J.H. GEUVERS and R.P. NEDERPELT

INTRODUCTION

Besides the pure λ-calculus introduced in the previous section, in which there is no stratification in the terms, there are also systems of typed λ-calculus in which the set of terms is subdivided into sets of terms of a common type. There are many different ways of attaching types to λ-terms, depending, for example, on the set of types one wants to consider. In general, not all λ-terms have to be typable. Thus, the set of 'legal' expressions (relative to the type system) is only a specific subset of all λ-terms.

The fact that a term *s is of type* σ is usually denoted by $s : \sigma$. In some systems the term s is an untyped λ-term ($s \in \Lambda$). In these systems types are assigned to the terms of the (untyped) pure λ-calculus. In Barendregt [1992] this is called *typing à la Curry.* In other systems some type information is contained in the term s itself. This is usually achieved by attaching a type to each bound variable occurring in s directly behind a λ. One example of this is the identity on the type σ, which would be $\lambda x : \sigma . x$ or $\lambda x^{\sigma} . x$. Here the type of x is declared to be σ, so $\lambda x : \sigma . x$ is a function from σ to σ. Hence, the type of $\lambda x : \sigma . x$ is the class of functions from σ to σ. This type is usually denoted by $\sigma \to \sigma$. In Barendregt [1992] this is called *typing à la Church.* Sometimes it is useful to put even more type information in the terms than just the types of the bound variables. There are systems described in the literature in which applications are decorated with types as well. However, this is only relevant for more complicated systems which will not be dealt with here.

There are good reasons for assigning types to λ-terms. One is that the type of a term provides some information about the term because it is a partial specification of the algorithm that the term denotes. This also makes it possible to do a partial correctness check of a program by checking whether the λ-term can be typed and if so, to compare the type that has been found with the type (specification) that the term was intended to have. Another advantage of typing is that the type of a term can provide information on the manner in which it should be evaluated.

A completely different motivation for studying typed λ-calculi originates in logic: due to the so-called 'formulas-as-types' embedding, types in certain typed lambda calculi can be seen as formulas. Here, a term of a type represents a proof of the corresponding formula. Therefore, in this fashion the primary objects are the types and the terms are denotations for deductions in a logical system. This formulas-as-types embedding will be discussed in more detail later.

SIMPLY TYPED λ-CALCULUS

The simply typed λ-calculus is a relatively easy system of type assignment but it is interesting in that it illustrates the significant properties that a type system can have and also can serve to illustrate the difference between typing *à la Curry* and typing *à la Church.*

DEFINITION 33.1.: The *simply typed λ-calculus à la Curry*, $\lambda \rightarrow_{Cu}$, has as types the set Typ defined by

$$\begin{aligned} \mathsf{Con} &\subset \mathsf{Typ}, \\ \mathsf{TVar} &\subset \mathsf{Typ}, \\ \sigma, \tau \in \mathsf{Typ} &\Rightarrow (\sigma \rightarrow \tau) \in \mathsf{Typ}, \end{aligned}$$

in which Con is an arbitrary set of *type constants* (such base types as Int, Real and Bool) and TVar is a set of *type variables*. In the simply typed λ-calculus a term s is given a type σ under the assumption of specific types for the free variables that occur in s. A *context* of $\lambda \rightarrow_{Cu}$ is therefore defined as a finite sequence of *declarations* $x : \sigma$, in which x is a *term variable* and σ is a type. It is always assumed that all term variables in a context are different, thus a context always takes the form

$$x_1 : \sigma_1,\ \ldots,\ x_n : \sigma_n$$

with $n \geq 0$, all $x_i \in \mathsf{Var}$, all $\sigma_i \in \mathsf{Typ}$ and $x_i \equiv x_j \Rightarrow i = j$. The *derivation rules* for deriving *judgements* of the form

$$\Gamma \vdash s : \sigma,$$

in which Γ is a context, s a λ-term and σ a type are now as follows.

$$\text{(ctxt)} \quad \Gamma \vdash x : \sigma \qquad\qquad \text{if } x : \sigma \in \Gamma$$

$$(\lambda) \quad \frac{\Gamma, x : \sigma \vdash s : \tau}{\Gamma \vdash \lambda x.s : (\sigma \to \tau)}$$

$$\text{(app)} \quad \frac{\Gamma \vdash s : (\sigma \to \tau) \quad \Gamma \vdash t : \sigma}{\Gamma \vdash st : \tau}$$

We say that the term s is *typable* if there exists a context Γ and a type σ such that $\Gamma \vdash s : \sigma$.

NOTATION 33.2.: To reduce the number of brackets the conventions for denoting a type are as follows: $(\sigma_1 \to (\sigma_2 \to \cdots \to (\sigma_{n-1} \to \sigma_n) \cdots))$ is denoted by $\sigma_1 \to \sigma_2 \to \cdots \to \sigma_{n-1} \to \sigma_n$. In type notations the brackets associate to the right.

EXAMPLES 33.3.:

1. The following is a complete derivation of the fact that $\vdash \lambda xy.x : \sigma \to \tau \to \sigma$ in $\lambda \to_{Cu}$ for arbitrary types σ and τ.

$$\frac{\dfrac{x : \sigma,\ y : \tau \vdash x : \sigma}{x : \sigma \vdash \lambda y.x : \tau \to \sigma}}{\vdash \lambda xy.x : \sigma \to \tau \to \sigma}$$

2. $x : \sigma \to \tau \to \rho,\ y : \sigma \to \tau,\ z : \sigma \vdash xz(yz) : \rho$.

The version à la Church of the simply typed λ-calculus differs from $\lambda \to_{Cu}$ in its derivation rules. Moreover, the terms are taken from a variant Λ_{Typ} of Λ in which the abstractions are decorated with the type of the abstracted variable:

$$\Lambda_{\mathsf{Typ}} ::= \mathsf{Var} \mid (\lambda \mathsf{Var} \colon \mathsf{Typ}\ .\ \Lambda_{\mathsf{Typ}}) \mid (\Lambda_{\mathsf{Typ}} \Lambda_{\mathsf{Typ}}).$$

As in the case of $\lambda \to_{Cu}$, the system links a type to terms by means of the derivation of judgements.

DEFINITION 33.4.: The *simply typed λ-calculus à la Church*, $\lambda \to_{Ch}$, has as types the same set Typ of $\lambda \to_{Cu}$. Likewise, the notions of *declaration* and *context* are the same as for $\lambda \to_{Cu}$. The set of terms is Λ_{Typ}.

The derivation rules are as follows.

$$\text{(ctxt)} \quad \Gamma \vdash x : \sigma \qquad\qquad \text{if } x : \sigma \in \Gamma$$

$$(\lambda) \quad \frac{\Gamma, x : \sigma \vdash s : \tau}{\Gamma \vdash \lambda x{:}\sigma.s : \sigma \to \tau}$$

$$\text{(app)} \quad \frac{\Gamma \vdash s : \sigma \to \tau \quad \Gamma \vdash t : \sigma}{\Gamma \vdash st : \tau}$$

To distinguish derivability in $\lambda\!\to_{Cu}$ from that in $\lambda\!\to_{Ch}$, $\vdash_{Cu}$ and $\vdash_{Ch}$ are sometimes used for $\lambda\!\to_{Cu}$ and $\lambda\!\to_{Ch}$ respectively. In the following, when $\lambda\!\to$ is written it will refer to either $\lambda\!\to_{Cu}$ or $\lambda\!\to_{Ch}$. It is not difficult to adapt the examples above to the $\lambda\!\to_{Ch}$ case.

The differences between $\lambda\!\to_{Cu}$ and $\lambda\!\to_{Ch}$ may seem minimal, but even this minimal difference is enough to analyse the crucial features. In the system of polymorphic λ-calculus that will be introduced later these small differences play a much more crucial role. Below, a number of significant properties are dealt with, after which precise connections are made between the systems. Before attempting to prove these properties a number of less complex lemmas are given below.

LEMMA 33.5. (*substitution*): For both $\lambda\!\to_{Cu}$ and $\lambda\!\to_{Ch}$,

If $\Gamma_1, y : \tau, \Gamma_2 \vdash s : \sigma$ and $\Gamma_1 \vdash t : \tau$, then $\Gamma_1, \Gamma_2 \vdash s[y := t] : \sigma$.

LEMMA 33.6. (*generation*): For both $\lambda\!\to_{Cu}$ and $\lambda\!\to_{Ch}$,

1. If $\Gamma \vdash x : \sigma$, then $x : \sigma \in \Gamma$.

2a. If $\Gamma \vdash_{Cu} \lambda x.s : \tau$, then $\tau \equiv \sigma \to \rho$ for some σ and ρ with $\Gamma, x : \sigma \vdash_{Cu} s : \rho$.

2b. If $\Gamma \vdash_{Ch} \lambda x{:}\sigma.s : \tau$, then $\tau \equiv \sigma \to \rho$ for some ρ with $\Gamma, x : \sigma \vdash_{Ch} s : \rho$.

3. If $\Gamma \vdash st : \tau$, then $\Gamma \vdash s : \rho \to \tau$ for some ρ with $\Gamma \vdash t : \rho$.

The proofs of both Lemmas are by induction on derivations. (Lemma 33.5 is by induction on the derivation of $\Gamma_1, y : \tau, \Gamma_2 \vdash s : \sigma$.)

THEOREM 33.7. (*subject reduction*): For both $\lambda\!\to_{Cu}$ and $\lambda\!\to_{Ch}$,

If $\Gamma \vdash s : \sigma$ and $s \twoheadrightarrow_\beta t$, then $\Gamma \vdash t : \sigma$.

PROOF: Suppose $\Gamma \vdash s : \sigma$. By induction on the structure of the term s it can be proven that if $s \to_\beta t$, then $\Gamma \vdash t : \sigma$. (Note that this is sufficient.) Two cases can be distinguished: s is an abstraction or an application. (s cannot be a variable if $s \to_\beta t$.) We only consider the case that $\vdash$ is $\vdash_{Ch}$.

(λ) Say $s \equiv \lambda x : \tau . u$ with $t \equiv \lambda x : \tau . u'$ and $u \to_\beta u'$. Then by Lemma 33.6,

$$\Gamma, x : \tau \vdash u : \rho \text{ for some } \rho.$$

From the induction hypothesis it follows that $\Gamma, x : \tau \vdash u' : \rho$. Consequently, $\Gamma \vdash t : \sigma$.

(app) Say $s \equiv uv$. Then by Lemma 33.6, $\Gamma \vdash u : \tau \to \sigma$ and $\Gamma \vdash v : \tau$ for some τ. It is now possible to distinguish subcases according to whether the reduction takes place in u or v or whether uv is itself the redex that has been contracted. If the reduction is within u or v, the proof is completed by induction hypothesis. If uv is the redex, then $u \equiv \lambda x : \tau . w$ for some w and x and so $\Gamma, x : \tau \vdash w : \sigma$ by Lemma 33.6. The proof is completed by applying the Substitution Lemma 33.5. □

Subject reduction is clearly an important property: if a program s evaluates to a program t, this t should satisfy the same specifications as s. An important corollary of subject reduction is a typed version of the Church-Rosser property.

THEOREM 33.8. (*Church-Rosser*): For both $\lambda\!\to_{Cu}$ and $\lambda\!\to_{Ch}$,
If $\Gamma \vdash s : \sigma$ and $s \twoheadrightarrow_\beta t_1$, $s \twoheadrightarrow_\beta t_2$, then there is a t such that $\Gamma \vdash t : \sigma$ and $t_1 \twoheadrightarrow_\beta t$, $t_2 \twoheadrightarrow_\beta t$.

PROOF: For $\lambda\!\to_{Cu}$, the proof follows from the Church-Rosser Theorem for $\twoheadrightarrow_\beta$ for untyped terms (Theorem 32.34) and Subject Reduction: the first gives us a common reduct t of t_1 and t_2 and the second shows that this t is typable with type σ. For $\lambda\!\to_{Ch}$ the situation is only slightly more complicated. It is easy to verify that the Church-Rosser Theorem holds for the set Λ_{Typ}. The rest of the procedure is exactly the same as for $\lambda\!\to_{Cu}$. □

An important difference between $\lambda\!\to_{Cu}$ and $\lambda\!\to_{Ch}$ is that in the second system every typable term has a unique type, while in the first this is not the case. (An easy counterexample is $\lambda x.x$, which is of type $\sigma \to \sigma$ for every type σ.) This *uniqueness of types property* for $\lambda \to_{Ch}$ is not surprising, because in $\lambda \to_{Ch}$ there is a lot of type information in the terms.

THEOREM 33.9. (*uniqueness of types*): For $\lambda{\to}_{Ch}$,

$$\text{If } \Gamma \vdash s : \sigma \text{ and } \Gamma \vdash s : \tau, \text{ then } \sigma \equiv \tau.$$

PROOF: The proof is by induction on the structure of s, using the Generation Lemma 33.6. □

A positive feature of $\lambda{\to}_{Cu}$ is that, although there is no unique type for every term, there is a recursive function that decides for every untyped λ-term whether it is typable in $\lambda \to_{Cu}$, and if it is, computes a type which is most general in some sense. This can be very convenient for programming applications, because it means that a programmer does not have to provide any type information when writing the program.

DEFINITION 33.10.: A *type substitution* in $\lambda{\to}_{Cu}$ is a mapping that assigns types to type variables. A type substitution S is extended to a mapping from types to types. $S(\sigma)$ is written for the application of S to the type σ. If Γ is a context, $S(\Gamma)$ is obtained by applying S to all the types in Γ. The composition of two type substitutions S and T is denoted by $S \circ T$ (first T, then S).

DEFINITION 33.11.: Let t be an untyped λ-term. The *principal pair* of t, $pp(t)$, is a pair (Γ, σ) in which Γ is a context that contains declarations for all the free variables of t and σ is a type such that

1. $\Gamma \vdash_{Cu} t : \sigma$,
2. if $\Gamma' \vdash_{Cu} t : \sigma'$, then there is a type substitution S such that $S(\sigma) = \sigma'$ and $S(\Gamma) = \Gamma'$.

If t is a closed term, the principal pair of t is of the form $(\langle - \rangle, \sigma)$ ($\langle - \rangle$ is the empty context). In this case, σ is called the *principal type scheme* of t, $pts(t)$.

Note that a principal type scheme is always unique modulo renaming of type variables. For example, if $\alpha \to (\beta \to \alpha) \to \alpha$ is a principal type scheme of t, then $\beta \to (\gamma \to \beta) \to \beta$ is likewise a principal type scheme of t. (But note that $\alpha \to (\alpha \to \alpha) \to \alpha$ is not and neither is $\gamma \to (\beta \to \alpha) \to \gamma$.) So, when it is stated that a type τ is *the* principal type scheme of a term t, it is meant that τ is a representative of the equivalence class (modulo renaming of type variables) of principal type schemes of t.

The crucial theorem for $\lambda \to_{Cu}$ is that principal pairs and principal type schemes exist.

THEOREM 33.12.: For every term t that is typable in $\lambda \to_{Cu}$, the principal pair of t, $pp(t)$, exists. Moreover, there is a recursive function that, given (a code of) an untyped λ-term, decides whether that term is typable in $\lambda \to_{Cu}$ and if so, returns its principal pair.

The proof of the theorem is not really difficult, but still too involved to be given here. For details, we refer to Hindley and Seldin [1986]. The idea is contained in some examples of terms and their pts which we give along with a detailed construction of the principal type scheme for the term $t \equiv \lambda y.y(y\mathsf{I})$.

EXAMPLES 33.13.:

1. The principal type scheme of I is $\alpha \to \alpha$ for $\alpha \in \mathsf{TVar}$.
2. The principal type scheme of K is $\alpha \to \beta \to \alpha$. As an exercise, show that $\alpha \to \alpha \to \alpha$ is not correct as the principal type scheme for K.
3. The principal type scheme of S is $(\alpha \to \beta \to \gamma) \to (\alpha \to \beta) \to \alpha \to \gamma$.

EXAMPLE 33.14.: We compute the *pts* of $t := \lambda y.y(y\mathsf{I})$. The principal type of I is $\alpha \to \alpha$ and hence the type of y should be something like $(\alpha \to \alpha) \to \sigma$ for some σ. (The type σ will be specified further later.) This implies that the type of $y\mathsf{I}$ is σ and from the application of y to $y\mathsf{I}$ we generate the type equality $\sigma = \alpha \to \alpha$. This equality has to be solved in order to find a *pts* for t. The obvious solution is to take $\sigma \equiv \alpha \to \alpha$.

(Note that there are no other requirements for the type σ that prevent this choice.) Thus, the type for $y(y\mathsf{I})$ will be $\alpha \to \alpha$ and, finally, the type of t turns out to be $((\alpha \to \alpha) \to \alpha \to \alpha) \to \alpha \to \alpha$. Intuitively, it will be clear that this type is the principal type scheme.

One can apply the same informal reasoning to show that a term is not typable in $\lambda{\to}_{Cu}$. Consider, for example, the term $\lambda x.xx$. One has to find a type σ for which $\sigma \equiv \sigma \to \tau$. Such a type does not exist and hence $\lambda x.xx$ cannot be typed in $\lambda{\to}_{Cu}$.

The basis of an algorithm for finding the principal type scheme is an algorithm for finding a *most general unifier* of two types. When trying to type a term in $\lambda{\to}_{Cu}$ one generates a set of equality-conditions $E = \{\sigma_1 = \tau_1, \ldots, \sigma_n = \tau_n\}$ that have to be *solved*, where solving the equality-conditions means finding a type substitution S such that $S(\sigma_i) = S(\tau_i)$ for all $1 \leq i \leq n$. Such a type substitution that solves E is called a *unifier of* E, and S is the *most general unifier of* E if for each other unifier T of E there is a substitution R such that $T = R \circ S$. (See section 25.) The crucial fact is now that there is an algorithm that decides for every set of equality conditions E whether a unifier of E exists and, if so, it gives the most general unifier of E. (See section 25.)

EXAMPLE 33.15.: The principal pair of $\lambda z.xy(x(zy)y)$ is

$$(x : \alpha \to \alpha \to \alpha,\ y : \alpha,\ (\alpha \to \alpha) \to \alpha)).$$

As corollaries to Theorem 33.12, we find that *typability*, *type assignment* and *type checking* are decidable in $\lambda{\to}_{Cu}$.

DEFINITION 33.16.:

1. *Typability* in $\lambda{\to}_{Cu}$ is the problem of deciding for $s \in \Lambda$ whether there is a context Γ and a type σ such that $\Gamma \vdash_{Cu} s : \sigma$.

2. *Type assignment* in $\lambda{\to}_{Cu}$ is the problem of deciding for a context Γ and $s \in \Lambda$ whether there is a type σ such that $\Gamma \vdash_{Cu} s : \sigma$.

3. *Type checking* in $\lambda{\to}_{Cu}$ is the problem of deciding for a context Γ, type σ and $s \in \Lambda$ whether $\Gamma \vdash_{Cu} s : \sigma$ holds.

The typability problem is sometimes denoted by $? \vdash s : ?$, the type assignment problem by $\Gamma \vdash s : ?$ and the type checking problem by $\Gamma \vdash s : \sigma\ ?$.

THEOREM 33.17.: Typability, type assignment and type checking are decidable in $\lambda{\to}_{Cu}$.

PROOF: The decidability of typability follows immediately from Theorem 33.12. If s and Γ are given, we compute the principal pair of s, say it is (Γ', σ), and check whether Γ is a substitution instance of Γ'. This yields a decision procedure for type assignment. If s, Γ and σ are given, we again compute the principal pair, say it is (Γ', σ'), and check whether Γ, σ is a substitution instance of Γ', σ'. □

The formal connection between $\lambda{\to}_{Cu}$ and $\lambda{\to}_{Ch}$ is given by an *erasure map* $| - |$, which removes all types in the terms of $\lambda{\to}_{Ch}$.

DEFINITION 33.18.: The *erasure map* $| - |$ from Λ_{Typ}, the terms of $\lambda{\to}_{Ch}$, to Λ, the terms of $\lambda{\to}_{Cu}$, is defined as follows:

$$\begin{aligned} |x| &:= x, \\ |\lambda x{:}\sigma.s| &:= \lambda x.|s|, \\ |st| &:= |s||t|. \end{aligned}$$

LEMMA 33.19.: 1. If $\Gamma \vdash_{Ch} s : \sigma$, then $\Gamma \vdash_{Cu} |s| : \sigma$.
2. If $\Gamma \vdash_{Cu} s : \sigma$, then for some t, $\Gamma \vdash_{Ch} t : \sigma$ and $|t| \equiv s$.

PROOF: By induction on the structure of s, using the Generation Lemma 33.6. □

This lemma gives a way of deciding the typability, type assignment and type checking for $\lambda{\to}_{Ch}$ via the decidability of the same problems for $\lambda{\to}_{Cu}$. However, there is no need to go via $\lambda{\to}_{Cu}$: given the extra type information in the terms of $\lambda{\to}_{Ch}$, the decidability problems are easier.

Another important property of the simply typed λ-calculus is that all terms are normalizable (even strongly normalizable). Together with the Church-Rosser property, this implies that the equality between typable terms in the simply typed λ-calculus is decidable. (Let s and t

be typable terms in $\lambda\!\rightarrow_{Cu}$. We can compute their normal forms, $nf(s)$ and $nf(t)$. By the Church-Rosser property we know that $s =_\beta t$ iff $nf(s) \equiv nf(t)$. The latter relation is (trivially) decidable.) It is clear that the (strong) normalization for $\lambda\!\rightarrow_{Cu}$ follows from the (strong) normalization for $\lambda\!\rightarrow_{Ch}$ and vice versa. To give some indication, we treat a proof of normalization for $\lambda\!\rightarrow_{Ch}$.
The idea of the normalization proof is to define a notion of 'depth' of a redex for every redex in a term t. This depth is a natural number. Then the 'level' of the term t is defined as the pair (d, n), where d is the maximal depth of all redexes in t and n is the number of redexes of depth d. The terms are now ordered by the *lexicographical ordering* on the levels, that is,

$$(d, n) <_l (d', n') := d < d' \text{ or } (d = d' \text{ and } n < n').$$

The normalization then follows from defining a reduction strategy that reduces the level of a term.

DEFINITION 33.20.:

1. The *rank* of a type of the simply typed λ-calculus is defined as follows:
   - $\text{rank}(\sigma) := 1$ for σ a type variable or type constant,
   - $\text{rank}(\sigma \rightarrow \tau) := \text{rank}(\sigma) + \text{rank}(\tau)$.
2. The *depth* of a redex $(\lambda x\!:\!\sigma.s)t$ is defined as the rank of the type of $\lambda x\!:\!\sigma.s$.
3. The *level* of a term s is defined as the pair (d, n), where d is the maximal depth of all redexes in s and n is the number of all redexes in s of depth d.

Note that the notion of rank is relative to a context Γ in which the term $\lambda x\!:\!\sigma.s$ is typed: the rank of the type of $\lambda x\!:\!\alpha.yx$ is 2 if $y : \alpha \rightarrow \alpha$ and 3 if $y : \alpha \rightarrow \alpha \rightarrow \alpha$. The dependency on the context is not mentioned explicitly.

In the following definition we need the notion of *rightmost* redex, which is defined similarly to that of the leftmost redex (see the previous section).

DEFINITION 33.21.: The reduction strategy F on well-typed terms is defined as follows: $F(s) := t$ if t is obtained from s by contracting the rightmost redex of maximal depth in s.

THEOREM 33.22.: The reduction strategy F computes the normal form for the typable terms of $\lambda\!\rightarrow_{Ch}$.

PROOF: It suffices to show that, if $F(s) = t$, then level$(t) <_l$ level(s), where $<_l$ is the lexicographical ordering. Now suppose that level$(t) = (d, n)$ and $(\lambda x : \sigma.u)v$ is the rightmost redex of depth d in s. The desired result follows from an analysis of the ways in which the contraction of this redex can create new redexes. (If no new redex is created, then the level of t is certainly smaller than the level of s.) For details about this analysis, we refer to Barendregt [1992]. □

We have already mentioned the normalization for $\lambda\!\rightarrow_{Cu}$ and the decidability of equality between typable terms as corollaries of the normalization for $\lambda\!\rightarrow_{Ch}$. Another important consequence of normalization is that it gives a handle on the number of closed terms of a given type, as shown by the following lemma. (The combinators c_n are given in Definition 32.11).

LEMMA 33.23. (*Corollary of normalization*): Let α be an arbitrary type variable. If $\vdash s : (\alpha \rightarrow \alpha) \rightarrow \alpha \rightarrow \alpha$ in $\lambda\!\rightarrow_{Cu}$, then $s \twoheadrightarrow_\beta c_n$ for some $n \in \mathbb{N}$ or $s \twoheadrightarrow_\beta \mathsf{I}$. (And if $\vdash s : (\alpha \rightarrow \alpha) \rightarrow \alpha \rightarrow \alpha$ in $\lambda\!\rightarrow_{Ch}$, then $s \twoheadrightarrow_\beta \lambda f : \alpha \rightarrow \alpha \,.\, \lambda x : \alpha \,.\, f^n(x)$ for some $n \in \mathbb{N}$ or $s \twoheadrightarrow_\beta \lambda f : \alpha \rightarrow \alpha \,.\, f$.)

PROOF: If $\vdash s : (\alpha \rightarrow \alpha) \rightarrow \alpha \rightarrow \alpha$ in $\lambda\!\rightarrow_{Cu}$, then $s \twoheadrightarrow_\beta t$ with t in normal form. By induction on the derivation of $\vdash t : (\alpha \rightarrow \alpha) \rightarrow \alpha \rightarrow \alpha$, it can now be shown that t is of the required form. (Using the fact that a term in normal form has the shape $\lambda z_1 \cdots z_n.yu_1 \cdots u_k$ for some $n \geq 0$ and $u_1, \ldots, u_k$ terms in normal form; see Barendregt [1984].) □

The normalization of the simply typed λ-calculus is a desirable property because it makes the equality of typable terms decidable. (The equality of the *un*-typed λ-calculus is not decidable, see Fact 32.39.) There is, however, a general result (which we shall not mention here) implying that in such a normalizing system not all recursive functions are typable.

There are different ways of introducing the natural numbers into $\lambda\!\rightarrow$, for example, adding a constant type N and the constants $0 : N$ and $\mathsf{Suc} : N \rightarrow N$. This is not yet very interesting (the addition cannot be defined). An option is to add more constants for recursive functions, for instance, $+ : N \rightarrow N \rightarrow N$ and $\times : N \rightarrow N \rightarrow N$, but then again we feel the need to add extra reduction rules to make $+$ and Suc commute, such as $+x(\mathsf{Suc}\, y) \rightarrow \mathsf{Suc}(+xy)$, etc. It is clear that the addition of new reduction rules requires reconsidering the proofs of Church-Rosser and normalization.

Another way is to take $N = (o \rightarrow o) \rightarrow o \rightarrow o$, where o is a type. (This option is suggested by Lemma 33.23.) Define $0 := \lambda fx.x$, $\mathsf{Suc} := \lambda yfx.f(yfx)$. Then

$$\mathsf{add} := \lambda zyfx.yf(zfx) : N \rightarrow N \rightarrow N$$

defines addition. (See Exercise 33.5.)

POLYMORPHIC λ-CALCULUS

A calculus that is more powerful then $\lambda \rightarrow$, but in which reduction is still normalizing (and hence equality is decidable), is the polymorphic λ-calculus introduced by Girard [1972] and later independently reinvented by Reynolds [1974]. The main intuition behind it is that many functions, for example, the identity $\lambda x.x$, do not just work for one type but for all types in the same way. One could say that the type of $\lambda x.x$ is not $\alpha \rightarrow \alpha$, for arbitrary $\alpha \in \mathsf{Typ}$, but $\forall\alpha \,.\, \alpha \rightarrow \alpha$. Incorporation of this quantification over types in the syntax is advantageous in several respects. For one, it makes it possible to express provability in second order arithmetic in the typed lambda calculus (this was Girard's motivation). Reynolds used the quantification over types to capture the notion of explicit polymorphism known in programming languages.

The system of polymorphic λ-calculus, called λ2 following Barendregt [1992], also comes in a Curry-variant $\lambda2_{Cu}$ and a Church-variant $\lambda2_{Ch}$. We define them together.

DEFINITION 33.24.: The *polymorphic λ-calculus*, λ2, has as types the set Typ defined by

$$\begin{array}{rcl} \mathsf{Con} & \subseteq & \mathsf{Typ}, \\ \mathsf{TVar} & \subseteq & \mathsf{Typ}, \\ \sigma, \tau \in \mathsf{Typ} & \Rightarrow & (\sigma \to \tau) \in \mathsf{Typ}, \\ \alpha \in \mathsf{TVar}, \sigma \in \mathsf{Typ} & \Rightarrow & \forall \alpha . \sigma \in \mathsf{Typ}, \end{array}$$

where Con is an arbitrary set of *type constants* and TVar is a set of *type variables.* A *context* of $\lambda 2$ is a finite sequence of *declarations* $x : \sigma$, where x is a *term variable* and σ is a type. All term variables in a context are chosen to be different.

For $\lambda 2$ à la Curry, $\lambda 2_{Cu}$, the *derivation rules* for deriving judgements of the form

$$\Gamma \vdash s : \sigma,$$

where Γ is a context, $s \in \Lambda$ and σ a type, are the rules (ctxt), (λ) and (app) of Definition 33.1, plus the following two derivation rules:

$$(\forall\!-\!I) \quad \frac{\Gamma \vdash s : \tau}{\Gamma \vdash s : \forall \alpha . \tau} \qquad \text{if } \alpha \in \mathsf{TVar} \text{ and } \alpha \text{ not in } \Gamma,$$

$$(\forall\!-\!E) \quad \frac{\Gamma \vdash s : \forall \alpha . \tau}{\Gamma \vdash s : \tau[\alpha := \sigma]} \quad \text{if } \sigma \in \mathsf{Typ}.$$

For $\lambda 2$ à la Church, $\lambda 2_{Ch}$, Λ_{Typ} (see Definition 33.4) is extended with the sets $\lambda \mathsf{TVar}.\Lambda_{\mathsf{Typ}}$ and $\Lambda_{\mathsf{Typ}}\mathsf{Typ}$. The *derivation rules* in $\lambda 2_{Ch}$ for deriving judgements of the form

$$\Gamma \vdash s : \sigma,$$

where Γ is a context, $s \in \Lambda_{\mathsf{Typ}}$ and σ a type, are the rules (ctxt), (λ) and (app) of Definition 33.4, plus the two derivation rules:

$$(\forall\!-\!I) \quad \frac{\Gamma \vdash s : \tau}{\Gamma \vdash \lambda \alpha . s : \forall \alpha . \tau} \qquad \text{if } \alpha \in \mathsf{TVar} \text{ and } \alpha \text{ not in } \Gamma,$$

$$(\forall\!-\!E) \quad \frac{\Gamma \vdash s : \forall \alpha . \tau}{\Gamma \vdash s\sigma : \tau[\alpha := \sigma]} \quad \text{if } \sigma \in \mathsf{Typ}.$$

Derivability in $\lambda 2_{Cu}$ will be denoted by $\vdash_{2Cu}$, derivability in $\lambda 2_{Ch}$ by $\vdash_{2Ch}$.

As usual, we omit brackets in type notations by letting them associate to the right. The differences between the Curry version and the Church version of $\lambda 2$ are much more serious than for $\lambda \to$. For one thing, there are more reductions in $\lambda 2_{Ch}$ than in $\lambda 2_{Cu}$, owing to the extra abstractions over types. In $\lambda 2_{Ch}$, the polymorphic identity is $\lambda \alpha . \lambda x : \alpha . x$, of type $\forall\alpha\ .\ \alpha \to \alpha$, which can be applied to a type obtaining a redex $(\lambda\alpha.\lambda x:\alpha.x)\sigma$, which is of type $\sigma \to \sigma$. For most of the meta-theoretic properties, the two systems still behave the same, except for the decidability of typing.

EXAMPLES 33.25.:

1. The following is a complete derivation of the fact that $\vdash \lambda xy.x : \forall\alpha.(\alpha \to \forall\beta.(\beta \to \alpha))$ in $\lambda 2_{Cu}$.

$$\dfrac{\dfrac{\dfrac{\dfrac{x:\alpha,\ y:\beta \vdash x:\alpha}{x:\alpha \vdash \lambda y.x : \beta \to \alpha}}{x:\alpha \vdash \lambda y.x : \forall\beta.(\beta \to \alpha)}}{\vdash \lambda xy.x : \alpha \to \forall\beta.(\beta \to \alpha)}}{\vdash \lambda xy.x : \forall\alpha.(\alpha \to \forall\beta.(\beta \to \alpha))}$$

2. Also $\vdash \lambda xy.x : \forall\alpha\beta.(\alpha \to \beta \to \alpha)$ in $\lambda 2_{Cu}$.

3. The first typing of K $(\equiv \lambda xy.x)$ corresponds to $\vdash \lambda\alpha.\lambda x{:}\alpha.\lambda\beta.\lambda y{:}\beta.x : \forall\alpha.(\alpha \to \forall\beta.(\beta \to \alpha))$ in $\lambda 2_{Ch}$, the second to $\vdash \lambda\alpha\beta.\lambda x{:}\alpha.\lambda y{:}\beta.x : \forall\alpha\beta.(\alpha \to \beta \to \alpha)$.

The extension of $\lambda \to$ to $\lambda 2$ by adding a kind of 'universal quantifier over all types' allows the typing of terms in a polymorphic way (as in the example of the polymorphic identity, $\lambda x.x : \forall\alpha.(\alpha \to \alpha)$), but also allows much stronger features such as *self application*. This is shown by the following examples.

EXAMPLES 33.26.: In $\lambda 2_{Cu}$, the following holds:

1. $\lambda x.xx : (\forall\alpha.(\alpha \to \alpha)) \to \forall\alpha.(\alpha \to \alpha)$.

2. $\lambda x.x(xx)(xxx) : (\forall\alpha.\alpha) \to \forall\alpha.\alpha$.

In the previous example we used the fact that $\lambda 2$ is *impredicative*: a term s of type $\forall\alpha.(\alpha \to \alpha)$ is a function from σ to σ for any type σ, so in particular s is a function from $\forall\alpha.(\alpha \to \alpha)$ to $\forall\alpha.(\alpha \to \alpha)$, and hence such an s can be applied to itself. This is the case in the first example. In the second example we have: if $x : \forall\alpha.\alpha$, then $x : \alpha[\alpha := (\forall\alpha.\alpha) \to \forall\alpha.\alpha]$ by the rule $(\forall{-}E)$; that is, $x : (\forall\alpha.\alpha) \to \forall\alpha.\alpha$.

In general, one speaks of an impredicative definition if an object is defined by making use of a quantification over the collection of all objects that the defined object is part of. We see this in the first example: the type $\forall\alpha.(\alpha \to \alpha)$ contains a quantification over all types α, concerning in particular $\forall\alpha.(\alpha \to \alpha)$ itself. Impredicativity is sometimes viewed with suspicion by logicians. This is firstly due to the fact that it can give rise to contradictions, as in the case of the Russell set (see section 17) which is defined as the set of all sets that do not contain themselves as an element. Another reason for suspicion is that the meaning of the concept itself is not clear. The explanation of an inductive definition is that in every clause an object is explained in terms of its components, which are already well-understood at that point. In an impredicative definition, there is a quantification over a collection that is not yet well understood (because the object that is being defined is part of the collection).

The self application in $\lambda 2_{Cu}$, as shown in the Examples 33.26 and explained above, is much more explicitly dealt with in the system à la Church: there we have to provide the type of x before we can apply it to x itself. For example, if $x : \forall\alpha.(\alpha \to \alpha)$, then $x\sigma : \sigma \to \sigma$ for any type σ; so, in particular, $x(\forall\alpha.(\alpha \to \alpha)) : (\forall\alpha.(\alpha \to \alpha)) \to \forall\alpha.(\alpha \to \alpha)$, and hence $x(\forall\alpha.(\alpha \to \alpha))x : \forall\alpha.(\alpha \to \alpha)$.

EXAMPLES 33.27.: The examples in 33.26 have the following counterparts in $\lambda 2_{Ch}$. (Write $\bot$ for the type $\forall\alpha.\alpha$.)

1. A counterpart of $\lambda x.xx$ is $\lambda x : \forall\alpha.(\alpha \to \alpha)\ .\ x(\forall\alpha.(\alpha \to \alpha))x$ and has type $(\forall\alpha.(\alpha \to \alpha)) \to \forall\alpha.(\alpha \to \alpha)$.
2. Here is another way to 'decorate' the term $\lambda x.xx$ of $\lambda 2_{Cu}$ to make it of type $(\forall\alpha.(\alpha \to \alpha)) \to \forall\alpha.(\alpha \to \alpha)$ in $\lambda 2_{Ch}$:

   $\lambda x{:}\forall\alpha.(\alpha \to \alpha).\lambda\beta\ .\ x(\beta \to \beta)(x\beta) : (\forall\alpha.(\alpha \to \alpha)) \to \forall\beta.(\beta \to \beta)$.
3. For the term $\lambda x.x(xx)(xxx)$ of $\lambda 2_{Cu}$, we obtain:

   $\lambda x{:}\bot\ .\ x(\bot{\to}\bot{\to}\bot)(x(\bot{\to}\bot)x)(x(\bot{\to}\bot{\to}\bot)xx) : \bot{\to}\bot$.

The examples show that an untyped λ-term in $\lambda 2_{Cu}$ can give rise to a number of different terms in $\lambda 2_{Ch}$. This is also the case in $\lambda\!\to$, but there one only has to add types to the abstraction variables, occurring behind the λ's. In $\lambda 2_{Ch}$, one also has to add new λ-abstractions (over type variables) and new applications (of terms to types). This operation of adding types, type abstractions and type applications is usually called the *decoration* of the untyped term. The reverse operation is called *erasure*, which we already encountered in Definition 33.18 for the case of $\lambda\!\rightarrow$ and which is extended here to $\lambda 2$.

DEFINITION 33.28.: The *erasure map* $|-|$ from the terms of $\lambda 2_{Ch}$ to the terms of $\lambda 2_{Cu}$ is defined as follows:

$$\begin{aligned} |x| &:= x, \\ |\lambda x{:}\sigma.s| &:= \lambda x.|s|, \\ |\lambda\alpha.s| &:= |s|, \\ |st| &:= |s||t|, \\ |s\sigma| &:= |s|. \end{aligned}$$

Here it is understood that s and t are terms and σ is a type.

The connection between $\lambda 2_{Cu}$ and $\lambda 2_{Ch}$ is given by the following lemma, which is an immediate extension of Lemma 33.19.

LEMMA 33.29.: 1. If $\Gamma \vdash_{2Ch} s : \sigma$, then $\Gamma \vdash_{2Cu} |s| : \sigma$.
2. If $\Gamma \vdash_{2Cu} s : \sigma$, then for some t, $\Gamma \vdash_{2Ch} t : \sigma$ and $|t| \equiv s$.

The statement of this lemma can be made more precise by taking derivations into account. It can be shown that if $\Gamma \vdash_{2Cu} s : \sigma$, then there is a one-to-one correspondence between the set $\{t \mid \Gamma \vdash_{2Ch} t : \sigma \text{ and } |t| \equiv s\}$ and the set of derivations of $\Gamma \vdash_{2Cu} s : \sigma$. The same holds for $\lambda\!\rightarrow$, but it has slightly more interest here.

Without proof we list a few meta-theoretic properties of $\lambda 2$. Proofs are similar to the ones for $\lambda\!\rightarrow$.

THEOREM 33.30. (*subject reduction*): For both $\lambda 2_{Cu}$ and $\lambda 2_{Ch}$,
if $\Gamma \vdash s : \sigma$ and $s \twoheadrightarrow_\beta t$, then $\Gamma \vdash t : \sigma$.

THEOREM 33.31. (*Church-Rosser*): For both $\lambda 2_{Cu}$ and $\lambda 2_{Ch}$,
if $\Gamma \vdash s : \sigma$ and $s \twoheadrightarrow_\beta t_1$, $s \twoheadrightarrow_\beta t_2$, then for some t, $\Gamma \vdash t : \sigma$, $t_1 \twoheadrightarrow_\beta t$ and $t_2 \twoheadrightarrow_\beta t$.

Just as in the case for $\lambda\!\rightarrow_{Ch}$, every typable term in $\lambda 2_{Ch}$ has a unique type, while in $\lambda 2_{Cu}$ this is not the case.

THEOREM 33.32. (*uniqueness of types*): For $\lambda 2_{Ch}$,

$$\text{if } \Gamma \vdash s : \sigma \text{ and } \Gamma \vdash s : \tau, \text{ then } \sigma \equiv \tau.$$

Just as for $\lambda\rightarrow$, the (strong) normalization property holds for $\lambda 2$. The proof is much more complicated than for $\lambda \rightarrow$ and was first found by Girard. See Girard et al. [1989] for details. It is not obvious that if $\lambda 2_{Cu}$ is (strongly) normalizing, then $\lambda 2_{Ch}$ also is: owing to the new abstractions over types, there are more redexes in $\lambda 2_{Ch}$. It can, however, easily be shown that these new redexes cannot cause infinite reduction paths. We give the following without proof.

THEOREM 33.33.: Both $\lambda 2_{Ch}$ and $\lambda 2_{Cu}$ are strongly normalizing.

So, although it is possible to type $\lambda x.xx$, it is not possible to apply this term to itself in the polymorphic λ-calculus. (Recall that $(\lambda x.xx)(\lambda x.xx)$ has no normal form.)

The advantages of the normalization property have already been discussed for the simply typed λ-calculus. So have the drawbacks: it is not possible to define all recursive functions in $\lambda 2$ because of the decidable equality.

An issue that has been bothering researchers in typed λ-calculus for a long time is the decidability of typing in $\lambda 2_{Cu}$. It is not difficult to see that typability, type assignment and type checking are decidable for $\lambda 2_{Ch}$. (See Definition 33.16 for the definition of these notions.) For $\lambda 2_{Cu}$, the situation is much more complicated. Only very recently a tentative proof has been given that typability, type assignment and type checking are all undecidable for $\lambda 2_{Cu}$. From a programming language point of view, this undecidability of typing is bad. The compiler needs more information from the programmer to be able to do a type check.

Although not all recursive functions can be defined in $\lambda 2$, it is very powerful in representing data types and functions on them. For reasons of clarity, we work in $\lambda 2_{Ch}$.

DEFINITION 33.34.: The type of natural numbers N is defined as $N := \forall\alpha.((\alpha \to \alpha) \to \alpha \to \alpha)$. For N we have a term $Z : N$ defined by $Z := \lambda\alpha \,.\, \lambda f : \alpha \to \alpha \,.\, \lambda x : \alpha \,.\, x$ and a term $S : N \to N$ defined by $S := \lambda y : N \,.\, \lambda\alpha \,.\, \lambda f : \alpha \to \alpha \,.\, \lambda x : \alpha \,.\, f(y\alpha f x)$.

It is clear that the closed type N can serve very well as the data type of natural numbers. Then Z represents zero and S the successor function. (Cf. Exercises 33.5 and 33.6.) It can be shown that all closed terms of type N are β-equal to the identity or to a term of the form $S^n(Z)$. (See also Lemma 33.23.)

EXAMPLES 33.35.:

1. Addition on N can be defined by
   $\mathsf{add} := \lambda zy : N \,.\, \lambda\alpha \,.\, \lambda f : \alpha \to \alpha \,.\, \lambda x : \alpha \,.\, y\alpha f(z\alpha f x)$.
2. Multiplication on N can be defined by
   $\mathsf{mult} := \lambda zy : N \,.\, yN(\mathsf{add}\ z)Z$.

(Here $\lambda zy : N \,.\, t$ abbreviates $\lambda z : N \,.\, \lambda y : N \,.\, t$.)

It is also possible to define the product and disjoint sum between two types in $\lambda 2$. These constructions correspond to the definition of the connectives & and $\vee$ in terms of $\forall$ and $\supset$ in second order logic. (Following the so-called formulas-as-types embedding, $\lambda 2$ corresponds with second order propositional logic. See the following subsection.)

Besides the data type of natural numbers, there are many other inductive data types that can be represented by a closed type in $\lambda 2$. The general procedure behind this was first noted by Böhm and Berarduci [1985]. (See also Girard et al. [1989] for a treatment of data types in $\lambda 2$.) Here we only treat one extra example, the data type of lists over a base type σ.

EXAMPLE 33.36.: For σ a closed type of $\lambda 2_{Ch}$, the type of finite lists over σ, L_σ, is defined by

$$L_\sigma := \forall\alpha \,.\, \alpha \to (\alpha \to \sigma \to \alpha) \to \alpha.$$

The constructors 'nil' (the empty list) and 'cons' (adding an element to the head of a list) can be defined by

$$\begin{aligned} \mathsf{nil} &:= \lambda\alpha \,.\, \lambda x{:}\alpha \,.\, \lambda f{:}\alpha \to \sigma \to \alpha \,.\, x, \\ \mathsf{cons} &:= \lambda l : L_\sigma \,.\, \lambda t : \sigma \,.\, \lambda\alpha \,.\, \lambda x{:}\alpha \,.\, \lambda f : \alpha \to \sigma \to \alpha \,.\, f(l\alpha x f)t. \end{aligned}$$

It is interesting to see how the constructor 'head' : $L_\sigma \to \sigma$ that takes the first element of a list is defined. Note that 'head' has to be a total function because all terms in λ2 are normalizing. This means that 'head' also has to produce a term of type σ when applied to nil. For this case, we use an arbitrary term of type σ. So, let t be a (closed) term of type σ. Define head by

$$\mathsf{head} := \lambda l : L_\sigma \,.\, l\sigma t(\lambda z : \sigma \,.\, \lambda x : \sigma \,.\, x).$$

Then indeed $(\mathsf{head}\ \mathsf{nil}) = t$ and for all terms $l : L_\sigma$ and $t' : \sigma$, $(\mathsf{head}\ (\mathsf{cons}\ l\ t')) = t'$. (See Exercise 33.7.)

FORMULAS-AS-TYPES

In addition to the connection between programming languages and typed λ-calculus, there is also an interesting connection between logical systems and typed λ-calculi. This connection is established by the 'formulas-as-types' embedding from a system of logic into a typed λ-calculus. This connection was first formally described by Howard [1980] (the manuscript dates back to 1968), but the basic idea was already known to Curry. The idea of 'formulas-as-types' was independently invented, around 1970, by de Bruijn, in a slightly different form. (See de Bruyn [1980].) Therefore, the formulas-as-types embedding is often called the Curry-Howard isomorphism (or Curry-Howard-de Bruijn isomorphism), even though the embedding is not always a real isomorphism.

The basic idea behind formulas-as-types is that the formulas of the logic become types of a typed λ-calculus and, maybe even more important, that proofs of formulas in the logic become terms of the corresponding type in the calculus. This gives an operational meaning to proofs which are usually represented by derivations (deduction trees). In fact, this embedding gives a semantics of the derivations and so, at the same time, it gives a semantics of the logical connectives and the elimination and introduction rules. A philosophical aim of the formulas-as-types embedding was to explain the logical connectives in terms of what it means to have a proof of a formula that is built up with that connective. This corresponds with the intuitionistic way of looking at the connectives. Hence, the formulas-as-types embedding is best suited for interpreting systems of *constructive* logic into typed λ-calculus.

The formulas-as-types embedding has many nice consequences. One of them is that proofs become terms in the calculus, which makes manipulation with proof-terms possible. The typed λ-calculi have also been implemented and turn out to be powerful systems for interactive theorem proving. In dialogue with the machine a proof-term is constructed for a given type, where the machine takes care of the type checking. Another interesting application is that from a proof of a formula of the form $\forall x : \sigma\ \exists y : \tau . A(x, y)$ one can extract a program f in the form of a λ-term, that satisfies the specification $\forall x : \sigma . A(x, fx)$. For this it is essential that the logic is constructive, because only in constructive logic a proof of a formula $\forall x : \sigma\ \exists y : \tau . A(x, y)$ does yield for every $a : \sigma$ a *construction* of a $t(a) : \tau$ such that $A(a, t(a))$ holds.

For the systems $\lambda\!\rightarrow$ and $\lambda 2$ of the previous subsections it is not hard to find the logical systems that correspond with them by the formulas-as-types embedding. The system $\lambda\!\rightarrow$ corresponds with minimal first order propositional logic, i.e., the propositional logic with only the connective $\supset$. The system $\lambda 2$ corresponds to second-order propositional logic, with connectives $\supset$ and $\forall$ (ranging over all formulas). This system seems minimal, but all the connectives $\&$, $\vee$, $\exists$ and $\neg$ can be defined in it. We do not give a detailed description of the correspondence between these systems. For details see Barendregt [1992]. Instead we look at the interpretation of minimal first-order predicate logic into a typed λ-calculus ΛPRED. The minimal first-order predicate logic, called PRED, has as connectives $\supset$ and $\forall$ (ranging over basic domains). It is not difficult to extend the interpretation to the full system, but for our purposes it suffices to present the general idea. In our formalisation the logical language will also be presented in the manner of a typed λ-calculus.

DEFINITION 33.37:

1. The set of *functional types* of ΛPRED , Type^f, is described by the following abstract syntax.

$$\mathsf{Type}^f ::= \mathsf{Var}^{ty} \mid \mathsf{Type}^f \rightarrow \mathsf{Type}^f,$$

with Var^{ty} a countable set of *type variables*. The set of *predicate types* of ΛPRED, Type^p, consists of the expressions

$$\sigma_1 \rightarrow \sigma_2 \rightarrow \cdots \rightarrow \sigma_n \rightarrow \mathsf{Prop},$$

with $n \geq 0$ and all σ_i functional types. (Prop represents the type of all formulas.) The set of types of ΛPRED, Type, consists of the union of Type^f and Type^p.

2. The *object-terms* of the language of ΛPRED form a subset of the set of *pseudoterms* , T, which is generated by the following abstract syntax.

$$\mathsf{T} ::= \mathsf{Var}^{ob} \mid \mathsf{TT} \mid \lambda x : \mathsf{Type}^f \,.\, \mathsf{T} \mid \mathsf{T} \supset \mathsf{T} \mid \forall \mathsf{Var}^{ob} : \mathsf{Type}^f \,.\, \mathsf{T},$$

with Var^{ob} a countable set of *object-variables.* An object-term is of a certain type only under assumption of specific types (functional or predicate) for the free variables that occur in the term. That the object-term t *is of type* A *if* x_i *is of type* A_i *for* $1 \leq i \leq n$, is denoted by

$$x_1 : A_1,\ x_2 : A_2, \ldots,\ x_n : A_n \vdash t : A.$$

Here $x_1, \ldots, x_n$ are different object-variables and $A_1, \ldots, A_n$ are types. The sequence $x_1 : A_1,\ x_2 : A_2, \ldots,\ x_n : A_n$ is called an *object-context.* The rules for deriving these typing judgements are the following.

$$(\mathsf{var})\ \frac{}{\Gamma \vdash x : A} \qquad \text{if } x : A \text{ in } \Gamma$$

$$(\lambda - \mathsf{abs})\ \frac{\Gamma, x{:}A \vdash t : B}{\Gamma \vdash \lambda x{:}A \,.\, t : A \to B} \qquad \text{if } A \text{ is functional}$$

$$(\mathsf{app})\ \frac{\Gamma \vdash q : A \to B \quad \Gamma \vdash t : A}{\Gamma \vdash qt : B}$$

$$(\supset)\ \frac{\Gamma \vdash \phi : \mathsf{Prop} \quad \Gamma \vdash \psi : \mathsf{Prop}}{\Gamma \vdash \phi \supset \psi : \mathsf{Prop}}$$

$$(\forall)\ \frac{\Gamma, x : A \vdash \phi : \mathsf{Prop}}{\Gamma \vdash \forall x{:}A \,.\, \phi : \mathsf{Prop}} \qquad \text{if } A \text{ is functional}$$

3. The set of *proof-terms* is a subset of the set of *pseudoproofs*, P, generated by the following abstract syntax.

$$\mathsf{P} ::= \mathsf{Var}^{pr} \mid \mathsf{PP} \mid \mathsf{PT} \mid \lambda x : \mathsf{Type}^f \,.\, \mathsf{P} \mid \lambda x : \mathsf{T} \,.\, \mathsf{P},$$

where Var^{pr} is the set of *proof-variables.* The rules for generating statements of the form

$$x_1 : A_1,\ \ldots,\ x_n : A_n\ ;\ p_1 : \phi_1,\ \ldots,\ p_k : \phi_k \vdash s : A,$$

where the sequence $x_1 : A_1, \ldots, x_n : A_n$ is as in 2, $p_1, \ldots, p_k$ are different proof-variables and

$$x_1 : A_1, \ldots, x_n : A_n \vdash \phi_i : \mathsf{Prop} \text{ for } 1 \leq i \leq k,$$

are the following. (The sequence $p_1 : \phi_1, \ldots, p_k : \phi_k$ is called the *proof-context* as opposed to the object-context.) We use Γ as a meta-variable for an object-context and Δ for a proof-context, separated by a semi-colon.

$$\text{(axiom)} \quad \frac{}{\Gamma; \Delta \vdash p : \phi} \qquad \text{if } p : \phi \text{ in } \Delta,$$

$$(\supset\text{-in}) \quad \frac{\Gamma; \Delta, p : \phi \vdash s : \psi}{\Gamma; \Delta \vdash \lambda p : \phi \,.\, s : \phi \supset \psi}$$

$$(\supset\text{-el}) \quad \frac{\Gamma; \Delta \vdash s : \phi \supset \psi \quad \Gamma; \Delta \vdash t : \phi}{\Gamma; \Delta \vdash st : \psi}$$

$$(\forall\text{-in}) \quad \frac{\Gamma, x : A; \Delta \vdash s : \phi}{\Gamma; \Delta \vdash \lambda x{:}A \,.\, s : \forall x{:}A \,.\, \phi} \qquad \text{if } x \text{ not in } \Delta, A \text{ functional}$$

$$(\forall\text{-el}) \quad \frac{\Gamma; \Delta \vdash s : \forall x{:}A \,.\, \phi \quad \Gamma \vdash t : A}{\Gamma; \Delta \vdash st : \phi[x := t]}$$

$$\text{(conv)} \quad \frac{\Gamma; \Delta \vdash s : \phi \,, \quad \Gamma \vdash \psi : \mathsf{Prop}}{\Gamma; \Delta \vdash s : \psi} \qquad \text{if } \phi \simeq_\beta \psi.$$

The intention of the system should be clear: natural deduction proofs of the minimal first order predicate logic PRED are interpreted as typed λ-terms in ΛPRED. The language of PRED is also a typed λ-calculus and formalised in ΛPRED in a typing judgement obtained via derivations. Before discussing a formal correspondence between deductions in PRED and proof-terms in ΛPRED, we give two examples.

EXAMPLES 33.38.:

1. From the deduction in predicate logic

$$\cfrac{\cfrac{\forall x{:}A \,.\, (Px \supset Q)}{Px \supset Q} \qquad \cfrac{\forall x{:}A \,.\, Px}{Px}}{Q}$$

we obtain the judgement

$$P : A \to \mathsf{Prop},\ \ Q : \mathsf{Prop},\ \ x : A\ ;$$
$$p_1 : \forall x{:}A\ .\ (Px \supset Q),\ \ p_2 : \forall x{:}A\ .\ Px\ \vdash\ p_1x(p_2x) : Q.$$

Notice that the declaration of x is essential here for the construction of the proof.

2. From the deduction in predicate logic

$$\dfrac{\dfrac{\dfrac{\forall x : A\,.\,(Px \supset Q)}{Px \supset Q} \qquad \dfrac{\forall x : A\,.\,Px}{Px}}{Q}}{\forall x{:}A\ .\ Q}$$

we obtain the judgement

$$P : A \to \mathsf{Prop},\ \ Q : \mathsf{Prop}\ ;$$
$$p_1 : \forall x{:}A\ .\ (Px \supset Q),\ \ p_2 : \forall x{:}A\ .\ Px\ \vdash \lambda x : A\ .\ p_1x(p_2x) : \forall x{:}A\,.\,Q.$$

Here it is not necessary for the construction of the proof to declare x.

To a deduction of

$$\phi_1, \ldots, \phi_n \vdash \psi$$

in PRED, one can now associate an object-context Γ and a proof-term s such that

$$\Gamma\ ;\ p_1 : \phi_1,\ \ldots,\ p_n : \phi_n \vdash s : \psi$$

in ΛPRED. This can be done in such a way that s is a faithful representation of the deduction in PRED. That is, there is a one-to-one correspondence between deductions (in PRED) and proof-terms (in ΛPRED). Hence we can speak of the *formulas-as-types isomorphism* from PRED to ΛPRED. To achieve this one-to-one correspondence, the Γ that is constructed should assign types to all the free term-variables in the deduction. We do not give precise definitions and proofs here (see Geuvers [1993] for more details). From the examples it will be clear that sometimes we have to declare a term-variable x in the context Γ, even though this variable does not occur free in the conclusion or in any of the premises of the derivation.

The definitions that have been given above for the case of minimal first order predicate logic can easily be extended to second and higher order predicate logic, obtaining the typed λ-calculi ΛPRED2 and ΛPREDω.

For these systems too there is a one-to-one correspondence between deductions of the logic and proof-terms of the typed λ-calculus. One can consult Geuvers [1993] for details.

There are several ways in which the formulas-as-types isomorphism from a logic into a typed λ-calculus can be useful. First it can be a nice tool to study the logical system because deductions have become terms of the λ-calculus under the embedding, which makes them an easier subject for investigation. Related to this is the fact that the formulas-as-types embedding actually gives a semantics of proofs in (constructive) logic. One of the things that this semantics shows is the close connection between the rules for $\supset$ and the rules for $\forall$. Another way in which formulas-as-types can be useful is that the typed λ-calculus is better suited for implementation as an interactive theorem prover, even more so because the proofs are terms in the type system which gives a mechanism for storing them.

FURTHER DEVELOPMENTS

Typed λ-calculus has practical applications to both programming languages and automated theorem proving. Functional languages often have the simply typed λ-calculus as a theoretical base, usually extended with some kind of fixed point operator to be able to type more functions. An example of this is the language ML, which also has a mechanism for defining (inductive) data types. The formulas-as-types embedding provides a formalism in which objects and proofs can be manipulated in a similar way. This makes the proofs 'first class citizens', which makes it possible to operate with them inside the system. In this section we will say a bit more about the formulas-as-types approach and how it has been used to obtain a powerful logical framework.

One of the basic observations made by de Bruijn was that formulas and sets behave very much in the same way, as types. Both have 'inhabitants', which are seen as proofs in the case of a formula and as objects in the case of a set. Furthermore, a kind of function space can be formed for both, $A \to B$ representing the set of functions from A to B if both are sets and $A \to B$ representing the set of proofs of $A \supset B$ if both are formulas. This led de Bruijn to the idea that formulas and sets can be treated in a totally similar way. This is even more convincing if we look at a formula of the form $\forall x:A.B$, where A is a set and B a formula depending on x. (This dependency is usually denoted by writing $B(x)$ instead of B.) One natural interpretation of $\forall x:A.B(x)$ is the set of proofs of the proposition $\forall x:A.B$. But there is a second interpretation if we think of B as a *set* depending on x. If we proceed with our formulas-as-types analogy, $\forall x:A.B(x)$ should then be the set of functions f from A to $\bigcup_{a:A}B(a)$ such that for every $a:A$ we have $fa:B(a)$. This is therefore the set of type-dependent functions for which the image type depends on the input value. In mathematics, this set is denoted by

$$\Pi_{x:A}B(x).$$

In type theory a similar notation has become standard: $\Pi x:A.B$. We now have two kind of type-forming operations, the function space operator (given by the arrow $\rightarrow$) and the dependent function space operator (given by the Π). It is well known that a function space $A \rightarrow B$ is just a special case of a dependent function space $\Pi x:A.B$, namely where B does not depend on x. (In that case a function f of type $\Pi x:A.B$ takes inputs of type A to outputs of type B.) This implies that in the syntax of our (still to be defined) typed λ-calculus, we can do with just one type-forming operator, Π, and use the arrow, $\rightarrow$, as an abbreviation, writing $A \rightarrow B$ for $\Pi x:A.B$ in case x does not occur free in B. See Nederpelt [1990] for more details, e.g., about how to express mathematical notions and notions from programming languages in type systems.

Let us look at a simple system of typed λ-calculus that includes the ideas just mentioned. The system is a variant of the 'logical framework' LF, defined by researchers in Edinburgh, exploiting ideas from the Automath project (de Bruijn [1980]).

DEFINITION 33.39.: LF (Harper et al. [1987]) is a system for deriving judgements of the form

$$\Gamma \vdash A : B$$

where Γ is a *context* and A and B are terms, which are taken from the set of pseudoterms

$$\mathsf{T} ::= \mathsf{Var} \mid \mathbf{type} \mid \mathbf{kind} \mid \mathsf{TT} \mid \lambda x:\mathsf{T}.\mathsf{T} \mid \Pi x:\mathsf{T}.\mathsf{T}.$$

Note that T not only contains 'terms' but also the types! Both notions are treated similarly and are defined simultaneously. The derivation rules are the following. (**s** ranges over {**type, kind**}.)

(base) $\emptyset \vdash$

(ctxt) $\dfrac{\Gamma \vdash A : \mathbf{s}}{\Gamma, x : A \vdash}$ if x not in Γ

(ax) $\dfrac{\Gamma \vdash}{\Gamma \vdash \mathbf{type} : \mathbf{kind}}$

(proj) $\dfrac{\Gamma \vdash}{\Gamma \vdash x : A}$ if $x : A \in \Gamma$

(Π) $\dfrac{\Gamma, x : A \vdash B : \mathbf{s} \quad \Gamma \vdash A : \mathbf{type}}{\Gamma \vdash \Pi x{:}A.B : \mathbf{s}}$

(λ) $\dfrac{\Gamma, x : A \vdash B : C \quad \Gamma \vdash \Pi x{:}A.C : \mathbf{s}}{\Gamma \vdash \lambda x{:}A\,.\,B : \Pi x{:}A.C}$

(app) $\dfrac{\Gamma \vdash A{:}\Pi x{:}C.D \quad \Gamma \vdash B : C}{\Gamma \vdash AB : D[x := B]}$

(conv) $\dfrac{\Gamma \vdash A : B \quad \Gamma \vdash C : \mathbf{s}}{\Gamma \vdash A : C}$ if $C \simeq_{\beta\eta} B$

We use the convention of writing $A \to B$ for $\Pi x : A.B$ if $x \notin FV(B)$.

There are different ways of looking at this system. The first way, which remains closest to the intuitions discussed before, is to understand **type** as the universe of sets and formulas. Consider the (Π) rule. With this rule it is possible to define $A \to B$, which can be a function space (if A and B are understood as sets), but also an implication (if A and B are understood as formulas). Furthermore, if A is understood as a set and B as a formula, $\Pi x : A.B$ becomes a universally quantified formula. In all of these cases the **s** is always **type**. If **s** is **kind**, for example, if B is **type**, then $\Pi x : A.B$ is just $A \to \mathbf{type}$, which can be understood as the set of predicates on the set A. In this case one considers **type** as the universe of formulas, so $A \to \mathbf{type}$ represents the collection of formula-valued functions on the set A. Let $P : A \to \mathbf{type}$. Then the intuition is that, for $x : A$, the predicate P holds in x if Px is true (i.e., there is a term of type Px) and P does not hold in x if Px is not true (i.e., there is no term of type Px). Note that one can form $A \to A \to \mathbf{type}$ in the same way, which is the set of binary relations on A. The best way to understand the intuition behind the system is to consider an example which shows how to do first order predicate logic in the system.

EXAMPLE 33.40.: Look at the following context, representing a part of arithmetic. (For convenience we write the binary relations $=$ and $+$ *between* the arguments.)

$$\begin{aligned}\Gamma \ := \ & N:\mathbf{type},\ 0:N,\ S:N\to N,\ +:N\to N\to N,\\ & =:N\to N\to \mathbf{type},\\ & a_1:\Pi x,y{:}N.((x=y)\to(y=x)),\\ & a_2:\Pi x,y{:}N.(x+y=y+x),\\ & a_3:\Pi x,y,z{:}N.((x=y)\to(y=z)\to(x=z)),\\ & a_4:\Pi x{:}N.(x+0=x),\\ & a_5:\Pi x,y{:}N.(x+(Sy)=S(x+y)),\\ & a_6:\Pi f:N\to N\,.\,\Pi x,y{:}N\,.\,((x=y)\to(fx=fy)).\end{aligned}$$

In this context one can find a term of type $\Pi x,y:N.((Sx)+y=S(x+y))$. (See Exercise 33.9.) $\Pi x,y:N.A$ abbreviates $\Pi x:N.\Pi y:N.A$.

The system LF has the connectives $\supset$ and $\forall$ on the type level, both represented by the Π type operator. However, this does not yet make it into a logical *framework*; the interpretation sketched above only gives us minimal first order predicate logic. (It is possible to define a formal embedding from ΛPRED to LF to make this interpretation precise.)

The second way of looking at the system is much more powerful and is also the way the LF system was intended to be used. In this view the types in **type** are considered as sets only, except for one special type 'prop', which acts as the set of all formulas. Then formulas are just terms of type prop and in the context there are declarations

$$\supset: \mathsf{prop}\to\mathsf{prop}\to\mathsf{prop},\ \vee:\mathsf{prop}\to\mathsf{prop}\to\mathsf{prop}$$

enabling the generation of formulas. To generate a universally quantified formula, say over the type N, one declares $\forall:(N\to\mathsf{prop})\to\mathsf{prop}$. (Think of $\forall P$ as $\forall x:N.Px$.) Now prop : **type** and the formulas are terms of type prop, hence one cannot have that $t:A$ for $A:\mathsf{prop}$. (It can be shown in general that if $B:\mathbf{type}$ and $A:B$, then A cannot have an inhabitant t.) The terms of type prop are just 'names' of formulas and to make them into types (representing 'real' formulas) with possible inhabitants (being proofs of the formulas), it is necessary to declare a T : $\mathsf{prop} \to \mathbf{type}$ in the context. This T takes the name of a formula into the type of its proofs. At this point the logical rules can be declared in the context. For example, $\supset\!-E : \Pi x, y : \mathsf{prop}\ .\ (T(x \supset y) \to Tx \to Ty)$. We look again at an example from arithmetic.

EXAMPLE 33.41.:

$$\begin{array}{rl} \Gamma := & \mathsf{prop} : \mathbf{type},\ T : \mathsf{prop} \to \mathbf{type}, \\ & \supset, \vee, \ \& \ : \mathsf{prop} \to \mathsf{prop} \to \mathsf{prop}, \\ & N : \mathbf{type},\ 0 : N,\ S : N \to N,\ + : N \to N \to N, \\ & = \ : N \to N \to \mathsf{prop},\ \forall, \exists : (N \to \mathsf{prop}) \to \mathsf{prop}, \\ & \supset\!-E : \Pi x, y{:}\mathsf{prop}.(T(x \supset y) \to Tx \to Ty), \\ & \supset\!-I : \Pi x, y{:}\mathsf{prop}.((Tx \to Ty) \to T(x \supset y)), \\ & \forall\!-E : \Pi P : N \to \mathsf{prop}\ .\ \Pi x{:}N\ .\ (T(\forall P) \to T(Px)), \\ & \forall\!-I : \Pi P : N \to \mathsf{prop}\ .\ ((\Pi x{:}N.T(Px)) \to T(\forall P)), \\ & a_1 : \Pi x, y{:}N.(T(x = y) \to T(y = x)), \\ & a_2 : \Pi x, y{:}N.T(x + y = y + x), \\ & a_3 : \Pi x, y, z{:}N.(T(x = y) \to T(y = z) \to T(x = z)), \\ & a_4 : \Pi x{:}N.T(x + 0 = x), \\ & a_5 : \Pi x, y{:}N.T(x + (Sy) = S(x + y)), \\ & a_6 : \Pi f : N \to N\ .\ \Pi x, y{:}N\ .\ (T(x = y) \to T(fx = fy)). \end{array}$$

Of course there are also declarations for the introduction and elimination rules of the other connectives, but they are not written. In this context one can find a term of type $\Pi x, y : N\ .\ T((Sx) + y = S(x + y))$. (See Exercise 33.10.)

The approach to LF that we have just sketched is very flexible because it uses only the built in mechanisms of the system to encode another formal system. The first approach uses the logic contained in the system already, but this logic is relatively poor. To see how flexible the second approach is we sketch how to interpret the untyped λ-calculus in a context.

EXAMPLE 33.42.: A context for the untyped λ-calculus is the following.

$$\begin{aligned}\Gamma \;:=\;& \Lambda : \mathbf{type},\ \mathsf{prop} : \mathbf{type},\ T : \mathsf{prop} \to \mathbf{type},\ = \,: \Lambda \to \Lambda \to \mathsf{prop},\\ & abs : (\Lambda \to \Lambda) \to \Lambda,\ app : \Lambda \to \Lambda \to \Lambda,\\ & \beta : \Pi f : \Lambda \to \Lambda \,.\, \Pi x{:}\Lambda \,.\, T(app(abs\ f)x = fx),\\ & \xi : \Pi f, g : \Lambda \to \Lambda \,.\, (\Pi x{:}\Lambda \,.\, T(fx = gx)) \to T(abs\ f = abs\ g),\\ & c : \Pi x, y, p, q{:}\Lambda \,.\, (T(x = y) \to T(p = q) \to T(app\ p\, x = app\ q\, y)),\end{aligned}$$

plus some declarations to turn = into an equivalence relation.

From the examples it should be clear that a Logical Framework such as LF is very flexible in the sense that a lot of different formal systems can be interpreted in it. The same holds for de Bruijn's *Automath* systems on which LF is based. These systems have also been implemented. The Automath research project was specifically aimed at making practical use of these systems for proof verification. The idea is then that the user types in a λ-term and its type, which is then type checked by the computer. Due to the formulas-as-types embedding this means that the computer checks whether the proof (a term, say A) is indeed a proof of the given formula (a type, say B), so whether $A : B$ is derivable in the given context. It is clear that to make all this possible, the decidability of type checking is very important. Another point that has to be considered is whether the context that has been constructed to represent the logic is *adequate*. Therefore we have to make sure that it is not possible to prove more formulas in the framework (inside this specific context) than in the logic. This problem is not trivial, because in the framework there are generally a lot of types that have no meaning at all in the logic. For LF this problem is discussed in detail in Harper et al. [1987].

If one is interested in a specific formal system to work with on a computer, a different approach is possible. This entails defining a particular type system that represents this formal system in a rather direct way (e.g., ΛPRED that represents minimal first order predicate logic). A very successful example of this approach is the Calculus of Constructions (Coquand and Huet [1988], Coquand [1990]), which is a typed λ-calculus powerful enough to interpret higher order predicate logic. The system includes the polymorphic λ-calculus, which means that a large number of data types can be defined in the system (e.g., the polymorphic Church numerals and finite lists, as shown in the previous section). It also includes

LF, which is used in the first approach, as minimal first order predicate logic. Due to the higher order features of the Calculus of Constructions, this 'first order part' becomes a powerful higher order predicate logic.

The logic inside the Calculus of Constructions is constructive, which makes it possible to extract programs from proofs. The idea is that a proof of a formula of the form $\forall x:\sigma.\exists y:\tau.A(x,y)$ contains a function f of type $\sigma \rightarrow \tau$ such that $\forall x:\sigma.A(x,f(x))$ holds. In the implementation of the Calculus of Constructions it is possible to extract the function as an algorithm (to be precise, a λ-term). So, from a proof of a specification it is possible to extract an algorithm that satisfies the specification. This is discussed extensively in Mohring [1986], but see also Coquand and Huet [1988].

The normalization of the Calculus of Constructions is not easy to prove, which makes the proof of the decidability of typing also difficult. Another issue that has been studied in the literature is whether the embedding of higher order predicate logic into the Calculus of Constructions is adequate. To better understand the Calculus of Constructions and to have a good setting for studying these and other meta-theoretical issues, the so-called *cube of typed λ-calculi* has been introduced by Barendregt (see Barendregt [1992]). This cube provides a fine structure for the Calculus of Constructions by describing eight systems of typed λ-calculus, each of them corresponding to a vertex in the cube. The simplest system is $\lambda\!\rightarrow_{Ch}$ and the most complicated is the Calculus of Constructions. Going along the edge of the cube from one system to another represents an extension of the system. Any of the three directions represents one way of extension, so there are three possible immediate extensions of $\lambda\!\rightarrow_{Ch}$. The first way is to add polymorphism, which extends $\lambda\rightarrow_{Ch}$ to $\lambda 2_{Ch}$. The second is to add type dependency, which extends $\lambda\!\rightarrow_{Ch}$ to LF. The third is to add higher orders, which extends $\lambda\!\rightarrow_{Ch}$ to a system that we have not discussed here, since it is of limited interest. Combination of the mentioned extensions gives the other four vertices of the cube, with the Calculus of Constuctions as the extension with all three possibilities. Apart from describing the cube of typed λ-calculi, Barendregt [1992] also shows how this framework can be used to understand the Calculus of Constructions and how it can help prove things about it. The Calculus of Constructions has been implemented by several research groups, notably at INRIA in Paris (Dowek et al. [1991]) and at the University of

Edinburgh (where the implementation is called 'Lego', see Luo and Pollack [1992]). It has also been extended with extra type forming operators in order to define inductive types. For this extension see Dowek et al. [1991].

Another approach to defining a type system for doing mathematics is taken by Martin-Löf (see Martin-Löf [1975] and Martin-Löf [1984]). His Intuitionistic Type Theory, a first order dependent typed λ-calculus with extra type forming operators, provides a foundation for intuitionistic mathematics. Although it is not explicitly intended as a system for doing mathematics on a computer, it can be implemented and it is in fact used quite extensively (see Constable et al. [1986] and also Nordström [1990]).

EXERCISES

33.1. Give a closed term of type $(\alpha \to \gamma) \to \alpha \to \beta \to \gamma$ in $\lambda{\to}_{Cu}$ and $\lambda{\to}_{Ch}$.

33.2. Give closed terms of type $\beta \to ((\alpha \to \beta) \to \gamma) \to \gamma$ in $\lambda{\to}_{Cu}$ and $\lambda{\to}_{Ch}$.

33.3. Find closed terms s and t in $\lambda{\to}_{Cu}$ such that $s \twoheadrightarrow_\beta t$ but $\not\vdash_{Cu} s : pts(t)$. So, principal type schemes are not stable under β-expansion and it is not even guaranteed that the pts continues to be a type if we expand the term. Show that the reverse holds: If the principal type scheme of s is σ and $s \twoheadrightarrow_\beta t$, then $\vdash t : \sigma$.

33.4. (a) For σ and τ types of $\lambda 2_{Cu}$, one can form a type $\sigma \,\&\, \tau := \forall\alpha.(\sigma \to \tau \to \alpha) \to \alpha$ and closed terms $p_1 : (\sigma \,\&\, \tau) \to \sigma$, $p_2 : (\sigma \,\&\, \tau) \to \tau$ and $\mathsf{pair} : \sigma \to \tau \to \sigma \,\&\, \tau$ such that $p_1(\mathsf{pair} st) = s$ and $p_2(\mathsf{pair} st) = t$. Find such terms p_1, p_2 and pair. Show the indicated equalities.

(b) Similarly, one can form in $\lambda 2_{Cu}$ a type $\sigma \vee \tau := \forall\alpha.(\sigma \to \alpha) \to (\tau \to \alpha) \to \alpha$ and closed terms $\mathsf{in}_1 : \sigma \to (\sigma \vee \tau)$, $\mathsf{in}_2 : \tau \to (\sigma \vee \tau)$ and $\mathsf{case} : \forall\alpha.(\sigma \to \alpha) \to (\tau \to \alpha) \to (\sigma \vee \tau) \to \alpha$ such that $\mathsf{case} st(\mathsf{in}_1 u) = su$ and $\mathsf{case} st(\mathsf{in}_2 u) = tu$. Find the indicated terms and prove the equalities.

33.5. Recall the definition of the Church numerals in $\lambda{\rightarrow}_{Cu}$:
$N := (o \to o) \to o \to o$, with o a type constant and $0 := \lambda fx.x$, $\mathsf{Suc} := \lambda yfx.f(yfx)$.

(a) Show that $\mathsf{Suc}(\lambda fx.f^n(x)) = \lambda fx.f^{n+1}(x)$.

(b) Show that the function $\mathsf{add} := \lambda zyfx.yf(zfx)$ is of type $N \to N \to N$ in $\lambda{\rightarrow}_{Cu}$ and that it behaves as addition.

(c) Recall the definition of exponentiation on the Church numerals, as given in Exercise 32.11 on untyped λ-calculus: $\mathsf{exp} := \lambda xy.yx$. Give a reason why this term cannot be typed on the Church numerals in $\lambda{\rightarrow}_{Cu}$.

33.6. Let $N := \forall\alpha.((\alpha \to \alpha) \to \alpha \to \alpha)$ be the type of the polymorphic Church numerals and let add be the term $\lambda zyfx.yf(zfx)$, given in the previous exercise.

(a) Show that the function $\mathsf{mult} := \lambda zy.y(\mathsf{add}\, z)0$ is of type $N \to N \to N$ in $\lambda 2_{Cu}$ and that it behaves as multiplication.

(b) Show that exponentiation is λ-definable on the polymorphic Church numerals in $\lambda 2_{Cu}$.

33.7. Show that the constructors for lists, as defined in the section on polymorphic λ-calculus, indeed behave as desired. That is, check that for the terms cons, nil and head one has $(\mathsf{head\ nil}) = t$ and $(\mathsf{head}\ (\mathsf{cons}\ l\ t')) = t'$ for t' an arbitrary term of type σ.

33.8. Let σ be a type in $\Lambda\mathsf{PRED}$, R a binary relation on σ and A a proposition. Find a natural deduction proof of

$$(\forall x, y{:}\sigma\,.\,Rxy \to Ryx \to A) \to (\forall x{:}\sigma\,.\,Rxx \to A)$$

in first order predicate logic and the typed λ-term that this proof corresponds with in $\Lambda\mathsf{PRED}$.

33.9. Give a term of type $\Pi x, y{:}N\,.\,((Sx) + y = S(x + y))$ in LF in the context Γ, defined in Example 33.40.

33.10. Give a term of type $\Pi x, y{:}N\,.\,T((Sx) + y = S(x + y))$ in LF in the context Γ, defined in Example 33.41.

§34 EPISTEMIC LOGIC FOR DISTRIBUTED SYSTEMS:

BASIC EPISTEMIC LOGIC; APPLICATION TO DISTRIBUTED SYSTEMS; PROTOCOL VERIFICATION; COMMON KNOWLEDGE; THE HATTED PROFESSORS; COMMON KNOWLEDGE IN DISTRIBUTED SYSTEMS; THE CO-ORDINATED ATTACK PROBLEM; IMPLICIT KNOWLEDGE IN A GROUP; IMPLICIT KNOWLEDGE IN A DISTRIBUTED SYSTEM

J.-J. Ch. MEYER

As we have seen in 13, *epistemic logic* is a version of modal logic in which the box operator is read as "is known". More generally, we may consider a number of distinct subjects or agents, as they are called in computer science, and introduce a different box operator $\Box_i$ for each agent i. In this section we shall see how epistemic logic can be used in computer science, in particular for reasoning about agents in a distributed environment such as processors in a network. This line of research was initiated by Joe Halpern and fellow-workers, and most of what we discuss in this section is based on their work. First, some definitions from §13 will be reviewed. The references are given at the end of this section.

BASIC EPISTEMIC LOGIC

Basic epistemic logic is a propositional modal logic with operators $\Box_i$ ($i = 1, \ldots, m$), where m is the number of agents. Thus, the basic logic is classical propositional logic extended with the operators $\Box_1, \ldots, \Box_m$. To emphasize that, in an epistemic context, knowledge is referred to, these operators are written as $[K_1], \ldots, [K_m]$. The brackets [ ] are slightly non-standard, but are added to emphasize the $\Box$-character of the operators. Formulas are interpreted on *Kripke models* M of the form $\langle W, R_1, \ldots, R_m, \models\rangle$, where W is a non-empty set of possible worlds, R_i ($i = 1, \ldots, m$) are binary relations on W, the accessibility relations, and $\models$ is a truth assignment relation between possible worlds and the atomic formulas in the language. When epistemic logic is used in computer science, the relations are generally taken to be equivalence relations, which means that *de facto* $S5$ models are being considered. (See §13; a motivation is given in Exercise 34.9.)

Epistemic formulas are interpreted on $S5_{(m)}$-Kripke models as follows: Given a model $M = \langle W, R_1, \ldots, R_m, \models \rangle$, we define $M \models_w A$ (A holds at world w of the model M) for $w \in W$ and epistemic formula A as follows:

$$\begin{array}{ll} M \models_w P & := w \models P \ (P \text{ atomic}); \\ M \models_w A \,\&\, B & := M \models_w A \text{ and } M \models_w B; \\ M \models_w A \vee B & := M \models_w A \text{ or } M \models_w B; \\ M \models_w A \supset B & := \text{not } M \models_w A \text{ or } M \models_w B; \\ M \models_w \neg B & := \text{not } M \models_w B; \\ M \models_w [K_i]B & := \text{for all } w' \text{ in } W, \text{ if } wR_iw' \text{ then } M \models_{w'} B. \end{array}$$

M is a *model* of A (or A holds in M), $M \models A$, iff for all w in W, $M \models_w A$. A is *valid* in $S5_{(m)}$, denoted $\models_{S5(m)} A$, iff A holds in all $S5_{(m)}$-Kripke models.

The logic $S5_{(m)}$ can be axiomatised in a Hilbert-style proof system by the axioms:

(A1) All (or enough) instances of propositional tautologies
(A2) $([K_i]A \,\&\, [K_i](A \supset B)) \supset [K_i]B$
(A3) $[K_i]A \supset A$
(A4) $[K_i]A \supset [K_i][K_i]A$
(A5) $\neg[K_i]A \supset [K_i]\neg[K_i]A$

and the rules:

(R1) Modus Ponens: $A,\ A \supset B\ /\ B$
(R2) Necessitation: $A\ /\ [K_i]A$, i.e., if A is provable, then $[K_i]A$ is provable.

(A1) and (R1) state that our logic is an extension of classical propositional logic; (A2) states that the knowledge of an agent is closed under implication; (A3) says that knowledge is true, i.e., an assertion that is known is true; (A4) expresses that knowledge itself is known, i.e., the fact that an assertion is known is known itself as well. This property is called "*positive introspection*"; (A5) expresses that ignorance about a fact is known, i.e., the fact that an assertion is not known, is known by the agent. This, at least philosophically, more controversial property is called "*negative introspection*". In computer science it can often be defended by some finiteness condition with respect to the agent, such as a finite memory, so that the agent can detect something that is unknown to him/her/it when it is missing in his/her/its database. The necessitation rule (R2) should not be confused with an assertion $A \supset [K_i]A$, which is not valid. (R2) says that if A is (formally) provable (in $S5_{(m)}$), then $[K_i]A$ is provable (in $S5_{(m)}$).

A proof of the soundness and completeness of the system $S5_{(m)}$ with respect to the class of $S5_{(m)}$-Kripke models can be found in, for example, [HC84] and [MH93].

After this short introduction to some epistemic notions, we are ready to consider an application of these notions in the context of distributed systems.

APPLICATION TO DISTRIBUTED SYSTEMS

Consider a *distributed system* with processors $1, \ldots, m$ connected by a communication network. Every processor is a state machine. The local state of a processor is a function of the initial state, the messages received and possible internal actions. The precise way in which this function works is irrelevant for our present purposes. The *global state* s of the system is a vector of local states: $s = (s_1, \ldots, s_m)$. The set of global states is denoted by S.

We can now consider a *Kripke model* M *associated with this distributed system*: $M = \langle W,\ R_1, \ldots, R_m,\ \models \rangle$, where
$W = S = \{(s_1, \ldots, s_m) | s_i \text{ is a local state of processor } i\}$,
$R_i = \{(s, s') | s_i = s'_i\}$ for $i = 1 \ldots, m$ (i.e., sR_is' iff $s_i = s'_i$), and
$\models$ is a truth assignment relation.

The set W thus contains all possible global states. One can imagine primitive propositions in this context to include assertions such as: for the (local) variable y used by processor i it holds that $y = 0$. Again, the precise form of these primitive propositions is irrelevant for our present purposes. The justification of the definition of the relations R_i is the following: processor i is supposed to know in which local state it is itself, but it is ignorant regarding the local states of the other parallel processors. Thus, it considers all global states as possible provided these (global states) agree on the local state i itself is in. Note that the R_i ($i = 1, \ldots, m$) are equivalence relations and hence the Kripke models associated with distributed systems are $S5_{(m)}$-models.

This 'concrete' model will be discussed later in connection with the notion of common knowledge. Below are some remarks concerning possible refinements of the above model. In this model two global states of a system are considered equivalent for agent i if they have the same i-local state. This is very much related to what is called *observational equivalence* in the realm of concurrency semantics: as far as agent i can see, there is no difference between global states which have the same i-local state. Actually, it is a rather simple and perhaps even naive notion of observational equivalence. One can imagine more sophisticated options for the equivalence relation between states.

In general, states may be replaced by so-called *points* (r, t) consisting of an *execution* r of the system (i.e., a description of what the system is subsequently doing and will do in the future) together with a time stamp t according to a global clock of the system. (Halpern c.s. uses the term *run* instead of execution, but as this term is rather overloaded, we prefer the term execution here.) In subsequent subsections we shall make the notion of an execution more precise, according to our needs, but for the moment we just assume the notion of execution to be primitive and without further structure: an execution is some description of the system's behaviour. We furthermore suppose that t takes values from some time domain T. Given some set $\mathcal{R}$ of possible executions, we can now consider a Kripke model $\langle\{(r,t)|r \in \mathcal{R},\ t \in T\},\ \cong_1, \ldots, \cong_m,\ \models\rangle$, where the $\cong_i$ are equivalence relations induced by some *view* of agent i deciding what has to be regarded equivalent, and $\models$ is a truth assignment relation. The view of i gives a criterion according to which the agent (processor) i considers points as 'the same', or in other words, which points are regarded possible given the point i is at.

Formally, the view of agent i is a function v_i from points to observations that associates some observations made by i with a given point. Now two points are equivalent on the basis of a view v_i as defined by: $(r,t) \cong_i (r',t')$ iff $v_i(r,t) = v_i(r',t')$. The observation associated with a point (r,t) may be very rich: in the extreme case it may be the point itself (so-called point knowledge), i.e., $v_i(r,t) = (r,t)$. In this extreme case, the $\cong_i$-equivalence classes are all singletons, so that the agent can see exactly what point it is at, and, consequently, the agent's knowledge will be the largest possible.

On the other hand, it may be very poor – in the extreme case, it may again be some constant, i.e., $v_i(r,t) = C$ for all points (r,t), for some constant C in the domain of observations. In this extreme we have that the relation $\cong_i$ will consist of only one big equivalence class (every point is equivalent to any other point) so that the agent cannot distinguish between possible points at all, and hence its knowledge will be minimal. As an intermediate case, the view of agent i may just involve considering the local state of i, which is the case that we considered in the first instance above. However, the view may also be more complex, involving the history of the process (and so also the execution r) in a non-trivial way, without amounting to rendering the complete point itself. For example, i's view may concern the global history of the process or just its own local history (the actions that it has executed itself), but also its location in a program (cf. [vHM92]). The general approach of Halpern & Moses covers all these possibilities. The primitive and complex propositions are now interpreted in the usual way, for example:

$$\mathcal{R} \times T \models_{(r,t)} [K_i]A \text{ iff } \mathcal{R} \times T \models_{(r',t')} A \text{ for all } (r',t') \in \mathcal{R} \times T \text{ such that } (r,t) \cong_i (r',t').$$

This general model can be used to reason about (knowledge in/of) distributed processes. For more details about applications using this more advanced model, we refer to [HM84a, 90], [ChM85, 86], [FV86], [FI86], [KT86], [HF89]. Although the above framework is generally used for an interleaving approach to concurrency, it can also be adapted for modelling *"true" concurrency* (cf., e.g., [LMRT90]). To do this, the points in Halpern & Moses' approach are replaced by pairs (r,l) of partially ordered sets (posets) of actions, where r models the causality relation between the actions of the process concerned, and l is a sub-poset of r describing to what extent the process is advanced at the current moment (cf. [vHM92]).

PROTOCOL VERIFICATION

After our more general considerations on the use of epistemic logic in the context of distributed systems, we consider in this subsection the possibilities to use epistemic logic in a more concrete area, viz., that of *protocol verification.* [HZ87] have demonstrated how an epistemic analysis of protocols for sending messages in a computer network more or less naturally leads to 'alternating bit'-like algorithms.

One reason for researchers such as Halpern to occupy themselves with epistemic logic was the application to *protocols*, programs that are able to guarantee the transmission of messages over a distributed system (network). Several more or less complicated protocols have been proposed in the literature. A very well-known one is the "*alternating bit protocol*", in which messages are alternately ended with either a 0 or a 1. Of course, it is of the greatest interest to know whether these protocols are adequate in the sense that they indeed guarantee a perfect transmission of messages. However, *protocol verification*, proving the correctness of protocols, is a non-trivial enterprise. Halpern c.s. had the idea to use the notion of knowledge to gain some insight into the behaviour of protocols, which change the knowledge that is present at the nodes in a network by sending messages. In this way, Halpern & Zuck investigated a number of protocols. A simple case is discussed here to get an impression of their approach.

We are given two processors, a sender S and a receiver R. S has an input tape with an infinite sequence X of data elements: $X = <x_0, x_1, x_2, \ldots>$ with $x_i \in \{0,1\}$.

S reads these data elements and sends them to R, which writes the data received on an output tape. However, the network may be faulty in the sense that reception of the data is NOT guaranteed: *deletion errors* may occur. The goal is to give a protocol that guarantees the following things:

(a) *safety* : at any moment the sequence of data elements written by R is a prefix of X.

(b) *liveness* : every $x_i \in X$ will eventually be written by R (provided the communication medium satisfies certain *fairness* properties).

The problem above is called the *sequence transmission* problem.

On the basis of the properties of knowledge possessed by S and R, the following protocols are obtained, which are written as an algorithm, where we still employ programming-alien elements such as knowledge operators (later it will be shown how to eliminate these and obtain actual programs in a PASCAL-like language):

SOLUTION 1: PROTOCOL I

PROTOCOL for S:

```
i := 0;
while   true do
        begin read x_i;
           send x_i until K_S K_R(x_i);
           send "K_S K_R(x_i)" until K_S K_R K_S K_R(x_i);
           i := i + 1
        end
```

PROTOCOL for R:

```
when   K_R(x_0) set i := 0;
while  true do
       begin write x_i;
          send "K_R(x_i)" until K_R K_S K_R(x_i);
          send "K_R K_S K_R(x_i)" until K_R(x_{i+1});
          i := i + 1
       end
```

Here "A" (for an assertion A) stands for the message that a state has been reached in which A holds. Intuitively, the protocols are clear: after reading a data element x, S sends it (i.e., x) to R until it (i.e., S) knows that x has been received by R. Then S sends an acknowledgement that it (S) knows that R has received x until it (S) knows that R has received this acknowledgement. On the other hand, when R receives a data element x, it writes it on its tape, and sends an acknowledgement to S until it (R) knows that this acknowledgement has been received by S. Then it (R) sends a further acknowledgement stating that it (R) knows that S has received the first acknowledgement until it (R) receives a new data element (indicating that the sender has also received the second acknowledgement).

Protocol I is put in terms of the knowledge operators K_S and K_R. These are not available in ordinary programming languages. We can eliminate the knowledge operators by using acknowledgements ack, ack^2, ack^3, as follows:

```
PROTOCOL for S:
    while  true do
           begin read y;
             send y until ack is received;
             send ack^2 until ack^3 is received;
           end

PROTOCOL for R:
    when   receive y write y
    while  true do
           begin send ack until ack^2 is received;
             send ack^3 until y is received;
             write y
           end
```

Halpern & Zuck [HZ87] proved that Protocol I solves the sequence transmission problem in the sense that it guarantees safety, and – provided it is guaranteed by the medium that messages from S to R and from R to S eventually arrive – also liveness.

One may also construct extensions of Protocol I such that the above theorem also holds if in addition to "deletion errors", "*mutation errors*" (data elements may get mutilated, 0 may become 1 or vice versa) and "*insertion errors*" (undesired data elements are inserted) occur.

An interesting question is whether the protocol really needs *knowledge of depth 4*. This appears not to be the case when the alphabet is extended to $\{0, 0', 1, 1', ack, ack'\}$, i.e., when we have at our disposal two "coloured" versions of $\{0, 1, ack\}$. The protocol then becomes simpler:

SOLUTION 2: PROTOCOL II

```
PROTOCOL for S:
    i := 0;
    while  true  do
           begin read x_i;
             send x_i; "K_S K_R(x_{i-1})" until K_S K_R(x_i);
             i := i + 1
           end
```

PROTOCOL for R:

```
when  K_R(x_0) set i := 0;
while true do
      begin write x_i;
        send "K_R(x_i)" until K_R(x_{i+1});
        i := i + 1
      end
```

By means of the coloured alphabet, we can implement Protocol II as follows, using

$$var(m, i) = \begin{cases} m, \text{ if } i \text{ is even} \\ m', \text{ if } i \text{ is odd.} \end{cases}$$

PROTOCOL for S:

```
i := 0;
while  true do
       begin read x_i;
         send var(x_i, i) until var(ack, i) is received;
         i := i + 1
       end
```

PROTOCOL for R:

```
when  receive x_0 set i := 0;
while true do
      begin write x_i;
        send var(ack, i) until x_{i+1} is received;
        i := i + 1
      end
```

Halpern & Zuck [HZ87] proved that Protocol II guarantees safety and liveness as well, and also that this protocol can be extended such that the file is protected against errors other than just "deletion errors".

Although now S in Protocol II does not wait until knowledge of depth 4 is obtained before sending x_{i+1}, it can still be proved that S acquires knowledge of depth 4.

Protocol II is closely related to the *alternating bit protocol.* To obtain this protocol, one has to simulate the 'coloured' messages by the use of alternating closing bits:

$$var(m,i) = \begin{cases} m_0, \text{ if } i \text{ is even} \\ m_1, \text{ otherwise.} \end{cases}$$

Halpern & Zuck [HZ87] claim that their epistemic analysis is applicable to a whole class of related protocols that appear in the literature.

COMMON KNOWLEDGE

Basic epistemic logic may be extended with more powerful operators. We introduce two new operators into the language: $[E]A$ for '*everyone knows that A*' and $[C]A$ for '*it is common knowledge that A*'. Informally, $[E]A = [K_1]A \ \& \ \dots \ \& \ [K_m]A$ and $[C]A = [E]A \ \& \ [E][E]A \ \& \ [E][E][E]A \ \& \ \dots$ (ad infinitum). Although the former formula gives a perfectly fine definition of the $[E]$-operator, the latter formula can only be understood intuitively, for we do not consider infinite conjunctions in our language. First a proper semantics will be given for these new operators, and next a sound and complete axiomatisation.

DEFINITION 34.1.: Let the Kripke model $M = \langle W,\ R_1, \dots, R_m,\ \models \rangle$ be given. Let $R_E = R_1 \cup \dots \cup R_m$. Thus, wR_Ew' iff wR_1w' or ... or wR_mw'. Furthermore, let $R_C = (R_E)^*$, the reflexive-transitive closure of the relation R_E. Now we define:
$M \models_w [E]B :=$ for all w' in W, if wR_Ew' then $M \models_{w'} B$;
$M \models_w [C]B :=$ for all w' in W, if wR_Cw' then $M \models_{w'} B$.

To obtain an axiomatization, the following axioms are added to the system $S5_{(m)}$:

(A6) $[E]A \sim [K_1]A \ \& \ \dots \ \& \ [K_m]A$
(A7) $[C]A \supset A$
(A8) $[C]A \supset [E][C]A$
(A9) $([C]A \ \& \ [C](A \supset B)) \supset [C]B$
(A10) $[C](A \supset [E]A) \supset (A \supset [C]A)$
(R3) $A \ / \ [C]A$ (necessitation for $[C]$), i.e.,
if A is provable, then $[C]A$ is provable.

The resulting system is called $S5C_{(m)}$. This system can be shown to be sound and complete with respect to $S5C_{(m)}$-models, i.e., models of type $M = \langle W,\ R_1, \dots, R_m,\ R_E,\ R_C,\ \models \rangle$, where $R_E = R_1 \cup \dots \cup R_m$, and $R_C = (R_E)^*$ (cf. [MH93]).

EXAMPLE: THE DEAN AND THE HATTED PROFESSORS ([Leh84])

Consider the following situation. The dean of a computer science faculty of an anonymous university announces that owing to a stroke of unexpected financial luck (this, by the way, rules out every Dutch university for the story), he plans to increase the salaries of the three professors in the department. However, quite in line with his character, he would like to distribute the salary increases by means of a little game. He asks all three professors to his room and shows them five hats: two white ones and three black ones. He tells them that at noon he will turn off the light and place a hat on the head of each of the three professors, and that the two remaining hats will be destroyed. Then, so he tells them, he will come back in an hour's time, switch on the light, and ask whether one of the professors can tell him what colour his hat is. A wrong answer means unemployment, a correct answer means a pay increase. If nobody speaks, he will come again after another hour and ask the same question again. This will go on until some professor speaks up. After this announcement, he leaves the room, and asks his secretary to book a flight for him to go to a conference on epistemic logic that same day. He says that he will be free to leave at 2:01 p.m. How does the dean know that he need not go back to his professors at 3:00 p.m.?

We can analyse this situation by means of epistemic logic. Let A_j be the assertion that there are precisely j white hats (and thus $3-j$ black hats) involved. After the dean's announcement, it is common knowledge that there are either 3 black hats and no white ones, 2 black hats and 1 white one, or 1 black hat and 2 white ones, i.e., it holds that $[C](A_0 \vee A_1 \vee A_2)$, and hence also $[E](A_0 \vee A_1 \vee A_2)$. Moreover, it is common knowledge that in the case of A_2, the person with the black hat can determine the colour of his hat, since he sees two white hats already, and there aren't any more.

So if A_2 is the case, the puzzle is solved by 1:00 p.m. Since every professor knows this, and knows that everyone else knows this too (because it is common knowledge!), if A_2 is not the case, then it holds at 1:00 p.m. that $E(\neg A_2)$, and so $[E](A_0 \vee A_1)$.

Now suppose that A_1 is the case. The two professors wearing black hats see one white hat. Since they know that A_2 is not the case, they know that they themselves cannot be wearing white hats. In this case they can state at 2:00 p.m. that they have black hats, and the puzzle is solved again.

If none of the professors have spoken up yet by 2:00 p.m., everyone will know that neither A_2 nor A_1 is the case, so they may conclude that A_0 must be the case, i.e., $[E]A_0$ holds. So, at 2:01 p.m. they can conclude that all three of them have black hats, and they can prepare themselves mentally to state so at 3:00 p.m. However, the dean is no fool either and does not want to waste any time on this issue any more. Knowing this procedure, including the fact that the three professors will all speak up at 3:00 p.m., he can leave already at 2:01, leaving his secretary the message that she can inform the three professors of a (minor) pay rise.

COMMON KNOWLEDGE IN DISTRIBUTED SYSTEMS

As before, we consider a *distributed system* consisting of m processors $1, \ldots, m$. We consider the *Kripke model* $M = \langle S,\ R_1, \ldots, R_m,\ \models\rangle$ *associated with this distributed system*, with the set S of global system states $s = (s_1, \ldots, s_m)$ where the s_i $(i = 1, \ldots, m)$ are local states of the processors $1, \ldots, m$. The accessibility relations R_i are defined as before: sR_is' iff $s_i = s'_i$.

This associated Kripke model does not contain any information about the actual state transformations that the system executes or is subject to. The actual process is determined by the structure of the processors, the way they are programmed and the protocols by which they communicate. All these issues are beyond the scope of epistemic logic. Although epistemic logic cannot express anything about the way in which a process comes about, it can say something about the knowledge that is available at the processors in the system during an execution of the system. In general, the notion of an execution through a Kripke model is defined as follows:

DEFINITION 34.2.: An *execution* in a Kripke model M, with set W of worlds, is a (finite or infinite) sequence of states $w^{(1)}, w^{(2)}, \ldots$ with $w^{(i)} \in W$ $(i = 1, 2, \ldots)$. Such an execution is denoted as:

$$w^{(1)} \Rightarrow w^{(2)} \Rightarrow \ \ldots .$$

Thus an execution in a Kripke model is just a sequence of states in that model, which should be interpreted as a walk through the model as a result of the execution of processes by the processors of the system. This is, of course, rather abstract. In fact, in the theory of distributed programming this idea is often made much more concrete by assuming some programming language, by means of which one may describe the execution of processes by the processors. The execution of each statement in such a programming language, which several or all processors may do in parallel, is then specified by a so-called (possibly labelled) transition system ([HP79], [Plo81]), which is a formal system that describes the exact way in which executions of statements change the state of the system. Here we need not go to this more detailed level, since the properties we will discuss in this section can be described already adequately on the abstract level.

One may now ask whether something can be said about the course of common knowledge during a certain execution. In particular we may pose this question in the context of Kripke models associated with distributed systems. Is it possible to devise a distributed system such that common knowledge *increases* during some execution? That is to say, for some A, $[C]A$ does not hold in an initial part $w^{(1)} \Rightarrow w^{(2)} \Rightarrow \ldots \Rightarrow w^{(k)}$ of the execution (i.e., $(M, w^{(i)}) \not\models [C]A$ for $1 \leq i \leq k$), while $[C]A$ holds after this initial part and continues to hold (i.e., $(M, w^{(i)}) \models [C]A$ for $i > k$). Somewhat surprisingly, this is *not* the case if we do not impose further restrictions on the reachability of the states of the system besides the ones we prescribed before, as we shall see presently. First a proposition and a definition are given.

PROPOSITION 34.3.: Let w be a world in Kripke model M. Let K_w be the 'upward cone' of w, i.e., $K_w = \{w' \in W \mid wR_C w'\}$. Then: if $[C]A$ holds in w (i.e., $M \models_w [C]A$), then $[C]A$ holds in the whole of K_w. (See Exercise 34.6.)

DEFINITION 34.4.: Let $M = \langle W, R_1, \ldots, R_m, \models \rangle$ and R_C be defined as in Definition 34.1. Then M is called *strongly connected* if for all $w, w' \in W$ it holds that $wR_C w'$.

PROPOSITION 34.5.: The Kripke model associated with a distributed system, as defined before, is strongly connected. (See Exercise 34.6.)

We now have the following general result:

THEOREM 34.6.: Let M be a strongly connected Kripke model. Suppose that for some world w it holds that $M \models_w [C]A$. Then $M \models [C]A$, i.e., for every world w, $M \models_w [C]A$.

PROOF: Directly from Proposition 34.3. □

COROLLARY 34.7.: Let M be the Kripke model associated with a distributed system with processors $1, \ldots, m$ as defined before. Suppose that $M \models_s [C]A$ for some state s. Then $M \models [C]A$, i.e., $M \models_s [C]A$ for every state s.

PROOF: Directly from Theorem 34.6 and Proposition 34.5. □

COROLLARY 34.8.: Let M be the Kripke model associated with a distributed system with processors $1, \ldots, m$ as defined before. Then common knowledge is constant through every execution of M.

The results on common knowledge in distributed systems are somewhat surprising and may even seem counterintuitive. We would expect common knowledge in distributed systems to increase by communication. The solution of this paradox is the fact that in practice not every global state of the system is reachable. In this case it is possible that the associated Kripke model loses the property of being strongly connected, and thus escapes the results obtained thus far. (See the proof of Proposition 34.5.) So now we have to consider Kripke models $M = \langle W, R_1, \ldots, R_m, \models \rangle$, where $W \subseteq S_1 \times \ldots \times S_m$ (rather than $W = S_1 \times \ldots \times S_m$). Although in this case it cannot be proved that in general C-knowledge is constant, we can still obtain a result of a somewhat more special nature. First we need a definition.

DEFINITION 34.9.: An execution $s^{(1)} \Rightarrow s^{(2)} \Rightarrow \ldots$ is called *non-simultaneous* if for every transition $s^{(k)} \Rightarrow s^{(k+1)}$ there exists a processor i, $1 \leq i \leq m$, with $s_i^{(k)} = s_i^{(k+1)}$.

THEOREM 34.10 ([HM84a]).: In non-simultaneous executions, common knowledge is constant.

PROOF: Suppose $s \Rightarrow s'$ with $s_i = s'_i$, and, consequently, sR_is', and suppose $M \models_{s'} [C]A$. Now it holds that

if $M \models_{s'} [C]A$, then $M \models_{s'} [E][C]A$ and hence $M \models_{s'} [K_i][C]A$.

R_i is an equivalence relation, so in particular it holds that: if sR_is', then $s'R_is$. From this it follows by means of the definition of the semantics of the $[K_i]$-operator: $M \models_s [C]A$. Thus any C-knowledge that is present in s' is also present in s. Analogously: any C-knowledge present in s is present in s' as well. Hence, C-knowledge is constant at the non-simultaneous transition $s \Rightarrow s'$. By induction it follows that C-knowledge is constant in any non-simultaneous execution. □

We may conclude that in a distributed system, in which not all global states are accessible, it is not the case that in general common knowledge is constant, although it must be constant along non-simultaneous executions. In other words, the only *possibility* for a change in common knowledge is via transitions that are simultaneous, in the sense of being not non-simultaneous.

There is more to be said about common knowledge in distributed systems. We already saw that special conditions have to be met to enable common knowledge to be changed. In practice it is even worse since in a network it is generally not guaranteed that messages come across. Communication channels may be faulty, so that messages get lost. Halpern & Moses show in [HM84a, 90] that under these circumstances it is even impossible in principle to attain common knowledge in such a network. In order to get a feeling for this result, we discuss the example of the Byzantine generals who want to launch a co-ordinated attack.

THE CO-ORDINATED ATTACK PROBLEM

Consider a situation of two army regiments on two hills on both sides of a valley in which a hostile army is situated. If the two regiments attacked simultaneously, they would conquer the enemy. If only one of the two regiments attacked, it would be defeated by the enemy. There is no initial battle plan available which the two regiments agree upon. Neither general of the regiments would decide to attack without knowing for sure that the other one will also attack. The general of one of the two regiments, let us call him A, now wants to co-ordinate a simultaneous attack of the two regiments. To this end he only may use a messenger who has to go through enemy territory to convey the message m ("attack!") to the other general, B. Unfortunately, not in the least for the messenger himself, his safe arrival is not guaranteed. In order to be certain that the message has been delivered (and that its recipient will attack), an *acknowledgement*, i.e., a message $[K_B]m$, has to be sent by the recipient, general B, to the sender, general A. However, this has to again be done by a messenger, which is not safe either. So, in order for general B to be sure about the simultaneous attack, he has to have an acknowledgement of the arrival of this acknowledgement. (Otherwise, it is *he* who might attack on his own, with dramatic consequences.) So a message about this, $[K_A][K_B]m$, has to be sent from A to B. But A will again need a confirmation of this by means of a message $[K_B][K_A][K_B]m$, *ad infinitum*. At any moment, however, there will be an assertion of the form "$[K_B][K_A][K_B]\ldots m$" or "$[K_A][K_B][K_A]\ldots m$" that is not (yet) true. This means that *common knowledge* of m, i.e., $[C]m$, cannot be attained (in finite time).

Unfortunately for the two generals, common knowledge of m ("$[C]m$") is necessary to be (both) sure that they will attack simultaneously. (Otherwise, there will always be doubt in the mind of one of them whether the last acknowledgement that has been sent has been received by the other one.)

Halpern & Moses ([HM84a, 90]) have proved this rigourously by means of a theorem that says on the one hand that "$[C]m$" is necessary for solving the "*Co-ordinated Attack Problem*", and on the other that C-knowledge cannot be attained in a distributed system in which communication is not guaranteed. The "Co-ordinated Attack Problem" is therefore fundamentally unsolvable. This was in fact already a folk theorem of operating systems theory, where similar types of co-ordinated attacks occur in the context of computers (cf. [Gra87]).

There is a bright spot in this problem, however. Halpern & Moses and Fischer & Immerman have shown that in order to solve some practical problems of this nature, certain weaker forms of common knowledge suffice which *are* attainable in practice. The interested reader is referred to [FI86] and [HM90]. The reader is also referred to [Mos92] for an overview of recent developments in this area.

IMPLICIT KNOWLEDGE IN A GROUP

There is yet another variant of knowledge: *implicit knowledge.* The intuition behind this notion is the following. We say that a group of agents has implicit knowledge of a fact if the knowledge of that fact is distributed over the members of that group. In a closed system a group of cooperating agents can only acquire knowledge that is already implicit in the group. This will be discussed in greater detail in the next subsection, where a so-called *conservation principle for implicit knowledge* will be treated. The idea of implicit knowledge is formalized by the following definition of the implicit knowledge operator $[I]$, which we add to the basic epistemic language.

DEFINITION 34.11.: Given a Kripke structure
$M = \langle W,\ R_1, \ldots, R_m,\ \models\rangle$. Let $R_I = R_1 \cap \ldots \cap R_m$. Then:

$$M \models_w [I]A \text{ iff } M \models_{w'} A \text{ for all } w' \text{ such that } wR_Iw'.$$

Thus, if all agents could pool their knowledge together, they would only consider those worlds as possible which are epistemic alternatives for each of them in isolation. One can summarize this intuition by using the reading of $[I]A$ as: *a wise man knows A*, as opposed to that of $[C]A$: *any fool knows A*. Indeed: if some agent in the group knows that w' is not an epistemic alternative for him, then the 'wise man' knows this as well. The 'wise man' only considers those states as possible that are considered possible by all agents.

From a formal point of view, the operator $[I]$ is in a way (but not in the technical modal logic sense) the 'dual' of the operator $[E]$: the former is defined by means of the intersection of the accessibility relations, whereas the latter is defined on the basis of their union.

Somewhat unexpectedly, then, the axiomatization of the $[I]$-operator is much more involved that that of its 'dual' $[E]$. First we note that, obviously, implicit knowledge is only interesting for $m \geq 2$ agents: for $m = 1$, $[I]A$ coincides with $[K]A$. Axiomatisation of implicit knowledge for the case $m = 1$ is therefore trivial. For $m \geq 2$, however, the question of axiomatisation is quite another matter. We cannot use a clear-cut correspondence result as in the case of the $[E]$-operator, since – contrary to the union of accessibility relations – there is no axiom that corresponds directly to the intersection of accessibility relations (cf. [vdHM92b], Theorem 2.17). But, again somewhat surprisingly, we can gain from the fact that in these modal logics we cannot distinguish models in which we have relations that are intersections of the accessibility relations associated with the (knowledge of the) agents involved, from models in which we have relations that are *contained within* these intersections.

And with respect to the latter weaker property, we *do* have a corresponding axiom, viz.

(A11) $\quad [K_i]A \supset [I]A \qquad (i = 1, \ldots, m)$.

It is easy to show that this axiom corresponds with the property that $R_I \subseteq R_i$, for $1 \leq i \leq m$, where R_I stands for the accessibility relation associated with the modal operator $[I]$. We shall call models of the type $M = \langle W, R_1, \ldots, R_m, R_I, \models \rangle$ with $R_I = R_1 \cap \ldots \cap R_m$ $S5I_{(m)}$-models.

If we add (A11) to the system $S5_{(m)}$, and furthermore add versions of the axioms (A2) - (A5) for $[I]$ rather than $[K_i]$, we obtain a sound and complete system with respect to $S5I_{(m)}$-models (cf. [HM85], [vdHM92b]). So, in retrospect it can be said that the axiomatization of implicit knowledge itself is rather simple, although the proof of completeness is much harder than one would perhaps expect at first glance. The system consisting of the axioms and rules of $S5_{(m)}$ together with (A11) is called $S5I_{(m)}$.

From the intuitive description of implicit knowledge we would expect that the following rule holds:

$$\text{(R4)} \qquad \frac{(A_1 \,\&\, \ldots \,\&\, A_m) \supset B}{([K_1]A_1 \,\&\, \ldots \,\&\, [K_m]A_m) \supset [I]B}$$

The intuition behind this rule is elaborated upon by means of an example. Let $m = 3$ and assume B to be Fermat's Last Theorem. We furthermore assume that three mathematical agents 1, 2, 3 work on this old and famous problem in order to discover a proof of this theorem. Suppose that the actual situation is such that B follows from three lemmas: A_1, A_2 and A_3. Mathematician 1 has proved A_1 — so he knows A_1. Likewise mathematicians 2 and 3 proved and know A_2 and A_3, respectively. Thus, in this situation we have $[K_1]A_1, [K_2]A_2, [K_3]A_3$, i.e., $[K_1]A_1 \;\&\; [K_2]A_2 \;\&\; [K_3]A_3$. If the mathematicians contacted each other at a mathematical congress, they would be able to gather their partial knowledge and thereby know (a proof of) B.

Rule (R4) indeed appears to be deducible from (A2), (A11) and propositional logic. Conversely, (A11) is also deducible from (R4) and the other axioms. So we can replace (A11) by (R4) and obtain an alternative complete axiomatisation of implicit knowledge.

IMPLICIT KNOWLEDGE IN A DISTRIBUTED SYSTEM

In this subsection we shall see how the notion of implicit knowledge can be applied in distributed systems with unreliable message exchange in order to characterize attainable states of knowledge and to obtain a 'message logic'. This application was developed by R. Fagin & M. Vardi. We only give a rough sketch of the idea. The reader interested in the details is referred to [FV86].

The following assumptions are made concerning the distributed system:

1. The environment of the distributed system ('Nature') does not change.
2. There is synchronous communication (in rounds) between the agents (processors) in the system.
3. The communication is unreliable: messages may get lost, i.e., never received.
4. If a message is received, it is received in the *same* round as it was sent.
5. Messages concerning the past may be very diverse: e.g. "I knew A just after round r", "I sent message M to agent j in round r", "I received precisely the set Θ of messages from agent j in round r", "Every message that I sent in round r to agent j I still know to be true". Finite Boolean combinations of these messages are also allowed. On the other hand, to avoid all kinds of serious technical problems (see [FV86]), messages concerning the future are restricted to a special type: "If A holds in round r, then in round r the only messages I will send to agent j are in the set Θ", where A is a message about the past. Messages are assumed

to be honest in the sense that they are true when they are sent. For messages about the future, this means that promises that are conveyed in these messages must be kept, and are fulfilled in the future.

6. At time $t = 0$ the agents receive information about their environment (Nature). At time $t > 0$ the only information received is through the communication network of the system.

Suppose W is the set of states the system may be in, and let $\mathcal{M}$ stand for the set of messages that can be sent (and received) by the agents in the system. A state is viewed as a function assigning truth values to the primitive propositions. A *k-round execution* is given by a tuple $(w,\ T,\ \text{sent},\ \text{received})$, where $w \in W$ is a state describing 'Nature'; $T:\ \{1,\ldots,m\} \rightarrow 2^W$, $T(i)$ denotes agent i's initial information about 'Nature', we assume that $w \in T(i)$ for all agents i; sent: $\{1,\ldots,m\} \times \{1,\ldots,m\} \times \{1,\ldots,k\} \rightarrow 2^{\mathcal{M}}$, $\text{sent}(i,j,r)$ denotes the set of messages sent by i to j in round r; received: $\{1,\ldots,m\} \times \{1,\ldots,m\} \times \{1,\ldots,k\} \rightarrow 2^{\mathcal{M}}$, $\text{received}(i,j,r)$ denotes the set of messages received by i from j in round r. In line with the above, we assume that $\text{sent}(i,j,0) = \text{received}(i,j,0) = \emptyset$, and $\text{received}(i,j,k+1) \subseteq \text{sent}(j,i,k+1)$ for all $k \geq 0$.

Let $R = (w,\ T,\ \text{sent},\ \text{received})$ and $R' = (w',\ T',\ \text{sent}',\ \text{received}')$ be two k-round executions. Now we define $R \approx_i R'$ iff

1. $T(i) = T'(i)$;
2. $\forall 1 \leq j \leq m\, \forall 1 \leq r \leq k\ [\ \text{sent}(i,j,r) = \text{sent}'(i,j,r)]$;
3. $\forall 1 \leq j \leq m\, \forall 1 \leq r \leq k\ [\ \text{received}(i,j,r) = \text{received}'(i,j,r)]$.

This means that for agent i, two executions are indistinguishable iff they concern the same sets of possible worlds (expressing uncertainty about the status quo of 'Nature'), and the same histories with respect to sending and receiving. If agent i is actually in execution R above, then he also considers execution R' above as (equally) possible.

LEMMA 34.12.: Let $R = (w,\ T,\ \text{sent},\ \text{received})$ and $R' = (w',\ T',\ \text{sent}',\ \text{received}')$ be two 0-rounds. Then: $R \approx_i R'$ iff $T(i) = T'(i)$.

PROOF: Immediate, since conditions 2 and 3 of $R \approx_i R'$ are vacuous for $k = 0$. □

Epistemic formulas are now interpreted on k-round executions $R =$ (w, T, sent, received) as follows:

$R \models P$:= $w \models P$ (P atomic);
$R \models A \,\&\, B$:= $R \models A$ and $R \models B$;
$R \models A \vee B$:= $R \models A$ or $R \models B$;
$R \models A \supset B$:= not $R \models A$ or $R \models B$;
$R \models \neg B$:= not $R \models B$;
$R \models [K_i]B$:= for all R', if $R \approx_i R'$ then $R' \models B$;
$R \models [I]B$:= for all R', if $R \approx_I R'$ then $R' \models B$, where $\approx_I = \approx_1 \cap \ldots \cap \approx_m$.

Note that we have: $R \models [I]B$ iff

for all R', if $R \approx_i R'$ for all $i \in \{1, \ldots, m\}$, then $R' \models B$.

If R is a k-round execution and $r \leq k$, then we say that A is implicit knowledge after r rounds of R if $R' \models [I]A$, where R' is the r-round prefix of R, i.e., the execution which one obtains from R by considering only r rounds (thus restricting the functions sent and received to $\{1, \ldots, r\}$ in their third argument).

One can now show the so-called *conservation principle for implicit knowledge*:

THEOREM 34.13.: Let A be a purely propositional formula (i.e., a formula without modal operators) and let R be a k-round execution. Assume that $0 \leq r \leq k$ and $0 \leq s \leq k$. Then A is implicit knowledge after r rounds of R iff A is implicit knowledge after s rounds of R.

PROOF: It is sufficient to prove that A is implicit knowledge after r rounds of R iff A is implicit knowledge after 0 rounds of R : Let $R =$ (w, T, sent, received) and let R_0 be the 0-round prefix of R. If A is implicit knowledge after the 0-th round of R, then for all 0-round executions $R' =$ (w', T', sent$'$, received$'$), if $R_0 \approx_i R'$ (i.e., $T(i) = T'(i)$) for all $i \in \{1, \ldots, m\}$, then $R' \models A$ (i.e., $w' \models A$). This implies that for all k-round executions $R =$ (w, T, sent, received) and $R' =$ (w', T', sent$'$, received$'$), if $T(i) = T'(i)$ for all $i \in \{1, \ldots, m\}$, then $R' \models A$. Now consider R_r, which is the r-round prefix of R. A is implicit knowledge after r rounds of R iff $R_r \models [I]A$, i.e., for all r-round executions $R' =$ (w', T', sent$'$, received$'$), if $R' \approx_i R_r$ for all $i \in \{1, \ldots, m\}$, then $R' \models A$. Suppose

$R' \approx_i R_r$ for all i. Then $T'(i) = T(i)$ for all i (and some further conditions about the sent and receive functions we are not interested in now). By the above assertion we now immediately have that $R' \models A$.

Conversely, suppose that A is *not* implicit knowledge after the 0-th round of R. Then there is some 0-round execution $R' = (w', T', \text{sent}', \text{received}') \approx_i R_0$ for all i and such that $R' \not\models A$, i.e., $w' \not\models A$. By definition, $w' \in T'(i)$ for all i. Since $R' \approx_i R_0$ for all i, we know that $T'(i) = T(i)$ for all i, i.e., $T' = T$. Now consider the r-round execution $R^* = (w', T, \text{sent}_r, \text{received}_r)$ where sent_r and received_r are the restrictions of the functions sent and received of R to r rounds. For this R^* it holds that $R^* \not\models A$, since $w' \not\models A$. Moreover, with respect to the r-round prefix $R_r = (w, T, \text{sent}_r, \text{received}_r)$ of R it holds that $R^* \approx_i R_r$ for all i. Thus, there is some r-round execution $R^* \approx_i R_r$ for all i and such that $R^* \not\models A$. Hence, $R_r \not\models [I]A$. □

This theorem means that in an execution, implicit knowledge about 'Nature' remains constant after the initialisation at the 0-th round: the best an agent can attain by communication in a closed system is the knowledge that is implicit in the system at the initialisation and remains constant afterwards.

Note that the conservation principle does not hold for arbitrary formulas: for example, consider $A = [K_1]P$, where P is a primitive proposition. Suppose that in an execution agent 1 initially (in round 0) does not gain any information about the truth of P in Nature, but learns that P is true in round 1 because of a message from agent 2. Now we have that A is false and therefore not implicit knowledge after round 0, whereas A is true, and even known by agent 1 after round 1 (it holds that $[K_1]A$), and thus A is implicit knowledge after round 1. Implicit knowledge about knowledge of agents is therefore not constant.

Lastly, we discuss how the logic $S5I_{(m)}$ can be extended to get a "*message logic*" $ML_{(m)}$.

Define a *state formula* to be a formula that completely describes a state of the system. To be able to do this using finite formulas, we need to assume that the set of primitive propositions is finite. So, for example, if we have two primitive propositions P and Q, there are precisely 4 state formulas: $P \mathbin{\&} Q$, $P \mathbin{\&} \neg Q$, $\neg P \mathbin{\&} Q$ and $\neg P \mathbin{\&} \neg Q$.

Now we need to add the following axiom to the system $S5I_{(m)}$:

(ML) $\quad [I]\neg A \supset ([K_1]\neg A \vee \ldots \vee [K_m]\neg A)$,

where A is a state formula. The axiom (ML) says that if it is implicit knowledge that a state is impossible, then the stronger formula is true that some agent knows that the state is impossible. That is, if by pooling their knowledge the agents can rule out a state, then some agent can do this already on his/her own.

That the axiom (ML) is not generally sound for arbitrary formulas, even for primitive formulas, is easily seen in the following example. Assume that P and Q constitute the primitive propositions and assume $m = 2$. Consider a 0-round execution, in which both P and Q are false (in 'Nature'), agent 1 learns (via $T(1)$) that either both P and Q are true or they are both false, and agent 2 learns (via $T(2)$) that Q is false. Now $\neg P$ is implicit knowledge, since this is implied by the shared knowledge of agents 1 and 2. However, neither agent 1 nor agent 2 knows $\neg P$. So the axiom (ML) does not hold for $A = P$ in this case.

The soundness of (ML) for state formulas A for an arbitrary k-round execution $R = (w,\, T,\, \text{sent},\, \text{received})$ is sketched as follows: Suppose that for every agent i it holds that i does not know that $\neg A$, i.e., $\neg[K_i]\neg A$ holds for all $1 \leq i \leq m$. This says that every agent i considers the state w_A described by A possible.

Formally, we have that $\forall i\, \exists R'_i[R'_i \approx_i R \mathbin{\&} R'_i \models A]$.

Say that $R'_i = (w'_i,\, T'_i,\, \text{sent}'_i,\, \text{received}'_i)$, where any sent'_i and $\text{received}'_i$ suffice (provided honesty of messages is satisfied) as well as any T'_i, provided that for every i, $w'_i \in T'_i(i) = T(i)$. Moreover, since A is purely propositional and $R'_i \models A$, we have that $w'_i \models A$ (for every i). Furthermore, since A is a state formula, uniquely determining a state, $w'_i = w_A$ for every i. Now take the k-round execution $R^* = (w_A,\, T,\, \text{sent},\, \text{received})$. Then for every i, $R^* \approx_i R'_i$ and hence $R^* \approx_i R$ since also $R'_i \approx_i R$. Moreover, $R^* \not\models \neg A$ (since $w_A \models A$ and thus $R^* \models A$). So now there is a k-round execution R^* such that $R^* \approx_i R$ for all i, and $R^* \not\models \neg A$. This implies that $R \models \neg[I]\neg A$, i.e., $\neg A$ is no implicit knowledge after k rounds of R.

Fagin & Vardi also show that the logic $ML_{(m)}$ is complete for $m \geq 2$, but this goes beyond our brief discussion. See [FV86]. There is a related paper [FHV92], which may be regarded as a follow-up in some respects of [FV86]. Inspired by the results of the latter paper, in [FHV92] it is investigated from a somewhat more general perspective how properties of knowledge may depend subtly on assumptions made about a system (such as (non)deterministic behaviour, whether processors never forget, and whether or not the environment of the system affects the state transitions of processors) and the assertion language used for describing knowledge (such as the nature of the primitive propositions and the (un)availability of a distributed knowledge operator). As we have already seen in this section, these assumptions may confine the set of attainable knowledge situations, and Fagin, Halpern & Vardi showed that, even in the case in which the external notion of knowledge is taken to be plainly $S5_{(m)}$, the logic $S5_{(m)}$ may no longer be complete when such assumptions about the system are considered. For various cases (such as the case in which the environment determines the initial states, processors do not forget, process state transitions are independent of the initial environment, and the primitive propositions characterize the initial environment) extensions of $S5_{(m)}$ that are complete axiomatizations are provided. The results from [FV86] regarding the message logic $ML_{(m)}$ that we have discussed above can now be seen as a special case of this more general approach: $ML_{(m)}$ is such an extension of $S5_{(m)}$ that is complete in the situation that we have treated in this section. We refer to [FHV92] for details.

REFERENCES

[ChM85, 86] M. Chandy & J. Misra, How Processes Learn, *Proc. 4th ACM PODC*, 1985, pp. 204-214; revised version in *Distributed Computing* 1(1), 1986, pp. 40-52.

[FHV92] R. Fagin, J.Y. Halpern & M.Y. Vardi, What Can Machines Know? On the Properties of Knowledge in Distributed Systems, *J. Assoc. Comput. Mach.* 39 (2), 1992, pp. 328-376.

[FI86] M.J. Fischer & N. Immerman, Foundations of Knowledge for Distributed Systems, *Proc. 1st TARK* (J.Y. Halpern, ed.), Morgan Kaufmann, Los Altos, 1986, pp. 171-185.

[FV86] R. Fagin & M.Y. Vardi, Knowledge and Implicit Knowledge in a Distributed Environment, *Proc. 1st TARK* (J.Y. Halpern, ed.), Morgan Kaufmann, Los Altos, 1986, pp. 187-206.

[Gra87] J. Gray, Notes on Database Operating Systems, *IBM Res. Rep.* RJ 2188, IBM, 1987.

[HC84] G.E. Hughes & M.J. Cresswell, *A Companion to Modal Logic*, Methuen, London, 1984.

[HF89] J.Y. Halpern & R. Fagin, Modelling Knowledge and Action in Distributed Systems, *Distributed Computing* 3, 1989, pp. 159-177.

[HM84a, 90] J.Y. Halpern & Y.O. Moses, Knowledge and Common Knowledge in a Distributed Environment, *Proc. 3rd ACM PODC*, 1984a, pp. 50-61; revised version in *J. ACM*, 37(3), 1990, pp. 549-587.

[HM85] J.Y. Halpern & Y.O. Moses, A Guide to the Modal Logics of Knowledge and Belief, *Proc. 9th IJCAI*, 1985, pp. 480-490.

[vdHM92b] W. van der Hoek & J.-J. Ch. Meyer, Making Some Issues of Implicit Knowledge Explicit, *Int. J. of Foundations of Computer Science* 3(2), 1992, pp. 193-223.

[vHM92] M. van Hulst & J.-J. Ch. Meyer, A Taxonomy of Knowledge in Distributed Systems, *Proceedings of CSN92* (J.L.G. Dietz, ed.), SION, Utrecht, 1992, pp. 42-53.

[HP79] M. Hennessy & G.D. Plotkin, Full Abstraction for a Simple Programming Language, in *Proc. 8th MFCS* (J. Becvar, ed.), Lect. Notes in Comp. Sci. 74, Springer, Berlin, 1979, pp. 108-120.

[HZ87] J.Y. Halpern & L.D. Zuck, A Little Knowledge Goes a Long Way: Simple Knowledge-Based Derivations and Correctness Proofs for a Family of Protocols, in *Proc. 6th ACM PODC*, 1987, pp. 269-280.

[KT86] S. Katz & G. Taubenfeld, What Processes Know: Definitions and Proof Methods, in *Proc. ACM PODC'86*, 1986, pp. 249-262.

[Leh84] D.J. Lehmann, Knowledge, Common Knowledge, and Related Puzzles, *Proc. 3rd ACM PODC*, 1984, pp. 62-67.

[LMRT90] K. Lodaya, M. Mukund, R. Ramanujam & P.S. Thiagarajan, Models and Logics for True Concurrency, SPIC Science Foundation, School of Mathematics, Techn. Report TCS-90-3, Madras, 1990.

[MH93] J.-J. Ch. Meyer & W. van der Hoek, *Epistemic Logic for AI and Computer Science*, Cambridge University Press, 1994.

[Mos92] Y. Moses, Knowledge and Communication, *Proc. 4th TARK* (Y. Moses, ed.), Morgan Kaufmann, San Mateo, 1992, pp. 1-14.

[Plo81] G.D. Plotkin, A Structural Approach to Operational Semantics, Report DAIMI FN-19, Comp. Sci. Dept., Aarhus University, 1981.

EXERCISES

34.1. Prove the soundness of the system $S5_{(m)}$.

34.2. Check whether the following formulas are valid with respect to $S5_{(1)}$-Kripke models (P, Q are atomic, we omit the index 1 from $[K_1]$): (a) $[K]P \vee [K]\neg P$,
(b) $\neg[K]P \vee \neg[K]\neg P$,
(c) $\neg P \vee [K]\neg[K]\neg P$,
(d) $([K]P \;\&\; [K]Q) \supset [K](P \supset Q)$,
(e) $P \;\&\; [K]\neg[K]Q \;\&\; ([K]\neg[K]\neg P \supset [K]Q)$.

34.3. Prove the soundness of the system $S5C_{(m)}$.

34.4. Show that the following two assertions are provable in the system $S5C_{(m)}$: (a) $[C]A \supset [C][C]A$ and (b) $\neg[C]A \supset [C]\neg[C]A$.

34.5. In the situation of the dean and the hatted professors, analyse what would have happened if the dean had not made a public announcement about the procedure, but had asked each professor to come to his room separately showing each the 5 hats and telling exactly the same story as in the original situation: at noon he will turn off the light, and place a hat on their heads, etc.

34.6. Prove Propositions 34.3 and 34.5.

34.7. Show the soundness of axiom (A11) with respect to $S5I_{(m)}$-models.

34.8. Show that in $S5I_{(m)}$ (R4) is a derived theorem.

34.9. A *knowledge representation system* (*k.r.s.*) is a pair (W, A) where W is a non-empty set (of objects or possible worlds) and A is a set of attributes. An *attribute* is a function f from W to $\{0,1\}$; $f(w) = 1$ is to be read as: the attribute f applies to w. The idea is that an agent tries to distinguish the objects in W by means of the attributes in A.
Suppose there are m agents $1,\ldots,m$ with knowledge representation systems (W, A_i) for $i = 1,\ldots,m$.
For $w, w' \in W$ we define w ind(A_i) w' := for every f in A_i, $f(w) = f(w')$; i.e., agent i cannot distinguish between w and w' by means of the attributes in A_i; in other words, w and w' are *indiscernable* for agent i. Let $R_i :=$ ind(A_i).
(See C.M. Rauszer, *Rough Logic for Multi-Agent Systems.*)

i) Prove that R_i is an equivalence relation, i.e., R_i is reflexive, transitive and symmetric.
Consequently, each R_i induces a *partition* of W, i.e., a collection of disjoint subsets of W, the union of which equals W. (See section 15.) Let $\mathcal{E}_i$ be the partition of W induced by R_i. $\mathcal{E}_i$ is called the *knowledge base* of agent i.

ii) Let $W = \{\alpha, \beta, \gamma, \delta\}$, $A_1 = \{a, b, c\}$ and $A_2 = \{b, c, d, e\}$. Suppose the attributes for agent 1 and 2 are defined respectively as follows.

	a	b	c
α	0	1	1
β	0	1	1
γ	1	1	1
δ	1	1	1

	b	c	d	e
α	1	1	1	0
β	1	1	1	0
γ	1	1	1	0
δ	1	1	0	1

Determine $R_1 = \text{ind}(A_1)$ and $R_2 = \text{ind}(A_2)$. Next determine $\mathcal{E}_1$ and $\mathcal{E}_2$, the knowledge base of agent 1 and 2 respectively.

34.10. Let $M = \langle W,\ R_1,\ R_2,\ \models\rangle$ be the following Kripke model. $W = \{\alpha, \beta, \gamma, \delta\}$; R_1 and R_2 are defined as in exercise 34.9. ii); by definition, $\alpha \models P$ and $\delta \models P$.
Let $p_i(P)$ express the probability that agent i assigns to P.

i) Check that $M \models_\alpha p_1(P) = 1/2$, $M \models_\beta p_1(P) = 1/2$, $M \models_\gamma p_1(P) = 1/2$ and $M \models_\delta p_1(P) = 1/2$.
Consequently, $M \models_\alpha C(p_1(P) = 1/2)$.

ii) Check that $M \models_\alpha p_2(P) = 1/3$, $M \models_\beta p_2(P) = 1/3$, $M \models_\gamma p_2(P) = 1/3$ and $M \models_\delta p_2(P) = 1$.
Consequently, $M \models_\alpha K_1(p_2(P) = 1/3)$ but $M \not\models_\alpha K_2K_1(p_2(P) = 1/3)$. Also not $M \models_\alpha C(p_2(P) = 1/3)$.

Hint: It will be helpful to draw a picture of the Kripke model involved.
Note that – strictly speaking – expressions such as $p_1(P) = 1/2$ are not formulas. However, we can extend the language of epistemic logic in such a way that these expressions can be considered in an adequate way. See H.C.M. de Swart, ***Linking Kripke semantics and game theory; a logical approach to Aumann's theorem.***

34.11. i) Let $M = \langle W,\ R_1,\ R_2,\ \models\rangle$ be a Kripke model with R_1, R_2 equivalence relations. Let $R_i(w) := \{w' \in W \mid wR_iw'\}$, $i = 1, 2$, and $R_C(w) := \{w' \in W \mid wR_Cw'\}$.
Prove that $\{R_i(w) \mid w \in R_C(w_0)\}$ is a *partition* of $R_C(w_0)$, i.e., a) $\cup\{R_i(w) \mid w \in R_C(w_0)\} = R_C(w_0)$, and
b) if $R_i(w_1) \neq R_i(w_2)$, then $R_i(w_1) \cap R_i(w_2) = \emptyset$.

ii) Let $M = \langle W,\ R_1,\ R_2,\ \models,\ \mu\rangle$ be a Kripke model with measure μ, i.e., $\langle W,\ R_1,\ R_2,\ \models\rangle$ is a Kripke model for $S5$ and $\mu : P(W) \to \mathbf{R}$ is a function that assigns to every subset V of W a real number $\mu(V)$, called the *measure* of V, such that

(*) if $V_1 \subseteq V_2 \subseteq W$, then $\mu(V_1) \leq \mu(V_2)$, and
(**) if $V_1 \cap V_2 = \emptyset$, then $\mu(V_1 \cup V_2) = \mu(V_1) + \mu(V_2)$.
For instance, if W is finite, μ, defined by $\mu(V)$ = the number of elements of V, satisfies (*) and (**). (See Exercise 34.10.)
Let $p_i(A)$ express the probability that agent i assigns to A.

$$M \models_w p_i(A) = q_i := \mu(X_A/R_i(w)) = q_i,$$

where $X_A := \{w' \in W \mid M \models_{w'} A\}$ and $\mu(X_A/R_i(w))$ is short for $\mu(X_A \cap R_i(w))/\mu(R_i(w))$. (Cf. Exercise 34.10.)
Prove *Aumann's theorem* (1976): If $M \models_w C(p_1(A) = q_1)$ and $M \models_w C(p_2(A) = q_2)$, then $q_1 = q_2$. In words, if it is common knowledge at w that the probability of A given agent 1's information is q_1 and that the probability of A given agent 2's information is q_2, then $q_1 = q_2$. Roughly speaking, Aumann's result says that rational agents cannot agree to disagree about the probability of A. **Hint**: Assuming that $M \models_w C(p_i(A) = q_i)$, $i = 1, 2$, prove that $\mu(X_A \cap R_C(w)) = q_i \cdot \mu(R_C(w))$, using i) and (**).

34.12. Let (W, A_i) be a knowledge representation system for agent i. And let $\mathcal{E}_i$ be the corresponding knowledge base of agent i. (See Exercise 34.9.) $\mathcal{E}_i$ is a subbasis of a topological space $(W, \mathcal{D}_i)$, i.e., $(W, \mathcal{D}_i)$ is a *topological space* if $\mathcal{D}_i$ consists of all unions of finite intersections of the elements in $\mathcal{E}_i$, $\emptyset$ and W itself. The elements of $\mathcal{D}_i$ are called *open* sets. For $X \subseteq W$ let $K_i(X)$:= the greatest open set contained in X. $K_i(X)$ is called the *positive knowledge* of agent i: it contains those objects or possible worlds which are classified by agent i *without any doubt* as elements of X by means of the attributes available to agent i. Prove that K_i is an interior operator, i.e., 1) $K_i(X) \subseteq X$, 2) $K_i(X) \subseteq K_i(K_i(X))$, and 3) $W - K_i(W - X) \subseteq K_i(W - K_i(W - X))$. Note that the interior operator K_i corresponds to a knowledge operator K that satisfies the axioms for $S5$.
For $W = \{\alpha, \beta, \gamma, \delta\}$ and A_1 and A_2 defined as in Exercise 34.9., determine $K_1(\{\alpha\})$, $K_1(\{\alpha, \beta\})$, $K_1(\{\alpha, \beta, \gamma\})$ and $K_1(\{\alpha, \beta, \gamma, \delta\})$; also determine $K_2(\{\alpha\})$, $K_2(\{\alpha, \beta\})$, $K_2(\{\alpha, \beta, \gamma\})$, $K_2(\{\delta\})$ and $K_2(\{\alpha, \beta, \gamma, \delta\})$.
(See C.M. Rauszer, *Rough Logic for Multi-Agent Systems.*)

§35 VERIFICATION BY HAND USING LINEAR TIME TEMPORAL LOGIC:

INTRODUCTION; ABSTRACT PROGRAMS; LINEAR TIME TEMPORAL LOGIC; APPLICATION TO PROGRAM VERIFICATION; PROVING PROPERTIES OF MUTEX

R. KUIPER and W. PENCZEK

INTRODUCTION

In the realm of computer science, temporal logics are versions of modal logics (cf. §13, §23 in Volume I and Hughes and Cresswell [1968, 1984]) that enable one to describe the temporal order between states occurring in the execution of a program. Order in time rather than explicit timing is considered. This is a quite convenient level of abstraction for specifying concurrent systems. On the one hand, it is expressive enough to concisely formalise requirements of type "if a message is broadcasted an infinite number of times, then at some time it will be received" - requirements that are notoriously difficult to handle in more classical approaches like Hoare's logic (cf. Apt [1981]). On the other hand, it is abstract enough to avoid real time arithmetic of the sort "duration one plus duration two is less than twice duration three".

Temporal interpretation of the modal operators $\Box$ and $\Diamond$ as "always" and "eventual", respectively, dates back to Prior's *Time and modality* [1957], though no connection with computer science was made at the time.

The first tentative steps in this direction are suggested by Burstall in *Program proving as hand simulation with a little induction* [1974]. In *Formalization of algorithmic reasoning* Kröger [1975] presented an early example of a logic that, although not exactly using the modal operators, can be viewed as a temporal logic applied to computer science. *The temporal logic of programs* by Pnueli [1977] marks the start of the extensive developments in temporal logic as applied in computer science.

Currently, temporal logic is extensively used for specifying and proving properties of programs. Without any aspiration to an exhaustive coverage of the field, some broad trends can be discerned. Two different views of the flow of time are adopted, namely, the linear or the branching towards the future. Verification also is done in two ways, either by hand, i.e., using a proof system, or in an automated manner, i.e., using model checking. This and the next section offer a modest introduction to the application of temporal logic in computer science: the two different types of temporal logic will be presented, exemplified on the two different applications.

The present section starts by introducing a general notion of program and a mutual exclusion algorithm example; this is the basis of discussion in both sections. Linear time temporal logic (LTL) is then used as a framework for verification by hand, i.e., using a proof system. Major advances in this area were made by Manna and Pnueli [1991] and Lamport [1991]. The exposition here is mainly based on Manna and Pnueli's approach [1991, 1993].

In the next section, automatic verification, i.e., the use of model checking, is considered in the framework of the branching time Computation Tree Logic (CTL). The main source used here is the approach developed by Clarke, e.a., [1986] and Emerson [1990].

ABSTRACT PROGRAMS

To avoid irrelevant detail and bias towards concrete programming languages, an abstract notion of program is used in which many different programming concepts can be adequately represented.

Programs as transition systems

Programs are represented by *transition systems*. In this approach, a program P is defined by providing the following:

- $Y = \{y_1, \ldots, y_n\}$ - A finite set of *state variables*. These variables represent whatever is desired to be recorded about a program state, for example, not only data variables but location variables as well. A *state* w is a type consistent interpretation of Y such that each variable $y \in Y$ has a value over its domain. The set of all states is denoted by W.

- Θ - The *initial condition*. This assertion characterizes the initial states.

- T - A finite set of *transitions*. A partial *transition function* $\rho_t : W \rightarrow W$ is associated with each transition $t \in T$. The transition t is *enabled* at the state w (written $en_w(t)$) if there is a state w' such that $w' = \rho_t(w)$. The set of all states reachable from the initial states is denoted by W_R.
- A possibly empty set J of *just transitions*. Anticipating their role in the semantics, their intuitive meaning is that not executing such a transition, if it is continuously enabled during an execution, is disallowed.

Note that concurrency is not part of the programming framework. This is intentional: both logics discussed here represent concurrent transitions by interleaving, i.e., non-deterministic sequential execution. A more detailed notion of program than the logics can handle would not be useful.

The algorithm MUTEX

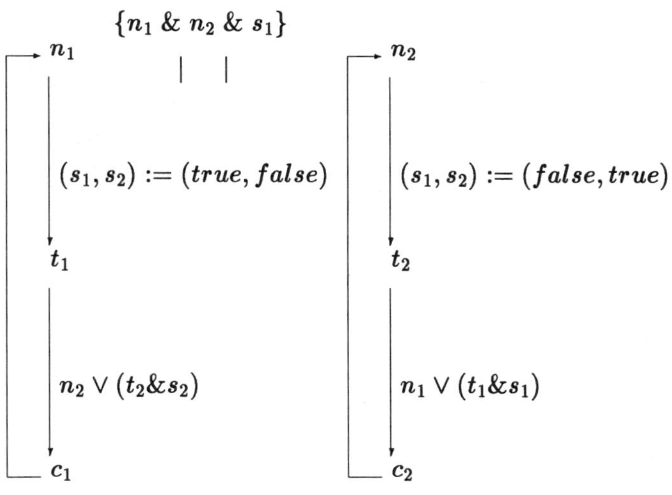

Figure 1: The algorithm MUTEX

As a running example, a simplification of Peterson's algorithm [1983] is used; the main simplification being that locations are assumed to be globally visible rather than encoded in shared variables. To aid intuition, the algorithm is first presented as a picture of two parallel processes P_1 and P_2, in Figure 1. Location names, e.g., n_1 for the *non-critical section* in P_1, are also used as propositions that are true if and only if control in a process resides at such location. On the edges guards, such as $n_2 \vee (t_2 \,\&\, s_2)$, now interpreted as a proposition, and assignments appear, the latter being indicated by ":=". A guard indicates the enabledness of a transition, an assignment its effect on the program variables. The starting state of the algorithm with control residing in n_1 and n_2, s_1 equal to *true* and s_2 equal to *false* is denoted by $\{n_1 \,\&\, n_2 \,\&\, s_1\}$.

The main purpose of the algorithm is to ensure *mutual exclusion*, i.e., that the processes are never in their *critical sections* c_1 or c_2 simultaneously. A secondary aim is to ensure *eventual access*, i.e., that each process, if it desires to, eventually enters its critical section. A process expresses such desire by being in its *trying section*, t_1 or t_2.

As a first approximation, consider an algorithm like the one presented in Figure 1 with, however, nothing written on the transitions except for guards $n_2 \vee t_2$ and $n_1 \vee t_1$ on the respective lower edges. A process can then, from its non-critical section, always freely enter its trying section. From there, it can progress to its critical section if and only if the other process is not in its critical section. It can then go back to its non-critical section. Obviously, this ensures mutual exclusion. A little less obviously, eventual access is not guaranteed. There are two scenarios that always deny, say, process P_2, access to its critical section.

1. P_1 never leaves its critical section.

2. P_1 leaves its critical section, but P_2, although already in t_2, is never quick enough in entering c_2: P_1 passes through n_1 and t_1 and enters c_1 again before P_2 does so.

There is only one simple and natural way to remedy the first possibility: just disallow a process to remain indefinitely in its critical section. The second possibility is a scheduling problem, and less easy to solve. Of course, one could postulate a scheduler taking care of this situation. However, this would require a very clever scheduler that would at least have to remember some details about each process' access history. This is not very realistic. Therefore, in the present exposition, the scheduler is kept simple and the burden is shifted to the algorithm.

The algorithm presented in Figure 1 uses shared signature propositions. Process P_1 has signature s_1 and P_2 has signature s_2. When a process enters the trying section, the corresponding signature is set to *true*, the other one to *false*. If both processes are in their trying section simultaneously, the signature corresponding to the process that arrived there later has value *true*. As the guards on the outgoing edges show, in that case, quite naturally, the other process has priority of access. The next time round, the roles are reversed. Initially, both processes are in their non-critical sections, n_1 and n_2, and the value of s_1 is arbitrarily set to true; it is updated as soon as a process makes a move.

The only assumption necessary to ensure eventual access is that no process stays indefinitely in t_1, c_1, t_2 or c_2 while the outgoing edge is enabled.

Transition system for MUTEX

The concurrent algorithm MUTEX, as shown in Figure 1, can be represented as a transition system as follows. The convention used is that the position of concurrent control in the two components is denoted by the names of the positions, interpreted as Boolean variables that are true if and only if control is at that position. Transitions are denoted by the names of their source and target node.

- $Y = \{n_1, t_1, c_1, n_2, t_2, c_2, s_1, s_2\}$ is the set of state variables. s_1 and s_2 are data variables whereas the others are location variables. Each variable can have just two values: *true* or *false*. Therefore, we write n_1 rather than $n_1 = true$, and $\neg n_1$ rather than $n_1 = false$. For writing the states of the program, we adopt another convention. Only propositions which are *true* are mentioned; for example, the state $\{n_1, n_2, s_1, \neg s_2, \neg t_1, \neg t_2, \neg c_1, \neg c_2\}$ is denoted by $\{n_1, n_2, s_1\}$.
- $\Theta = n_1 \ \& \ n_2 \ \& \ s_1 \ \& \ \neg c_1 \ \& \ \neg c_2 \ \& \ \neg t_1 \ \& \ \neg t_2 \ \& \ \neg s_2$ is the initial condition.
- $T = \{n_1t_1, t_1c_1, c_1n_1, n_2t_2, t_2c_2, c_2n_2\}$ is the set of transitions. In the definition of the transition function, the star $*$ is used to denote any matching state elements.

  $\rho_{n_1t_1}(\{n_1, s_1, *\}) = \{t_1, s_1, *\}$. This should be read as: any state of the form $\{n_1, s_1, ...\}$ is transformed by the transition n_1t_1 to the

state of the form $\{t_1, s_1, ...\}$, where ... denotes the same elements.
$\rho_{n_2t_2}(\{n_2, s_2, *\}) = \{t_2, s_2, *\}$,
$\rho_{t_1c_1}(\{t_1, n_2, *\}) = \{c_1, n_2, *\}$, $\rho_{n_1t_1}(\{n_1, s_2, *\}) = \{t_1, s_1, *\}$,
$\rho_{t_1c_1}(\{t_1, t_2, s_2\}) = \{c_1, t_2, s_2\}$, $\rho_{c_1n_1}(\{c_1, *\}) = \{n_1, *\}$,
$\rho_{n_2t_2}(\{n_2, s_1, *\}) = \{t_2, s_2, *\}$, $\rho_{t_2c_2}(\{t_2, n_1, *\}) = \{c_2, n_1, *\}$,
$\rho_{t_2c_2}(\{t_1, t_2, s_1\}) = \{t_1, c_2, s_1\}$, $\rho_{c_2n_2}(\{c_2, *\}) = \{n_2, *\}$.

- $J = \{t_1c_1, c_1n_1, t_2c_2, c_2n_2\}$. There are 10 states in the set W_R.

LINEAR TIME TEMPORAL LOGIC

Linear time temporal logic, LTL, can be used to describe an execution of a program, i.e., a sequence of states in which the sequential order intuitively reflects time. For reasons of convenience, sequences are taken to be infinite.

LTL and modal logic

There is a close connection between LTL and modal logic with the operators $\Box$ and $\Diamond$ as introduced in §13 and §23 in Volume I.

Sequences of states can be represented as modal models in which worlds together with the valuation function represent the states and a linear, serial ("endless") and discrete relation R represents the sequential ordering. "Discrete" means that between any two worlds there are only finitely many other worlds. Interpreted over such models the intuitive meaning of the modal operator $\Box$ is "always" and that of $\Diamond$ "eventual".

Recall that the simplest modal logic K results from adding one axiom to classical propositional logic:

$$\Box(\phi \supset \psi) \supset (\Box\phi \supset \Box\psi)$$

and one inference rule:

$$\text{if} \quad \vdash \phi, \quad \text{then} \quad \vdash \Box\phi.$$

K is complete over (Kripke-) models with unrestricted relation R. (See section 13 in Volume I.)

The following axioms are added to restrict R to be weakly linear, serial and discrete. A relation is weakly linear if it is reflexive, transitive and connected (i.e., for each three worlds w, u, v, if wRu and wRv, then uRv or $u = v$ or vRu). The resulting logic is called $S4.3Dum$.

- $\Box\phi \supset \phi$ - reflexivity,
- $\Box\phi \supset \Box\Box\phi$ - transitivity,
- $\Box(\Box\phi \,\&\, \phi \supset \psi) \vee \Box(\Box\psi \,\&\, \psi \supset \phi)$ - connectedness,
- $\Box\phi \supset \Diamond\phi$ - seriality,
- $\Box(\Box(\phi \supset \Box\phi) \supset \phi) \supset (\Diamond\Box\phi \supset \phi)$ - discreteness.

If R were totally connected (each two worlds related), linearity would hold. Unfortunately, no characterizing axiom exists. However, $S4.3Dum$ is complete over linear, serial and discrete models as well.

To express state changes corresponding to one step in a computation, it is required that the immediate successor of a state can be referred to explicitly. Modal models such as those just described do not have this property, as the relation does not distinguish between the immediate successor of a world and one that is further away. This can be remedied by defining a new relation, R_X, that just relates each world to its immediate successor. Let $R' = R - id_W$, where id_W stands for the identity relation on W. R_X is then equal to $R' - R' \circ R'$. A new modal operator X is introduced to correspond to the relation R_X, i.e., X refers to the immediate successor of a world, providing the desired additional expressive power. X is pronounced "next". In this manner, LTL can be understood as a multi-modal logic, with modalities corresponding to two relations, R and R_X. (See Goldblatt's *Logics of Time and Computation* [1992].)

It would be possible to conduct the formal definition of LTL, and in the next section CTL, along these lines. However, in the literature usually advantage is taken of the sequence character of the LTL models. This is the approach adopted in the further exposition (see clause S3 in Definition 35.1).

Syntax of Propositional LTL

Apart from the next operator, some additional expressiveness turns out to be desirable for program specification and verification. This prompted the introduction of the binary U, intuitively stating that some property is preserved until some other property is fulfilled. To facilitate the comparison with CTL in the next section, G is used instead of $\Box$ and F instead of $\Diamond$.

The set of LTL *formulas* is defined inductively, given a set PV (of Propositional Variables, including special ones *true* and *false*).

S1. every member of PV is a formula;

S2. if ϕ and ψ are formulas, then so are $\neg\phi$ and $\phi \vee \psi$;

S3. if ϕ, ψ are formulas, then so are $X\phi$ and $(\phi U \psi)$.

The connectives & and $\supset$ are defined in the standard way. The operators F (eventually) and G (always) can be derived as follows:

- $F\phi := trueU\phi$;
- $G\phi := \neg F \neg \phi$.

Semantics of Propositional LTL

DEFINITION 35.1.: Let $M = \langle p, V \rangle$ be a *model*, i.e., $p = (w_0, w_1, \ldots)$ and V assigns to every w_i a set of propositional variables. We denote the suffix $(w_i, w_{i+1}, \ldots)$ of p by p_i.

S1. $w_i \models q$ iff $q \in V(w_i)$, for $q \in PV$;

S2. $w_i \models \neg\phi$ iff not $w_i \models \phi$;

$w_i \models \phi \vee \psi$ iff $w_i \models \phi$ or $w_i \models \psi$;

S3. $w_i \models X\phi$ iff $w_{i+1} \models \phi$;

$w_i \models (\phi U \psi)$ iff there is $k \geq i$ such that $w_k \models \psi$ and for all j, $i \leq j < k$, $w_j \models \phi$.

A LTL formula ϕ is said to be *true in* a model M (written $M \models \phi$) iff $w_0 \models \phi$. And ϕ is *valid* iff for every model M, $M \models \phi$. This notion of validity is called *anchored validity*.

Proof system for Propositional LTL

The following proof system is sound and complete for Propositional LTL (see Goldblatt [1992]).

DEFINITION 35.2. (Axioms and proof rules for Propositional LTL):

A1. $G(\phi \supset \psi) \supset (G\phi \supset G\psi)$

A2. $X(\neg\phi) \sim \neg X\phi$

A3. $X(\phi \supset \psi) \supset (X\phi \supset X\psi)$

A4. $G\phi \supset \phi \,\&\, XG\phi$

A5. $G(\phi \supset X\phi) \supset (\phi \supset G\phi)$

A6. $\phi U\psi \sim \psi \vee (\phi \,\&\, X(\phi U\psi))$

R1. If ϕ is an instance of a propositional tautology, then $\vdash \phi$.

R2. If $\vdash \phi$ and $\vdash \phi \supset \psi$, then $\vdash \psi$.

R3. If $\vdash \phi$, then $\vdash G\phi$.

Next, LTL is extended to its first order version.

DEFINITION 35.3. (Syntax of First Order LTL):

Basic symbols:

- flexible variables $x, y, z \in Flex$ and vectors of these,
- (flexible) propositional variables $p, q \in Prop$,
- rigid variables $u, v, w \in Rig$,
- $Var = Flex \cup Rig$,
- function symbols $F \in Func$,
- predicate symbols $P \in Pred$.

Classical operators:

- Boolean connectives $\neg, \vee$ and derived operators like $\&$ and $\supset$,

- first order quantifier $\forall$ and, derived, $\exists$, (quantifiers are applied to rigid variables only);
- equality predicate $=$.

Modal operators:

- U (until), X (next) and derived ones like G (always) and F (eventually).

Formulas are then constructed from these in the obvious manner.

Semantics of First Order LTL

DEFINITION 35.4. (First Order LTL Model):
$M = \langle p, D, V \rangle$ is a first order LTL *model* if

- $p = (w_0, w_1, \ldots)$ is a sequence of worlds,
- D is a non-empty set, called the domain,
- $V = (I, V_{Rig}, V_{Prop}, V_{Flex})$, where
  - I is the interpretation of function and predicate symbols such that $I(F) : (D^n \rightarrow D)$ for any n-ary function symbol F, and $I(P) : D^n \rightarrow \{true, false\}$ for any n-ary predicate symbol P;
  - $V_{Rig} : Rig \rightarrow D$;
  - $V_{Prop} : \{w_0, w_1, \ldots\} \rightarrow (Prop \rightarrow \{true, false\})$;
  - $V_{Flex} : \{w_0, w_1, \ldots\} \rightarrow (Flex \rightarrow D)$.

DEFINITION 35.5. (First Order LTL Evaluation):
Evaluation of terms. Let the value of a term t in a model M at a world w be denoted by $t|_w^M$.

- For t a flexible proposition q or flexible variable x or rigid variable u: $q|_w^M = V_{Prop}(w)(q)$ or $x|_w^M = V_{Flex}(w)(x)$ or $u|_w^M = V_{Rig}(u)$,
- For a k-ary function F: $F(t_1, \ldots, t_k)|_w^M = I(F)(t_1|_w^M, \ldots, t_k|_w^M)$.

Evaluation of formulas. Let the value of a formula ϕ in a model M at the world w_i be denoted by $M, w_i \models \phi$.

- For a flexible proposition q: $M, w_i \models q$ iff $V_{Prop}(w_i)(q) = true$.
- For a k-ary predicate symbol P: $M, w_i \models P(t_1, \ldots, t_k)$ iff $I(P)(t_1|^M_{w_i}, \ldots, t_k|^M_{w_i}) = true$.
- For a next time application: $M, w_i \models X\phi$ iff $M, w_{i+1} \models \phi$.
- For an until application: $M, w_i \models (\phi U \psi)$ iff there is $k \geq i$ such that $M, w_k \models \psi$ and for all j, $i \leq j < k$, $M, w_j \models \phi$.
- "Eventual" and "always" are derived operators:

  $F\phi = (trueU\phi)$, $G\phi = \neg F \neg \phi$.
- For application of the universal quantifier: $M, w \models (\forall u)\phi$ iff for every model M' which is just like M except possibly that

  $V_{Rig}(u) \neq V'_{Rig}(u)$, $M', w \models \phi$,
- The existential quantifier is a derived operator: $\exists u[\phi] = \neg \forall u[\neg \phi]$.
- Boolean connectives and the equality predicate are interpreted as usual.

First order LTL is incomplete

At this point, one would expect to find a sound and complete proof system for first order LTL. However, it turns out that there is no finite complete proof system for first order LTL. At first sight, this may seem surprising because of the following, too superficial, considerations.

First order modal logic has a finite complete proof system (see section 23 in Volume I). One might be led to think that this is because it is a "combination" of two other logics that both have finite complete proof systems, namely, (non-modal) first order logic and propositional modal logic. However, as has just been shown, propositional LTL also has a finite proof system and the claim is that the first order extension no longer admits a finite proof system. Apparently, the idea that finite axiomatizability is preserved over combining logics is a fallacy. It will be argued that it is precisely the combination of properties of both logics involved that in the present situation precludes the existence of a finite proof system. The exposition is based on the work of Szalas [1986]; an infinite complete proof system for a slightly different variant of the logic has been given by Szalas [1987].

THEOREM 35.6.: Finite proof systems for first order LTL that are sound and complete do not exist.

PROOF: We show that there is no finite and complete axiomatization of first-order linear time temporal logic provided the language contains a zero-argument function symbol 0, an unary function symbol s and two-argument function symbols $+$ and $*$. The idea is to prove incompleteness by defining a finite set of formulas FN such that the set of its semantic consequences cannot be deduced from any finite set of axioms and inference rules for first order LTL. The following set of formulas FN is used:

N1. $\forall u\ [0 \neq s(u)]$,

N2. $\forall u \forall w\ [s(u) = s(w) \supset u = w]$,

N3. $\forall u\ [u + 0 = u]$,

N4. $\forall u \forall w\ [u + s(w) = s(u + w)]$,

N5. $\forall u\ [u * 0 = 0]$,

N6. $\forall u \forall w\ [u * s(w) = u * w + u]$,

N7. $\forall u\ [z = 0\ \&\ G(z = u \supset X(z = s(u)))]$.

N1 - N6 are the Peano axioms. N7 says that at the beginning state $z = 0$ and for each state if the value of z is equal to u, then the value of z at the successor state is $s(u)$.

One can show that there is a model $M = \langle p, D, V \rangle$ of the set FN, in which the interpretation of $0, s, +$ and $*$ is the standard interpretation (see section 24 in Volume I).

Next, one can prove that if $M = \langle p, D, V \rangle$ is a model of FN, then its interpretation of function and predicate symbols I (which is part of V) is isomorphic to the standard interpretation $\mathcal{N}$ of the natural numbers (see §24 in Volume I). N7 is needed to show that for every element d of D there exists a natural number i such that $d = s^i(0)$.

Note that the truth of a formula ϕ in the standard model $\mathcal{N}$ of natural numbers (written $\mathcal{N} \models \phi$) is equivalent to the validity of the first-order LTL formula N1 & ... & N7 $\supset \phi$.

Suppose that we are given a finite, sound and complete proof system of first-order LTL. This means that $\mathcal{N} \models \phi$ iff $\vdash$ N1 & ... & N7 $\supset \phi$. Consequently, each formula ϕ true in $\mathcal{N}$ can be deduced from the finite proof system and the finite set of formulas FN. This is a contradiction with Gödel's incompleteness theorem (see section 24 in Volume I). □

It can now be seen why the combination of first order logic and temporal logic leads to a logic that is not finitely axiomatizable: precisely the combined use of a next operator and predicates, as is most apparent in formula N7, allows one to mimick arithmetic in the logic.

To see that the above incompleteness result has real significance for proving properties of programs, consider the program given in Example 35.9.

APPLICATION TO PROGRAM VERIFICATION

LTL program semantics

Intuitively, an execution of a program can be viewed as a sequence of states where each program transition changes a state to its successor. In this setting concurrent steps by parallel components are represented by interleaving. Sequences can easily be made infinite by, for instance, repeating the last state in the case of finite computations.

An LTL model greatly resembles such an infinite state sequence. Therefore, the semantics of a program is given directly as a temporal formula; no intermediate semantics is provided. The set of models satisfying this formula is then taken to represent the set of all possible executions of the program.

To express enabledness of a transition in the logic, it is assumed that the language contains the following assertion:

- $en(t)$ for $t \in T$ such that $w \models en(t)$ iff $en_w(t)$.

To express the effect of transitions on the values of the state variables $y_1, \ldots, y_n$, the partial transition function $\rho_t : W \to W$, defined on states,

is rewritten as a partial function $f_t : D^n \rightarrow D^n$, defined on values of the vector $\vec{y} = <y_1, \ldots, y_n>$.
$f_t(<d_1, \ldots, d_n>) = <d'_1, \ldots, d'_n>$ iff

$$\rho_t(\{(y_1, d_1), \ldots, (y_n, d_n)\}) = \{(y_1, d'_1), \ldots, (y_n, d'_n)\}.$$

Two abbreviations are defined that enable the writing of LTL program semantics in a concise manner. The first one expresses the fact that a transition is performed as well as its effect.

- *taken*$(t) := \exists\ \vec{u}\ [\ \vec{y} = \vec{u}\ \ \&\ X(\ \vec{y} = f_t(\vec{u}))]$. By convention, $\vec{y} = f_t(\vec{u})$ evaluates to *false* where f_t is not defined. Thus, if t is not enabled, *taken*(t) is *false*. (Remember that u is a rigid variable and y a flexible one.)

This formula provides some insight into the motivation for two of the extensions that have been made to propositional modal logic. The wish to express the state change in one transition is the most important reason to include the next time operator in the logic. The need to express the effect of a transition on a potentially infinite number of states is the main reason to include rigid variables and existential quantification over these.

The second abbreviation expresses that no state change occurs.

- *idle* $:= \exists\ \vec{u}\ [\ \vec{y} = \vec{u}\ \ \&\ X(\ \vec{y} = \vec{u})]$.

The LTL live program semantics can now easily be expressed in terms of the abbreviations, $\oplus$ denoting "exclusive or".

DEFINITION 35.7. (LTL *live semantics*): The LTL live semantics of a program P is defined as follows:

$$\chi_{PL} = \Theta\ \&\ G((\bigoplus_{t \in T} taken(t)) \vee (\neg \bigvee_{t \in T} en(t)\ \&\ idle))$$

The first conjunct is simply the initial condition to be satisfied by any starting state. The second conjunct consists of two parts: enforcing the interleaved execution of one transition at the time as long as such transitions are enabled and enforcing repetition of the same state if no transitions are enabled anymore. The semantics is called "live" because of this second conjunct, only enforcing some progress as long as this is possible.

Proof rules for first order LTL, the program part

When a *logic* is used to describe properties of *programs* that manipulate *data*, three different parts can be distinguished accordingly in the proof system (see Manna and Pnueli [1991]):

- the logic part - to deal with logical truth,
- the (data) domain part - to deal with domain properties, and
- the program part - to deal with program properties.

The program part is used to remove the references to the program from a formula, i.e., an expression of the form $P \vdash \phi$, where P is the program, is reduced to an expression of the form $\vdash \psi$, where ψ is a LTL-formula. The logic and domain part are then used to deduce the truth of the resulting formula ψ.

Because of the incompleteness incurred for first order logics over interpreted domains, only relative completeness can usually be achieved, i.e., after applying the program part, the truth value of some of the resulting formulas has to be given by an oracle. (In the case of first order temporal logic, the situation is even worse. As argued above, even if no interpreted domain is used, first order LTL already does not admit of a finite proof theory.)

The formulas to be decided by the oracle could now be temporal formulas, expressing not only properties of the program but a complicated mixture of these and properties of the data domain and the structure of the models. It turns out that in the present case (programs as transition systems and LTL) the use of the oracle can be limited to non-temporal formulas only: a program proof system can be given that enables reducing formulas to this form. The general ideas and a discussion of completeness issues can be found in Manna and Pnueli, *Completing the temporal picture*, [1991].

Properties of programs can be classified according to, very roughly and intuitively speaking, the intricateness of their temporal interdependencies. At the lower end of the scale are the invariance properties, just stating that some assertion holds in every state of an execution, i.e., of form $G\phi$. The next level consists of progress properties, of form $G(\phi \supset F\psi)$. For each level there are matching proof rules. A complete hierarchy of properties and the corresponding sets of proof rules are presented in Manna and Pnueli [1991]. We limit ourselves to somewhat simplified versions of the most relevant rules for just the two cases mentioned; the present exposition is based on Manna and Pnueli [1993].

There is one unfortunate aspect here. In order to keep this introduction as simple as possible, especially with regard to the connection with CTL made in the next section, we would have liked to base the whole exposition on the simple semantics given above in Definition 35.7. To our best knowledge, however, proof rules of the kind we are interested in are only given for a slightly more elaborate semantics.

The idea is as follows. On the one hand, this new semantics is less restrictive, namely, also allowing states to be repeated if there still are enabled transitions, so-called stuttering, as discussed by Lamport [1983] in *What good is temporal logic?*. On the other hand, it can be made more restrictive, in that a set of *just transitions* is provided as part of each program definition which are not allowed to remain unused if, from some point on, continuously enabled.

This semantics can also be given as an LTL formula - straightforwardly, by adding an idling transition and a conjunct enforcing justice. This leads to the following definition, assuming a set J of just transitions.

DEFINITION 35.8. (LTL *just semantics*):
The LTL just semantics of a program P is defined as follows:

$$\chi_{PJ} = \Theta \ \& \ G((\bigoplus_{t \in T} taken(t)) \oplus idle) \ \& \ \bigwedge_{t \in J} FGen(t) \supset GFtaken(t).$$

The first conjunct is simply the initial condition to be satisfied by any starting state. The second conjunct consists of two parts, disjuncts, enforcing the interleaved execution of one transition or enforcing repetition of the state respectively. The semantics is called "just" because of the third conjunct, called the *justice requirement*, which enforces that a just transition, whenever it becomes continuously enabled, cannot be ignored forever.

EXAMPLE 35.9.: With respect to completeness, consider program P:
$z := 0$; while true do $z := s(z)$ od.
Let $M = \langle p, D, V \rangle$ be the model representing the execution of the program P, i.e., satisfying the live semantics, where $p = (w_0, w_1, \ldots)$,

$D = \{0, s(0), s^2(0), \ldots\}$, $V_{Flex}(w_i)(z) = s^i(0)$ and I is the standard interpretation of $0, s, +$ and $*$. Then, $M \models$ N1 & ... & N7. Therefore, there is no finite proof system for programs from which each formula true in M (i.e., true for the program P) could be deduced.

Example: To get some feel for the differences between the two semantics, compare the following possibilities for the example program MUTEX.

- If the live semantics is used, one process in the example program can remain in its non-critical section forever, as the other can go on making moves, thus satisfying the liveness requirement. No process, however, can remain forever in its critical section, as then the other one would get stuck at its trying section and the liveness requirement be violated. From the latter observation, it can easily be seen that with the live semantics, the algorithm guarantees each process eventual access to its critical section from its trying section.

- Consider the just semantics with an empty justice set. No process is compelled to do any move, so nothing about eventual access can be guaranteed.

- In order to guarantee eventual access, it is necessary to make sure that a process is not denied access because the other process remains forever in its critical section, but also that the process, once it is in its trying section, does not forever ignore the possibility of entering its critical section. This can be achieved by putting the outgoing transitions from the trying and the critical sections into the justice set.

There is a subtle difference between the live semantics and the one with the nonempty justice set as defined above: the former allows remaining forever in either of the non-critical sections, but not in both; the latter also allows this quite natural possibility.

It is with respect to the just semantics that the following proof rules are constructed. A formula is always proved with respect to a program. Rather than incorporating (the name of) this program in the proof rules, it is mentioned just once and then left implicit.

Transition properties

The fact that the execution of a transition t in any state where a non-temporal assertion ϕ holds, if enabled, yields a state where an assertion ψ holds is written as $\{\phi\}t\{\psi\}$. The aim is to define and prove this formally.

Recall that the effect of executing a transition t can be expressed in LTL as:

- $taken(t) := \exists \vec{u}\ [\ \vec{y} = \vec{u}\ \ \&\ \ X(\ \vec{y} = f_t(\vec{u}))]$.

Therefore,

- $\{\phi\}t\{\psi\}$ can be defined as $(\phi\ \&\ taken(t)) \supset \psi$.

Because of the presence of the X, this is already a temporal formula, and therefore outside the scope of a non-temporal first order proof system. The temporal aspect of this kind of formula is so slight, however, that with some effort they can still be verified as non-temporal formulas; namely, by rewriting the transition function as a first order assertion using primed and unprimed versions of the state variables $\vec{y}$. The notation $\phi(\vec{y})$ is used as a reminder that $\vec{y}$ is the vector of (flexible) state variables ϕ addresses. $\phi[\vec{y'} / \vec{y}]$ denotes syntactic replacement of unprimed by primed variables. It can be easily seen that

- $\{\phi\}t\{\psi\}$ iff $\phi(\vec{y})\ \&\ \exists \vec{u}\ [\ \vec{y} = \vec{u}\ \ \&\ \ \vec{y'} = f_t(\vec{u})] \supset \psi[\vec{y'}\ /\ \vec{y}\ ]$.

As before, by convention $\vec{y'} = f_t(\vec{u})$ evaluates to *false* where f_t is not defined. Thus, if t is not enabled, $\{\phi\}t\{\psi\}$ is true.

A set T of transitions *yields* ψ *from* ϕ (written $\{\phi\}T\{\psi\}$) if $\{\phi\}t\{\psi\}$ holds for all transitions $t \in T$.

Invariance properties

Invariance properties are properties of form $G\phi$, where ϕ is a non-temporal formula. The intuition is that the property ϕ holds in every state throughout every execution of the program.

The basic proof rule for invariance is INV:

I1 $\Theta \supset \phi$

I2 $\{\phi\}T\{\phi\}$

$$\frac{}{G\phi}$$

The idea behind this rule is that I1 ensures that ϕ holds at the beginning of every computation, and that I2 ensures that ϕ remains true.

As usual in case of inductive proofs, it might be necessary to choose a stronger invariant assertion to enable the induction. To obtain then the original invariant again, the rule for *monotonicity* is introduced.

MON

M1 $\phi' \supset \phi$

M2 $G\phi'$

$$\frac{}{G\phi}$$

Response properties

Response properties are properties of the form $G(\phi \supset F\psi)$. In the special fortuitous cases where the desired property ψ is achieved by executing only one just transition in a world where ϕ holds, the following rule applies. Again, nontemporal assertions only are considered.

RESP

R1 $G(\phi \supset (\psi \vee \gamma))$

R2 $\{\gamma\}T\{\psi \vee \gamma\}$

R3 $\{\gamma\}t_h\{\psi\}$, for some $t_h \in J$

R4 $G(\gamma \supset en(t_h))$

$$\frac{}{G(\phi \supset F\psi)}$$

R1 and R2 ensure that from the start, if the desired property is not achieved "through luck", then at least the intermediate assertion γ is preserved. R3 ensures that there is a helpful just transition that yields ψ from γ. R4 ensures that this helpful transition is always enabled and hence must be taken at some point.

Generally, several such helpful steps are required to achieve ψ, demanding repeated application of RESP with several intermediate assertions γ_i. The following rule enables one to apply this strategy more concisely. Let γ_0 be ψ.

CHAIN

C1 $G(\phi \supset (\bigvee_i \gamma_i)),\ i \geq 0$

C2 $\{\gamma_i\}T\{\bigvee_{j\leq i}\gamma_j\},\ i > 0$

C3 $\{\gamma_i\}h_i\{\bigvee_{j<i}\gamma_j\}, h_i \in J,\ i > 0$

C4 $G(\gamma_i \supset en(h_i)),\ i > 0$

$G(\phi \supset F\psi)$

The idea behind this rule is that in the course of an execution intermediate assertions γ_i with progressively lower indices will hold.
C1 ensures that at least one such assertion follows from ϕ.
C2 ensures that no transition raises the index.
C3 ensures that from each intermediate assertion one with a lower index can be established by executing a just transition.
C4 ensures that such a transition is enabled and thus will eventually be taken because of justice.

PROVING PROPERTIES OF MUTEX

Recall the notation conventions introduced when the example program MUTEX was presented, especially the use of location names as propositions indicating the position of control in the respective processes.

Mutual exclusion

To prove: $G\neg(c_1\&c_2)$. Proof: By application of INV. The only relevant transitions to consider are t_ic_i ($i = 1, 2$), as only these might falsify the invariant. Without loss of generality, choose t_1c_1.

I1 $n_1\&n_2\&s_1 \supset \neg(c_1\&c_2)$, evident.

I2 $\{\neg(c_1\&c_2)\&t_1\&(n_2 \vee (t_2\&s_2))\}t_1\{\neg(c_1\&c_2)\}$, evident.

$G\neg(c_1\&c_2)$ then follows by INV.

Remark: If the condition on t_1c_1 had just been s_2 rather than $t_2\&s_2$, strengthening of the invariant would have been necessary, by $t_1\&s_2 \supset \neg c_2$. Similarly, for t_2c_2.

Eventual access

To prove: $G(t_1 \supset Fc_1)$. Proof: By application of CHAIN. The following intermediate assertions and corresponding helpful transitions are used.

$\gamma_3 = t_1 \& t_2 \& s_1$, $h_3 = t_2c_2$ (establishing γ_2);
$\gamma_2 = t_1 \& c_2 \& s_1$, $h_2 = c_2n_2$ (establishing γ_1);
$\gamma_1 = (t_1 \& n_2 \& s_1) \vee (t_1 \& t_2 \& s_2)$, $h_1 = t_1c_1$ (establishing γ_0);
$\gamma_0 = c_1$.

C1 holds because all reachable states with t_1 are present in the disjunction. INV can be used to derive that the states $t_1 \& n_2 \& s_2$ and $t_1 \& c_2 \& s_2$ are not reachable.

C2 holds as all enabled non-helpful transitions do not raise the index. In fact, only one such transition exists, namely for γ_1, where n_2t_2 does not give rise to a lower index but leaves γ_1 true.

C3 holds because the helpful transitions, as shown above, lead to a lower indexed intermediate assertion.

C4 holds because only h_3 and h_1 have non-trivial enabledness conditions – these are evidently satisfied.

$G(t_1 \supset Fc_1)$ then follows from CHAIN.

EXERCISES

35.1. List all reachable states of the algorithm MUTEX.

35.2. Prove that the modal logic K with the axiom:

- $\Box\psi \supset \psi$ is sound over models with R reflexive,
- $\Box\psi \supset \Box\Box\psi$ is sound over models with R transitive,
- $\Box\psi \supset \Diamond\psi$ is sound over models with R serial.

35.3. Prove that the proof system for propositional LTL is sound, i.e., each axiom is valid and the proof rules preserve the validity.

35.4. Prove that the following LTL formula is valid:

$$F(\varphi \,\&\, \psi) \supset (F\varphi \,\&\, F\psi).$$

35.5. Find a model falsifying the following LTL formula:
$$(F\varphi \,\&\, F\psi) \supset F(\varphi \,\&\, \psi).$$

35.6. Write the LTL live and the LTL just semantics of the program P in Example 35.9.

35.7. Prove that the formula N7 is not a semantic consequence of the LTL just semantics of the program P in Example 35.9.

35.8. Temporal logic is quite amenable to abstract and incremental specification. "Abstract" means that descriptions are in terms of the input/output interface, i.e., not using concepts like a queue if this is a strictly internal one. "Incremental" means that descriptions can be strengthened by extending rather than rewriting them. Consider specifying the following instantiations of a buffer in LTL. Use variables *in* and *out* with values m from a domain of messages to describe the input and output channel. Assume that messages are unique and inputted at most once.

- Describe a buffer that guarantees that any message that appears on the output has been inputted sometime earlier. This is sometimes called " absence of miracles" (Dijkstra).
- Strenghthen this discription to also guarantee that as long as the buffer is not empty, it goes on to produce output.
- Strengthen this further to also guarantee that every message that is put into the buffer will also appear on the output sometime later.
- Finally, provide a condition that guarantees that the output is in the same order as the input.

35.9. From the specification of the buffer in the previous exercise, especially the absence of miracles, it should be clear that it is awkward that LTL only possesses future-time operators.

- Define the past-time counterparts of the existing operators, adapting the semantic model where necessary.
- Rewrite the specifications from the previous exercise, using the new operators.

35.10. The difference between a live semantics and a just one can be viewed as a difference in scheduling properties, i.e., in what is guaranteed about eventual execution of certain enabled transitions. A scheduler corresponding to a just semantics is stronger than one for a live semantics in that it guarantees eventual execution of any just transition that becomes continuously enabled from some point on. Intuitively, this means that a scheduler has to go on trying to execute each just transition at arbitrary moments.

The present exercise considers even stronger schedulers that guarantee the eventual execution of certain transitions, called *fair* transitions, that are enabled an infinite number of times. Therefore, a fair scheduler not only has to keep on trying, but also has to make sure that it does not only try at the wrong moments, i.e., when the transition is disabled.

- Take advantage of the new scheduling regime to weaken the guards on the transitions of MUTEX in such a manner that justice does not guarantee eventual access, but fairness does. Hint: consider whether the signature variables s_1 and s_2 are still necessary.
- Adapt the Response and Chain rules to reflect the new scheduling regime.
- Prove eventual access, i.e., $G(t_1 \supset Fc_1)$, for the new version of MUTEX using the new Chain rule.

§36 AUTOMATED VERIFICATION USING BRANCHING TIME LOGIC:

COMPUTATION TREE LOGIC (CTL); APPLICATION TO PROGRAM VERIFICATION; THE MUTEX EXAMPLE

R. KUIPER and W. PENCZEK

COMPUTATION TREE LOGIC (CTL)

The descriptions given by LTL are strictly limited to sequences. Thus, only properties of single executions can be described. The natural extension would be a logic that at any state also describes possible branching. Consequently, properties relating executions could also be described, such as the possible existence of several alternative extensions to a sequence at some state. This is exactly what Computation Tree Logic*, CTL*, does. The CTL* part dealing with sequences is identical to LTL; it is therefore a (strict) extension of LTL (see Stirling [1987]). For the sake of efficiency, a restricted version of CTL* is used for model checking; it is called CTL. As this restriction affects not only parts of the extension but also LTL-like parts, CTL and LTL are not extensions of each other.

The language of CTL* is composed of state and path formulas. A world together with the valuation corresponds to a state in an execution. Therefore, formulas interpreted over worlds are called *state formulas*. Formulas interpreted over sequences of worlds are called *path formulas*.

Syntax of CTL*

Let PV be a set of Propositional Variables. The set of *state formulas* and the set of *path formulas* are defined inductively:

S1. every member of PV is a state formula,

S2. if ϕ and ψ are state formulas, then so are $\neg\phi$ and $\phi \,\&\, \psi$,

S3. if ϕ is a path formula, then $E\phi$ is a state formula,

P1. any state formula ϕ is also a path formula,

P2. if ϕ, ψ are path formulas, then so are $\phi \,\&\, \psi$ and $\neg\phi$,

P3. if ϕ, ψ are path formulas, then so are $X\phi$ and $(\phi U \psi)$.

E is a path quantifier with the intuitive meaning: there is a path. X is the next step operator and U denotes Until. G and F are derived operators as in the case of LTL. The universal path quantifier is defined as a derived operator as follows: $A\phi := \neg E \neg \phi$.

Semantics of CTL*

Let $M = \langle W, R, V \rangle$ be an anchored *model*, i.e., W is a non-empty set of worlds, $R \subseteq W \times W$ is a binary relation on W which is total (each world has at least one successor), $V : W \times PV \longrightarrow \{true, false\}$ is a valuation function and there is one *root* world, denoted by v_0.
A *forward full path* starting at w_0 is an infinite sequence $p = (w_0, w_1, \ldots)$ of worlds such that $w_i R w_{i+1}$ for each $i \geq 0$. We denote the suffix $(w_i, w_{i+1}, \ldots)$ of p by p_i.

S1. $w \models q$ iff $V(w, q) = true$, for $q \in PV$,

S2. $w \models \neg\phi$ iff not $w \models \phi$,

$w \models \phi \,\&\, \psi$ iff $w \models \phi$ and $w \models \psi$,

S3. $w \models E\phi$ iff $p \models \phi$ for some path p starting at w,

P1. $p \models \phi$ iff $w_0 \models \phi$, for any state formula ϕ,

P2. $p \models \phi \,\&\, \psi$ iff $p \models \phi$ and $p \models \psi$,

$p \models \neg\phi$ iff not $p \models \phi$,

P3. $p \models X\phi$ iff $p_1 \models \phi$,

$p \models (\phi U \psi)$ iff there is $i \geq 0$ such that $p_i \models \psi$ and for all j, $0 \leq j < i$, $p_j \models \phi$.

A CTL* state formula ϕ is said to be *true in* a model M (written $M \models \phi$) iff $M, v_0 \models \phi$.

Example: Consider two programs which can perform some action a followed by an action b or c. Let the first program fork at the root, the second at the state reached after executing the first action a. Let a state be denoted by the last action executed before it is reached. The execution sequences ab and ac are the same for both programs; hence, if the program behaviour is described in terms of a, b and c only, LTL cannot distinguish between them. However, the formula $AX(EXb \ \& \ EXc)$ distinguishes between the two.

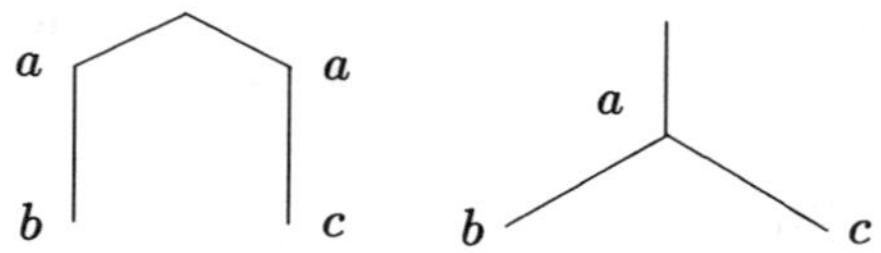

Considering that any LTL formula ϕ has a CTL* counterpart, namely, $A\phi$, the above example allows one to conclude that CTL* is a strict extension of LTL.

Computation Tree Logic, CTL

The language of CTL is the syntactic restriction of the language of CTL* such that only one single linear time operator (F, G, X or U) can follow a path quantifier (A or E). A formal syntax and semantics of CTL is given below.

Syntax of CTL

The set of CTL *formulas* is defined inductively:
S1. every member of PV is a formula,
S2. if ϕ and ψ are formulas, then so are $\neg\phi$ and $\phi \ \& \ \psi$,
S3′. if ϕ, ψ are formulas, then so are $A(\phi U \psi)$, $E(\phi U \psi)$ and $EX\phi$.
The semantics is the subset of the semantics of CTL* concerning CTL formulas. For example, the semantics of $A(\phi U \psi)$ is as follows:

> $w_0 \models A(\phi U \psi)$ iff for every forward full path $p = (w_0, w_1, \ldots)$ there is an $i \geq 0$ such that $w_i \models \psi$ and for all j, $0 \leq j < i$, $w_j \models \phi$,

The derived basic modalities are:

- $EF\phi := E(trueU\phi)$,
- $AF\phi := A(trueU\phi)$,
- $EG\phi := \neg AF\neg\phi$,
- $AG\phi := \neg EF\neg\phi$,

- $AX\phi := \neg EX\neg\phi$.

Example: The restriction of CTL* to CTL affects the LTL part in such a manner that CTL no longer contains LTL. For example, the LTL formula $GF\phi$ can be expressed in CTL* as the formula $AGF\phi$, but this is not a CTL formula.

Note that the earlier example of a CTL* formula that does not have an LTL counterpart, in fact, only concerns a CTL formula. Together with the last example, this allows the conclusion that inclusion does not hold either way for LTL and CTL.

APPLICATION TO PROGRAM VERIFICATION

CTL program semantics

We emphasize that model checking is for the CTL live semantics, while the proof system in the case of LTL is with respect to the LTL just semantics.

To ensure that programs are finite state, i.e., that their behaviour can be represented by finite state models, it is assumed that each data variable domain is finite. Moreover, for simplicity, it is assumed that all domains are equal to the set of Boolean values $\{true, false\}$. Therefore, a state w can be identified with the subset of the set Y of state variables containing those variables that are *true* in w. For defining CTL program semantics the set Y of state variables is taken to be a set PV of propositional variables. The states correspond to worlds.

The following auxiliary notions will be used:

- Let $\neg Y' = \{\neg y \mid y \in Y'\}$ be the set of the negated variables of $Y' \subseteq Y$.
- A function $\overline{} : 2^Y \longrightarrow 2^{Y \cup \neg Y}$ such that $\overline{Y'} = Y' \cup \neg(Y - Y')$, for any $Y' \subseteq Y$. The function $\overline{}$ extends each set of variables with the negations of those variables which are not in this set.
- The transition t is *enabled* at the state w (written $en_w(t)$) if there is a state w' such that $w' = \rho_t(w)$. It is assumed that the language

contains assertions $en(t)$ for $t \in T$ with the following semantic meaning: $w \models en(t)$ iff $en_w(t)$.

- Let $Q = \{\phi_1, \ldots, \phi_n\}$ be a set of formulas. Then:

$$\bigwedge Q := \phi_1 \& \ldots \& \phi_n \text{ and } \bigvee Q := \phi_1 \vee \ldots \vee \phi_n.$$

DEFINITION 36.1. (CTL live semantics): The CTL *live semantics* for a program P is defined as follows:

$$\chi_P \;=\; \Theta \;\&\; AG(\bigwedge_{w \subseteq Y} \{\bigwedge \overline{w} \supset$$

$$(1) \qquad [(\bigvee_{t \in T} en(t) \supset \bigwedge_{t \in En_w} EX(\bigwedge \overline{\rho_t(w)}) \;\&\; AX(\bigvee_{t \in En_w} \bigwedge \overline{\rho_t(w)})\,) \;\&$$

$$(2) \qquad (\neg \bigvee_{t \in T} en(t) \supset EX(\bigwedge \overline{w}) \;\&\; AX(\bigwedge \overline{w})\,)]\}),$$

$$(3) \qquad \text{where} \quad En_w = \{t \in T \mid en_w(t)\}.$$

(1) describes the successors of all states (or worlds) at which some transition t is enabled, whereas (2) reflects repetition of a state when no transition is enabled. (3) En_w defines the set of all transitions enabled at a state w. This formula does not determine a unique model; in fact, there are infinitely many CTL anchored models satisfying χ_P. It will be shown, however, that any two models M and M' of χ_P are CTL modally equivalent (i.e., for each formula ϕ, $M \models \phi$ iff $M' \models \phi$) and that there is a minimal proper model satisfying χ_P.

Firstly, we define the notion of bisimilar models. This is a criterion enabling one to state whether two CTL models are modally equivalent.

DEFINITION 36.2.: Two anchored models $M = \langle W, R, V\rangle$ and $M' = \langle W', R', V'\rangle$ with roots v_0 and v_0' are said to be *bisimilar* if there is a relation $Z \subseteq W \times W'$ such that the following conditions are satisfied:

- if $(w, w') \in Z$, then $V(w, y) = V'(w', y)$ for any $y \in Y$,
- $(v_0, v_0') \in Z$ and if $(w, w') \in Z$, then
- if wRw_1, then there exists $w_1' \in W'$ such that $w'R'w_1'$ and $(w_1, w_1') \in Z$,

- if $w'R'w'_1$, then there exists $w_1 \in W$ such that wRw_1 and $(w_1, w'_1) \in Z$.

The following lemma provides the criterion.

LEMMA 36.3.: Two finite CTL models are modally equivalent iff they are bisimilar.

The proof is by induction on the complexity of a formula. (See Browne et al. [1988]). □

Next we show how to define a model for a program P. We use the observation made about the states of a program P saying that each state w can be identified with the subset of Y containing those variables that are $true$ in w. We are interested only in the states that can be reached from the initial state.

DEFINITION 36.4.: The *model corresponding to a program P* is an ordered pair $M_P = \langle F_P, V_P \rangle$, where

- $F_P = (W_P, R_P)$ is a frame with W_P being the set of states in W reachable from the initial state satisfying Θ and wR_Pw' iff $\rho_t(w) = w'$ for some $t \in T$,
- $V_P : W_P \times Y \longrightarrow \{true, false\}$ is a valuation function such that $V_P(w, y) = true$ iff $y \in w$.

Finally, we prove that all models satisfying χ_P are modally equivalent. This is achieved by showing that each model for χ_P is bisimilar with the model M_P.

LEMMA 36.5.: All models satisfying χ_P are CTL modally equivalent.

PROOF: Let $M(\chi_P)$ denote the set of all anchored models satisfying χ_P. Let $M' = \langle W', R', V' \rangle \in M(\chi_P)$ and M_P be the model corresponding to the program P. It is then easy to check that the following condition holds:

- M' and M_P are bisimilar; the function $f : W' \rightarrow W_P$ such that $f(w') = w$ iff $V'(w') = V_P(w)$, defines bisimulation.

Since bisimilarity is symmetric and transitive, each two models in $M(\chi_P)$ are bisimilar and hence, by Lemma 36.3, they are modally equivalent. □

Note that bisimilarity between a model M' and the model M_P was defined in terms of a function. Since all worlds of M_P have different sets of variables *true* at them, there is no model M' which would have fewer worlds or arrows than M_P, which proves that M_P is a minimal model satisfying χ_P.

A model checking procedure can now be used to test whether ϕ is true in M_P. As all models for χ_P are modally equivalent, it is sufficient to check ϕ over one of them, say M_P. Minimality of M_P ensures that our checking is the most efficient.

Model checking for CTL

THEOREM 36.6.: There is a deterministic algorithm for determining whether a CTL formula ϕ is true at a world w in a finite model $M = \langle W, R, V\rangle$ of complexity $O(|\phi| \times (card(W) + card(R)))$.

PROOF: The algorithm shown here is designed so that when it terminates, each world w of M is labelled with the subformulas of ϕ which are true at w. It is presented informally, but it should be clear that an implementation in a suitable programming language is straightforward.

The algorithm operates iteratively. Starting with the atomic formulas, at each iteration it deals in principle with the subformulas that, following the syntax of CTL, can be built from those treated already. In principle, because the difficult to handle "for all paths" quantifier for the case $A(\phi_1 U \phi_2)$ is avoided, using the following equivalence:

$$A(\phi_1 U \phi_2) \sim \neg(E(\neg\phi_2 U(\neg\phi_1 \& \neg\phi_2)) \vee EG(\neg\phi_2)).$$

Thus, the cases are : ϕ is atomic or of form $\neg\phi_1$, $\phi_1 \& \phi_2$, $EX\phi_1$, $E(\phi_1 U \phi_2)$ or $EG\phi_1$. Only the last two cases are discussed as the others are straightforward.

To handle a formula of the form $\phi = E(\phi_1 U \phi_2)$, the algorithm first finds all worlds which are labelled with ϕ_2 and labels them with ϕ. Then it traverses the graph backwards, using the relation R^{-1}, and finds all worlds which can be reached by a path in which each world is labelled with ϕ_1. These worlds are labelled with ϕ. This step requires time $O(card(W) + card(R))$.

In the case $\phi = EG\phi_1$, first a structure $M' = \langle W', R', V' \rangle$ is built, where $W' = \{w \in W \mid M, w \models \phi_1\}$, $R' = R \cap (W' \times W')$ and $V' = V \lceil W'$, the restriction of V to W'. Next, the graph (W', R') is partitioned into maximal strongly connected components, i.e., maximal subgraphs in which there is a path of arrows between each two nodes. Those worlds which belong to maximal strongly connected components of size greater than 1 or with a self-loop are selected. Then the algorithm traverses the graph backwards and finds all worlds which can be reached by a path in which each world is labelled with ϕ_1. This step also requires time $O(card(W) + card(R))$. In order to handle an arbitrary CTL formula ϕ, the world-labelling algorithm is successively applied to the subformulas of ϕ, starting with the shortest and most deeply nested one. Let $|\phi|$ be the number of different subformulas. Since each pass takes time $O(card(W) + card(R))$, the algorithm requires time $O(|\phi| \times (card(W) + card(R)))$. □

THE MUTEX EXAMPLE

The model $M_P = \langle F_P, V_P \rangle$ corresponding to the simplified Peterson's algorithm P is defined as follows: $F_P = (W_P, R_P)$ such that

- $W_P = \{\{n_1, n_2, s_1\}, \{n_1, t_2, s_2\}, \{n_1, c_2, s_2\}, \{n_1, n_2, s_2\}, \{t_1, n_2, s_1\},$
  $\{c_1, n_2, s_1\}, \{t_1, t_2, s_1\}, \{t_1, t_2, s_2\}, \{t_1, c_2, s_1\}, \{c_1, t_2, s_2\}\}$;
  the root world is $\{n_1, n_2, s_1\}$.
- $R_P = \{(\{n_1, n_2, s_1\}, \{n_1, t_2, s_2\}), (\{n_1, n_2, s_1\}, \{t_1, n_2, s_1\}),$
  $(\{n_1, t_2, s_2\}, \{n_1, c_2, s_2\}), (\{t_1, t_2, s_1\}, \{t_1, c_2, s_1\}),$
  $(\{n_1, c_2, s_2\}, \{n_1, n_2, s_2\}), (\{t_1, n_2, s_1\}, \{c_1, n_2, s_1\}),$
  $(\{t_1, n_2, s_1\}, \{t_1, t_2, s_2\}), (\{n_1, t_2, s_2\}, \{t_1, t_2, s_1\}),$
  $(\{t_1, t_2, s_2\}, \{c_1, t_2, s_2\}), (\{c_1, t_2, s_2\}, \{n_1, t_2, s_2\}),$
  $(\{t_1, c_2, s_1\}, \{t_1, n_2, s_1\}), (\{c_1, n_2, s_1\}, \{n_1, n_2, s_1\}),$
  $(\{n_1, n_2, s_2\}, \{n_1, t_2, s_2\}), (\{n_1, n_2, s_2\}, \{t_1, n_2, s_1\}),$
  $(\{c_1, n_2, s_1\}, \{c_1, t_2, s_2\}), (\{n_1, c_2, s_2\}, \{t_1, c_2, s_1\})\}$.
- V_P is obvious.

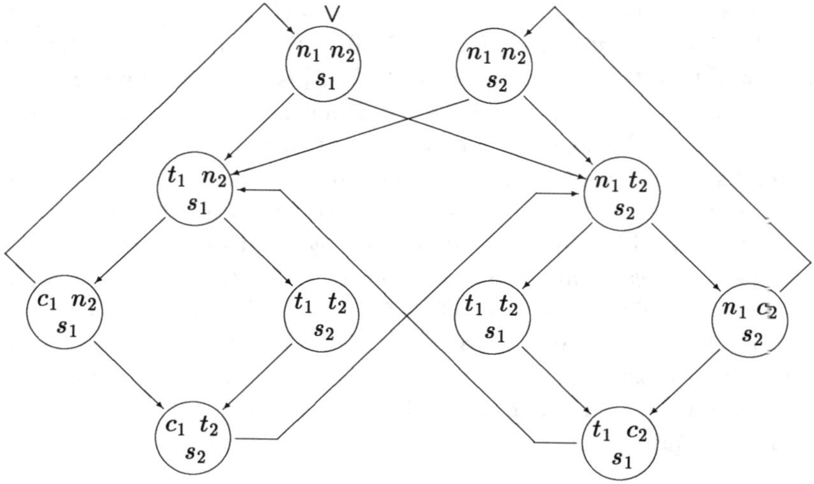

Figure 2: The diagram for model checking MUTEX

Model checking is now as follows.

- For mutual exclusion: $AG\neg(c_1\&c_2)$. This formula is equivalent to the formula $\neg E(trueU(c_1\&c_2))$ which does not contain the derived operators.

  - The labelling of the worlds in W_P with subformulas of the checked formula is performed according to the described model checking algorithm.
  - There are no worlds labelled $c_1 \& c_2$.
  - There are no worlds labelled $E(trueU(c_1 \& c_2))$.
  - All the worlds in W_P are labelled $\neg E(trueU(c_1 \& c_2))$.

  Therefore, $M_P \models AG\neg(c_1 \& c_2)$.

- For eventual access: $AG(t_1 \supset AFc_1)$. This formula is equivalent to the formula $\neg E(trueU(t_1\&EG\neg c_1))$, which does not contain the derived operators.

- The worlds $\{n_1, n_2, s_1\}, \{n_1, t_2, s_2\}, \{n_1, c_2, s_2\}, \{n_1, n_2, s_2\}, \{t_1, n_2, s_1\}, \{t_1, t_2, s_1\}, \{t_1, t_2, s_2\}, \{t_1, c_2, s_1\}$ are labelled $\neg c_1$.
- The worlds $\{n_1, n_2, s_1\}, \{n_1, t_2, s_2\}, \{n_1, c_2, s_2\}$, $\{n_1, n_2, s_2\}$ are labelled $EG\neg c_1$.
- There are no worlds labelled $t_1 \& EG\neg c_1$.
- There are no worlds labelled $E(trueU(t_1 \& EG\neg c_1))$.
- All the worlds are labelled $\neg E(trueU(t_1 \& EG\neg c_1))$.

Therefore $M_P \models AG(t_1 \supset AFc_1)$.

EXERCISES

36.1. Prove that each two models that are not LTL modally equivalent are not CTL modally equivalent.

36.2. Prove that the following CTL formula is valid: $(EG\psi \ \& \ AF\phi) \supset EF(\psi \ \& \ \phi)$.

36.3. Find a model falsifying the following CTL formula: $(EG\psi \ \& \ EF\phi) \supset EF(\psi \ \& \ \phi)$.

36.4. Write the live CTL semantics of the algorithm MUTEX.

36.5. Check whether the formula AFc_1 holds for MUTEX.

36.6. Build the model for MUTEX with an idle transition at each state. Note that this corresponds to the just LTL semantics without the justice requirement. Show by model checking that $AG\neg(c_1 \ \& \ c_2)$ still holds, but that $AG(t_1 \supset AFc_1)$ does not.

36.7. Modify the model checking algorithm to check CTL formulas over just paths, i.e., the paths satisfying the LTL formula $\bigwedge_{t \in J} FGen(t) \supset GFexe(t)$, where $exe(t)$ is an assertion saying that t was executed at a state (or world). Note that for the model in the previous exercise this corresponds to putting the justice requirement back to the LTL semantics.

36.8. Check whether the following formulas are true in the model for MUTEX in Exercise 36.6 over just paths, with $\mathcal{J} = \{t_1c_1, c_1n_1, t_2c_2, c_2n_2\}$: $AG\neg(c_1 \,\&\, c_2), AG(t_1 \supset AFc_1)$.

36.9. See exercise 35.10 for a brief description of the concept of fairness. The aim of the present exercise is to extend model checking to deal with fairness.

- Take advantage of the new scheduling regime to weaken the access condition of MUTEX.
- Adapt the model checking algorithm to the new scheduling regime.
- Model check eventual access, i.e., $AG(t_1 \supset AFc_1)$, for the new version of MUTEX using the new algorithm.

APPENDIX A TABLEAUX-BASED AUTOMATED THEOREM PROVER:

TABLEAUX REVISITED; FORMULAS AS PROLOG TERMS; PRELIMINARY PARTS OF THE PROGRAM; PROGRAMMING TABLEAUX RULES; CLAUSAL FORMS IN TABLEAUX; IMPROVING EFFICIENCY

E.A. de KOGEL and W.M.J. OPHELDERS

In section 11 (Volume I) the tableaux method is presented for propositional logic. It is extended to predicate logic in section 22. Section 22 also includes a procedure for systematically searching for a tableaux deduction of a given formula B from given premisses $A_1, \ldots, A_m$. In this appendix, such a search procedure is implemented in Prolog. In order to achieve implementation efficiently, we first extend the tableaux rules for the quantifiers with unification. The result is a theorem prover for classical predicate logic (without equality) that in many respects is better than resolution-based theorem provers.

TABLEAUX REVISITED

The Gentzen-type deductions found by using a systematic search procedure, such as the one described in §22, are generally unnecessarily long. As an example we show a part of the deduction of $\exists y \forall x[P(x) \supset Q(y)]$ from $\forall x \exists y[P(x) \supset Q(y)]$ in Figure 1. The entire sequent is not written out at every step, but only the new formulas that are introduced by an application of a rule. Every formula is given a number. If a rule is carried out that causes a split into two branches, formulas on the first branch are numbered with an extra .1, the second one is numbered with an extra .2. These branches are written above each other, rather than next to each other.

This tableau becomes very large, although a much shorter tableau proof exists, as we shall see in example A.2. One of the main reasons for the inefficiency of the systematic search procedure is that the $T\,\forall$ and $F\,\exists$ rules must be applied with respect to every a_i that occurs in the tableau, in this case resulting in many formulas of the form $F\ P(a_m) \supset Q(a_n)$ and of the form $T\ P(a_m) \supset Q(a_n)$, of which only a few actually contribute to the closure of branches.

initial		1	$T\ \forall x \exists y[P(x) \supset Q(y)]$
sequent		2	$F\ \exists y \forall x[P(x) \supset Q(y)]$
step 1a)			-
step 1b)	$T\forall(1)$	3	$T\ \exists y[P(a_1) \supset Q(y)]$
	$F\exists(2)$	4	$F\ \forall x[P(x) \supset Q(a_1)]$
step 2a)	$T\exists(3)$	5	$T\ P(a_1) \supset Q(a_2)$
	$F\forall(4)$	6	$F\ P(a_3) \supset Q(a_1)$
step 2b)	$F\supset(6)$	7	$TP(a_3)$
		8	$FQ(a_1)$
	$T\supset(5)$	9.1	$FP(a_1)$
	$T\forall(1)$	10.1	$T\ \exists y[P(a_2) \supset Q(y)]$
		11.1	$T\ \exists y[P(a_3) \supset Q(y)]$
	$F\exists(2)$	12.1	$F\ \forall x[P(x) \supset Q(a_2)]$
		13.1	$F\ \forall x[P(x) \supset Q(a_3)]$
step 3a)	$T\exists(10.1)$	14.1	$T\ P(a_2) \supset Q(a_4)$
	$T\exists(11.1)$	15.1	$T\ P(a_3) \supset Q(a_5)$
	$F\forall(12.1)$	16.1	$F\ P(a_6) \supset Q(a_2)$
	$F\forall(13.1)$	17.1	$F\ P(a_7) \supset Q(a_3)$
step 3b)	$F\supset(16.1)$	18.1	$TP(a_6)$
		19.1	$FQ(a_2)$
	$F\supset(17.1)$	20.1	$TP(a_7)$
		21.1	$FQ(a_3)$
	$T\supset(14.1)$	22.1.1	$FP(a_2)$
	$T\supset(15.1)$	23.1.1.1	$FP(a_3)$
	7 and 23.1.1.1 cause closure		
		23.1.2	$TQ(a_5)$
	$T\forall(1)$	24.1.1.2	$T\ \exists y[P(a_4) \supset Q(y)]$
		⋮	
		etcetera	

Figure 1: Part of a systematic tableau

The technique of unification can improve efficiency. When applying a $T\forall$ or $F\exists$ rule, it is hard to predict what term is the most suitable choice to apply it to. It is therefore better to fill in a placeholder for this term, and determine what term the placeholder refers to at a later stage. To this end, we introduce a set of so-called meta-variables, denoted by $X, Y, Z, X_1, Y_1, \ldots$, etc., that will serve as such placeholders.
The $T\forall$ and $F\exists$ rules are adapted as follows:

$$\frac{S,\ T\ \forall x[A(x)]}{S,\ TA(X),\ T\ \forall x[A(x)]} \qquad \frac{S,\ F\ \exists x[A(x)]}{S,\ FA(X),\ F\ \exists x[A(x)]}$$

In both cases, X should be a meta-variable that is new to the tableau. It should be noted that meta-variables are *not* part of our logic language, and we shall distinguish them from the variables of that language.

Now suppose that signed atoms TA and FB occur on a branch, and A and B can be unified, i.e., by replacing some meta-variables by well-chosen terms, we can make A and B equal. Then it seems we can close that branch. In fact, this is not necessarily true. For instance, assume that the meta-variable X was introduced by an application of the $T\forall$ rule, after which the $T\exists$ rule is applied with respect to a parameter a. Then we may never replace X by a term t in which a occurs. X is only a placeholder, and a valid tableau should result if we apply the $T\forall$ rule with respect to t in the first place. So, if a occurs in t, then the $T\exists$ rule can no longer be applied to a because a is no longer new to the branch. These considerations result in the following.

Condition: A meta-variable may never be instantiated with a term in which a 'newer' parameter occurs.

Thus, a branch is closed if signed atoms TA and FB occur in it, and A and B can be unified without violating this condition. Of course, the unifying substitution must be applied in the entire tableau before one tries to close other branches since wherever a meta-variable X occurs, it is a placeholder for one and the same term.

We shall implement the search procedure as it has just been sketched with meta-variables, conditions and unification, in Prolog.

FORMULAS AS PROLOG TERMS

In order to have a Prolog program which can manipulate formulas, we must represent these formulas as Prolog terms. The building blocks of

Prolog terms are *Prolog atoms* and *Prolog variables*. Parentheses, commas, etc., are also used.

- Prolog atoms are strings of letters, numbers and underscores _ starting with a lowercase letter. For instance, `abc`, `q_13` are Prolog atoms.
- Prolog variables are strings of letters, numbers and underscores starting with an uppercase letter. For instance, `X_13` and `Formula` are Prolog variables.

A Prolog interpreter will probably accept certain other classes of strings of characters as Prolog atoms and Prolog variables, but only the ones just defined will be needed.

DEFINITION A.1.: (Cf. the definition in §25) A *Prolog term* is either a Prolog variable, a Prolog atom, or an expression of the form $f(t_1, \ldots, t_n)$, in which f is a Prolog atom and $t_1, \ldots, t_n$ are Prolog terms. f is called the *functor* of $f(t_1, \ldots, t_n)$, in this case having arity n; $t_1, \ldots, t_n$ are the arguments of $f(t_1, \ldots, t_n)$. Examples of Prolog terms are `Formula`, `a_13`, `a_13(X,X)`, `abc(a_13,X,Y)`.

To represent the terms of our language by Prolog terms, we represent all variables, constants, function symbols and predicate symbols of our language by Prolog atoms. It is the responsibility of the user to make sure that in particular the set of variables and the set of constants are disjoint. To represent the connectives, we use the possibility of defining prefix and infix operators in Prolog. We represent $\neg$ by the unary prefix operator `not`, which is defined by executing the following goal:

```
:- op(100, fy, not).
```

After its execution, for every Prolog term t, the expression `not` t is internally identified with `not(`t`)`.

To represent &, $\vee$, $\supset$ and $\sim$, we define the binary infix operators `and`, `or`, `imp` and `iff`, respectively, by executing the following goals.

```
:- op(110, xfy, and).
:- op(120, xfy, or).
:- op(130, xfy, imp).
:- op(140, xfy, iff).
```

The meaning of the arguments of `op` is explained in exercise 25.12. Only a few examples will be given here. s `and` t is identified with `and(`s`,`t`)`. s `and` t `and` u is read as s `and (`t `and` u`)`. `not not` t is identified with `not(not(`t`))`. Hence, s `or` t `and` u is read as s `or (`t `and` u`)`, and `not` s `and` t is read as `(not` s`) and` t. In case of doubt, use parentheses.

We represent $\forall x[F]$ and $\exists x[F]$ by `all(`x'`,`F'`)` and `exist(`x'`,`F'`)`, respectively, if x' represents x and F' represents F.

For instance, the formula $\forall x \exists y[P(x) \supset Q(y)]$ can be represented by the Prolog term `all(x, exist(y, p(x) imp q(y)))`.

The signed formulas TG and FG are represented by the Prolog terms `t(`G'`)` and `f(`G'`)`, respectively, where G' is the representation of G as a Prolog term. A sequent is represented as a list of signed formulas.

Finally, we represent the *parameters*, i.e., the free variables to be introduced by the $T\exists$ and $F\forall$ rules, by the Prolog atoms `sk0`, `sk1`, `sk2`, ..., which therefore must not be used as constants or variables. Meta-variables will be represented by Prolog variables.

PRELIMINARY PARTS OF THE PROGRAM

Lists

Lists are very important in many Prolog programs. We shall define some frequently occurring predicates on lists. The `member` predicate is discussed in §25, the `del` and `conc` predicates are asked for in exercises 25.4 and 25.5.

```
member(H, [H|T]).
member(X, [H|T]) :- member(X, T).

del(X, [X|T], T).
del(X, [H|T], [H|S]) :- del(X, T, S).

conc([ ], X, X).
conc([H|X], Y, [H|Z]) :- conc(X, Y, Z).
```

Meta-Variables and Unification

As stated before, we represent the meta-variables used in the tableau method by Prolog variables because we can easily instantiate them with the terms we need when closing a branch.

As Prolog does not apply an occur check, and will certainly not take into account the conditions on parameters, we had to write our own unification program. Such a program explicitly controls the instantiation of Prolog variables, and in that respect it differs from the Prolog programs we have seen until now.

Prolog offers a few built-in predicates that help determine the instantiation state of terms. We shall use `var`, `nonvar`, `functor`, `arg`, `==` and `\==`.

`var`(t) will succeed if and only if t is a Prolog variable. To be more precise we should say: if the instantiation of t by the time of evaluation is a Prolog variable. However, for the sake of simplicity, we shall simply speak of t and mean its instantiation. Example:

```
:- X = a, var(X).
```

will fail because, although `X` is a Prolog variable, it is instantiated to a constant at the time of evaluation. On the other hand, although logically equivalent to the first example,

```
:- var(X), X = a.
```

will succeed, instantiating `X` to `a`, because `X` is not yet instantiated to `a` at the time of evaluation of `var(X)`.

`nonvar`(t) succeeds if `var`(t) fails. Hence, for every Prolog term t, either `var`(t) succeeds or `nonvar`(t) succeeds.

The ternary predicate `functor` can be used in two ways:

- If t is not a Prolog variable, hence, it is a Prolog atom g or a Prolog term of the form $g(t_1, \ldots, t_n)$, then upon calling `functor`(t, f, m), Prolog tries to match f and g, and m and n, respectively. The Prolog atom g is considered as a Prolog term with a nullary function symbol, hence, if $t = g$, then $n = 0$.

- If X is a Prolog variable, f is a Prolog atom, and n is a number, then upon calling `functor`(X, f, n), X is instantiated to a Prolog term of the form $f(X_1, \ldots, X_n)$, in which the $X_1, \ldots, X_n$ are fresh Prolog variables. This is the most general representation of a Prolog term with n-ary functor f, in the sense that every other such Prolog term can be obtained from it by further instantiations.

If t is not a Prolog variable, hence, t is of the form $f(t_1, \ldots, t_n)$, and m is a number such that $1 \leq m \leq n$, then on evaluation of $\texttt{arg}(m, t, s)$, Prolog tries to match s and the m-th argument of t, i.e., t_m.

Evaluation of s == t will succeed if and only if s and t are identical. For instance,

```
:- X = f(Y), X == f(Y).
```

will succeed. However,

```
:- X == f(Y), X = f(Y).
```

will fail because when `X == f(Y)` is evaluated, `X` is not yet instantiated to `f(Y)`.

We are now ready to discuss a simple program performing the occur check mentioned in §25. We do so quite extensively in order to provide more understanding of some of the concepts that we shall also use later on.

The goal `does_not_occur_in(`s`,`t`)` must succeed if s and t are Prolog terms and s does not occur in t. s does not occur in t if s and t are different and if t is not a Prolog variable, s does not occur in any of the arguments of t. We therefore always require that s \== t succeeds. As we saw before, either `var(`t`)` or `nonvar(`t`)` succeeds. In the latter case, t has the form $f(t_1, \ldots, t_n)$, and s must not occur in the arguments $t_1, \ldots, t_n$. To check this, we write the program `does_not_occur_in_args`, in which `does_not_occur_in_args(`m`,`s`,`t`)` succeeds if and only if

1) $m = 0$, or
2) $1 \leq m \leq n$ and s does not occur in $t_1, \ldots, t_m$, which can recursively be defined as `does_not_occur_in(`s`,`t_m`)` and `does_not_occur_in_args(`$m-1$`,`s`,`t`)`.

The program is quite straightforward:

```
does_not_occur_in(S, T) :-
  var(T),
  S \== T.
does_not_occur_in(S, T) :-
  nonvar(T),
  S \== T,
  functor(T, F, N),
  does_not_occur_in_args(N, S, T).

does_not_occur_in_args(0, S, T).
```

```
does_not_occur_in_args(M, S, T) :-
  M > 0,
  arg(M, T, TM),
  does_not_occur_in(S, TM),
  M_1 is M-1,
  does_not_occur_in_args(M_1, S, T).
```

Now we are ready to understand the procedure `unify` that matches its two arguments if the occur check succeeds. Either of the Prolog terms can be a Prolog variable or not, which yields four possibilities, each treated in a different clause.

The clauses for `unify_args` should be understood as analogous to the clauses for `does_not_occur_in_args` discussed above.

```
unify(X, Y) :-
  var(X),
  var(Y),
  X = Y.
unify(X, Y) :-
  var(X),
  nonvar(Y),
  does_not_occur_in(X, Y),
  X = Y.
unify(X, Y) :-
  nonvar(X),
  var(Y),
  does_not_occur_in(Y, X),
  X = Y.
unify(X, Y) :-
  nonvar(X),
  nonvar(Y),
  functor(X, F, N),
  functor(Y, F, N),
  unify_args(N, X, Y).

unify_args(0, X, Y).
unify_args(N, X, Y) :-
  N > 0,
```

```
   arg(N, X, A),
   arg(N, Y, B),
   unify(A, B),
   N_1 is N-1,
   unify_args(N_1, X, Y).
```

If x, y and t_x are Prolog terms, we can replace all occurrences of x in t_x by y, calling the result t_y. A program doing this can be used to program applications of the $T\forall$, $T\exists$, $F\exists$ and $F\forall$ rules. Some subtleties are involved here that will be discussed later.

The program is such that if TY is a Prolog variable not occurring in x, t or t_x, on evaluation of replace(x, y, t_x, TY), TY will be instantiated to t_y.

The structure is again similar to that of does_not_occur_in.

```
replace(X, Y, TX, Y) :-
  X == TX.
replace(X, Y, V, V) :-
  var(V),
  X \== V.
replace(X, Y, TX, TY) :-
  nonvar(TX),
  X \== TX,
  functor(TX, F, N),
  functor(TY, F, N),
  replace_args(N, X, Y, TX, TY).

replace_args(0, X, Y, TX, TY).
replace_args(N, X, Y, TX, TY) :-
  N > 0,
  arg(N, TX, AX),
  arg(N, TY, AY),
  replace(X, Y, AX, AY),
  N_1 is N-1,
  replace_args(N_1, X, Y, TX, TY).
```

Note the goals `functor(TX, F, N)`, `functor(TY, F, N)` in the third clause, illustrating the different ways to use `functor`. With the first call to `functor`, we determine the functor of `TX` and its arity, matching them with `F` and `N`, respectively, with the second call to `functor` we create a term with this same functor and arity, and match it with `TY`.

PROGRAMMING TABLEAUX RULES

By now the implementation of the rules is quite straightforward, and we shall write a program that even delivers tableau proofs, much like the one in Figure 1.

We must discuss a problem concerning the application of $T\forall$, $T\exists$, $F\exists$ and $F\forall$ rules first. We wish to use the predicate `replace` in the application of these rules. Application of the $T\forall$ rule to `t(all(`x`,`F`))` involves replacing all *free* occurrences of x in F by a new Prolog variable X. However, evaluation of `replace(`x`,`X`,`F`,`F_1`)` will construct F_1 from F by replacing *all* occurrences of x by X. Although it is possible to write a program that replaces only the free occurrences of x, we find it fair to demand that every occurrence of x in F is a free occurrence. This imposes the following restrictions on the formulas admitted as input.

- Prolog atoms that appear as bound variables may not occur in any other role as nullary functors, i.e., they may appear neither as constants nor as proposition symbols.
- No nested quantifications of the same variable are allowed. Hence, in `all(`x`,`F`)`, F may not contain subformulas of the form `all(`x`,`G`)` or `exist(`x`,`G`)`.

These requirements are rather fair, as every formula has an equivalent formula satisfying them, obtained by renaming bound variables if necessary.

The main predicate of our program is called `gtp`, which stands for Gentzen-type proof. It takes six arguments, so every call takes the form

`gtp(`*Sequent*`,` *Atoms*`,` Q`,` N`,` *Rstr*`,` *Proof*`).`

- A sequent, as discussed in §11, is here divided into two parts, *Sequent* and *Atoms*. *Sequent* is a list of signed formulas. *Atoms* is a list of signed atomic formulas. If we find a signed atomic formula F in *Sequent*, we try to find an atomic formula with the opposite sign in *Atoms*, and unify the atomic formulas. If this is impossible or after backtracking all possibilities are exhausted, then in the next call to `gtp` F is added to the *Atoms*. In this way the program will not try to unify the same pair of signed atomic formulas over and over again.

- Q stands for *quantifier depth*. Q is a natural number that indicates how often we may apply the $T\,\forall$ and $F\,\exists$ rules to a branch. This is necessary, since if $T\,\forall$ or $F\,\exists$ formulas are present, one could go on applying these rules forever. However, it is important that this process stops at some time, because otherwise it would never be possible to carry out any backtracking.

- N is the number of the parameter to be introduced in the next application of a $T\,\exists$ or $F\,\forall$ rule. At the application of such a rule, it is increased by 1.

- *Rstr* stands for restrictions. It is a list representing the order in which meta-variables and parameters are introduced to the branch. If we apply a $T\,\exists$ or $F\,\forall$ rule with respect to sk*n*, then the next call to `gtp` is made with the new list of restrictions [par(sk*n*) | *Rstr*]. If a meta-variable X is introduced, the next restriction list is [var(X) | *Rstr*]. Hence, items on the list are older if they appear further to the right, and the functors `par` and `var` indicate whether they were introduced as Skolem constants or as meta-variables. Note that as meta-variables may get instantiated, it is possible that var(t) appears on the list, while t is not a Prolog variable.

- *Proof* is a Prolog term from which we can easily construct a proof that can be understood by human readers. Each proof will be of one of the following three forms:

  - [F], if F is a signed atomic formula and the same atomic formula with the opposite sign occurs earlier in the tableau.
  - [F,L,P], which means that by applying a rule to the signed formula F, the signed formulas in the list L were added to the branch, and P is the remainder of the proof. This proof is

used if application of the rule does not cause splitting of the branch.

- [F,L_1,P_1,L_2,P_2]. This format is used if application of the appropriate rule to the signed formula F causes splitting of the branch. In that case L_1 (L_2) is the list of signed formulas added to the first (second) branch, and P_1 and P_2 are the respective remainders of the tableau in each branch.

We shall adopt a very simple search strategy, namely, applying the appropriate rule to the first signed formula in the sequent. When the first signed formula in the sequent is a `t all` or `f exist` formula, then after application of the appropriate rule we must keep it in the sequent. We shall then put it at the end of the sequent, to make sure that the other formulas get a chance of becoming the leftmost formula.

If gtp(*Sequent*, *Atoms*, *Q*, *N*, *Rstr*, *Proof*) succeeds, this means that there is a closed tableau with as its initial sequent $Sequent \cup Atoms$, and with proof *Proof*, found by searching according to the search strategy and with at most Q applications of the $T\,\forall$ and $F\,\exists$ rules in every branch.

We define separate predicates for tableaux rules that are much alike. They are divided as follows.

- The $T\,\neg$, $F\,\neg$, $T\,\&$, $F\,\vee$, $F\,\supset$-rules are similar, in the sense that they all add one or two new formulas to the branch and do not affect the restriction list. We shall call these rules simple. We shall write the predicate `simple`, such that simple(F, L) succeeds if and only if a simple rule is applicable to the signed formula F, and L is the list of formulas to be added to the branch. This is actually straightforward:

```
simple(t(not F), [f(F)]).
simple(f(not F), [t(F)]).
simple(t(F and G), [t(F),t(G)]).
simple(f(F or G),  [f(F),f(G)]).
simple(f(F imp G), [t(F),f(G)]).
```

Traditionally, the $T\neg$ and $F\neg$ rules are called *negation rules*, while the $T\,\&$, $F\vee$, $F\supset$-rules are called *alpha rules.*

- The $F\,\&$, $T\vee$, $T\supset$, $T\sim$ and $F\sim$-rules are similar, as they all cause a splitting of the branch and one or two formulas should be added to each branch. They do not affect the restrictions. Traditionally these rules are called *beta rules.* For these we write the predicate `beta`, where `beta(`F`, `L_1`, `L_2`)` succeeds if and only if a beta rule is applicable to the signed formula F, L_1 is the list of signed formulas to be added to one branch and L_2 is the list of signed formulas to be added to the other branch.

```
beta(f(F and G), [f(F)], [f(G)]).
beta(t(F or G),  [t(F)], [t(G)]).
beta(t(F imp G), [f(F)], [t(G)]).
beta(t(F iff G), [t(F),t(G)], [f(F),f(G)]).
beta(f(F iff G), [t(F),f(G)], [f(F),t(G)]).
```

- The $T\forall$ and $F\exists$ rules are similar, introducing a meta-variable that must be recorded in the restriction list. Traditionally, these rules are called *gamma rules.* We write the predicate `gamma` such that `gamma(`F`, `F_1`, `Y`)` succeeds if F is of the form `t(all(`x`,`G`))` or `f(exist(`x`,`G`))` and F_1 is `t(`G_1`)` or `f(`G_1`)`, respectively, where G_1 is the result of replacing all occurrences of x in G by a new meta-variable Y.

```
gamma(t(all(X,FX)), t(FY), Y) :-
  replace(X, Y, FX, FY).
gamma(f(exist(X,FX)), f(FY), Y) :-
  replace(X, Y, FX, FY).
```

- The $F\forall$ and $T\exists$ rules are similar, introducing a new parameter that must be recorded in the restriction list. Traditionally, these rules are called *delta rules.*

```
delta(f(all(X,FX)), f(FY), Y) :-
  replace(X, Y, FX, FY).
delta(t(exist(X,FX)), t(FY), Y) :-
  replace(X, Y, FX, FY).
```

The procedure **gtp** simply applies the appropriate rule to the leftmost signed formula in the sequent and constructs the appropriate new sequent(s) to pass to the next call to **gtp**.

```
gtp([F|S], A, Q, N, Rstr, [F,L,P]) :-
  simple(F, L),
  conc(L, S, S1),
  !,
  gtp(S1, A, Q, N, Rstr, P).

gtp([F|S], A, Q, N, Rstr, [F,L1,P1,L2,P2]) :-
  beta(F, L1, L2),
  conc(L1, S, S1),
  conc(L2, S, S2),
  !,
  gtp(S1, A, Q, N, Rstr, P1),
  gtp(S2, A, Q, N, Rstr, P2).

gtp([F|S], A, Q, N, Rstr, [F,[F1],P]) :-
  gamma(F, F1, Y),
  !,
  Q>0,
  Q_1 is Q-1,
  conc(S, [F], SF),
  !,
  gtp([F1|SF], A, Q_1, N, [var(Y)|Rstr], P).

gtp([F|S], A, Q, N, Rstr, [F,[F1],P]) :-
  delta(F, F1, Y),
  name(N, N_as_List),
```

```
    name(Y, [115,107|N_as_List]),
    N1 is N+1,
    !,
    gtp([F1|S], A, Q, N1, [par(Y)|Rstr], P).
```

In the last clause we use the built-in Prolog predicate `name` for the creation of new parameters. If a is a Prolog atom or a number, then the effect of evaluation of `name(`a`,`l`)` is that l is matched with the list of ASCII codes of the characters in a. For instance,

```
:- name(12, N_as_List).
```

leads to the matching `N_as_List = [49,50]`, since the ASCII code for 1 is 49, for 2 it is 50. If X is a Prolog variable and l is a list of ASCII codes, then, upon calling `name(`X`,`l`)`, X will be matched to the corresponding Prolog atom. Hence,

```
:- name(Y, [115,107,49,50]).
```

leads to the matching `Y = sk12`.

In the third clause we use that the Prolog interpreter will instantiate `Y` with a new Prolog variable, and thus will effectively introduce a new meta-variable.

If F is the leftmost signed formula of the sequent and the goals `simple(`F`, ...)`, `beta(`F`, ...)`, `gamma(`F`, ...)` and `delta(`F`, ...)` all fail, then we may assume that F is a signed atomic formula. In that case, we try to unify F with an atomic formula with the opposite sign in the list of signed atomic formulas. If all possibilities to do so are exhausted, then we go on with the remainder of the sequent and add F to the list of signed atomic formulas.

```
gtp([t(F)|S], A, Q, N, Rstr, [t(F)]) :-
    member(f(G), A),
    unify(F, G),
    check_restrictions(Rstr).

gtp([f(F)|S], A, Q, N, Rstr, [f(F)]) :-
    member(t(G), A),
    unify(F, G),
    check_restrictions(Rstr).
```

```
gtp([F|S], A, Q, N, Rstr, P) :-
  gtp(S, [F|A], Q, N, Rstr, P).
```

Note the importance of the `!`'s in the clauses for compound formulas. For instance, if we made the call `gtp([t(p and q)|`S`], ...)`, this would result in the new call `gtp([t(p),t(q)|`S`], ...)`. It is possible that evaluation of this last goal fails. If `!` is now left out of the first clause of `gtp`, then after backtracking, the program assumes that `t(p and q)` is a signed atomic formula. As this is obviously not our intention, we need the `!` (`cut`) to prevent the program from backtracking.

In the last clauses we use a procedure `check_restrictions` after every unification to find out whether forbidden instantiations are made. We still have to define this procedure. This is done in a straightforward way: whenever `par(sk`n`)` occurs in the list of restrictions, we check for every occurrence of `var(`t`)` to its right whether `sk`n does not occur in t.

```
check_restrictions([ ]).
check_restrictions([par(SkN)|Rstr]) :-
  not_in_instantiations(SkN, Rstr),
  check_restrictions(Rstr).
check_restrictions([var(T)|Rstr]) :-
  check_restrictions(Rstr).

not_in_instantiations(SkN, [ ]).
not_in_instantiations(SkN, [par(SkM)|Rstr]) :-
  not_in_instantiations(SkN, Rstr).
not_in_instantiations(SkN, [var(T)|Rstr]) :-
  does_not_occur_in(SkN, T),
  not_in_instantiations(SkN, Rstr).
```

Lastly, we must provide a 'user interface' which allows the user to make a call to the program without typing more than is necessary, and obtain a proof, if one is found. Therefore, we write the predicate `tableau`, taking only the initial sequent and the quantifier depth as arguments.

```
tableau(S, Q) :-
  write('The initial sequent is:'), nl,
  write_sequent(S),
  gtp(S, [ ], Q, 0, [ ], Proof),
  write_proof([ ], Proof).
```

The initial sequent is immediately written out. To do that, we use the built-in predicates `write` and `nl`. Evaluation of `write(`X`)` always succeeds and indeed writes X to the screen. Evaluation of `nl` also always succeeds, ensuring that the next text output to the screen will start on a new line.

Writing a sequent to the screen is then straightforward:

```
write_sequent([ ]).
write_sequent([F|S]) :-
  write('  '), write(F), nl,
  write_sequent(S).
```

Note that we have two spaces written in front of every signed formula. This enhances readability of the proof.

In putting out proofs, we shall also number the branches. This is necessary since after each splitting of a branch, two new successors are created. As it is hard to present those branches next to each other, we shall have to write them above each other. Branches are numbered with lists of 1's and 2's. If, for instance, the branch `[1,2,1]` splits, then the first branch is numbered `[1,2,1,1]` and the second `[1,2,1,2]`. This list of 1's and 2's is called `B_Ind` in the program.

```
write_proof(B_Ind, [F]) :-
  write('closure by '), write(F), nl.

write_proof(B_Ind, [F,L,P]) :-
  write('apply rule to '), write(F), nl,
  write_sequent(L),
  write_proof(B_Ind, P).
```

```
write_proof(B_Ind, [F,L1,P1,L2,P2]) :-
  write('apply rule to '), write(F), nl,
  conc(B_Ind, [1], B_Ind1),
    write('branch '), write(B_Ind1), nl,
    write_sequent(L1),
    write_proof(B_Ind1, P1),
  conc(B_Ind, [2], B_Ind2),
    write('branch '), write(B_Ind2), nl,
    write_sequent(L2),
    write_proof(B_Ind2, P2).
```

EXAMPLE A.2.: We consider the problem of deducing $\exists y \forall x[P(x) \supset Q(y)]$ from $\forall x \exists y[P(x) \supset Q(y)]$ (see also Figure 1). To do this, we start with $\{T\ \forall x \exists y[P(x) \supset Q(y)], F\ \exists y \forall x[P(x) \supset Q(y)]\}$. The least quantifier depth for which we obtain a proof is 4. Evaluation of

```
:- tableau([t(all(x, exist(y, p(x) imp q(y)))),
            f(exist(y, all(x, p(x) imp q(y))))], 4).
```

yields the following output.

```
The initial sequent is:
  t(all(x, exist(y, p(x) imp q(y))))
  f(exist(y, all(x, p(x) imp q(y))))
apply rule to t(all(x, exist(y, p(x) imp q(y))))
  t(exist(y, p(_528) imp q(y)))
apply rule to t(exist(y, p(_528) imp q(y)))
  t(p(_528) imp q(sk0))
apply rule to t(p(_528) imp q(sk0))
branch [1]
  f(p(_528))
apply rule to f(exist(y, all(x, p(x) imp q(y))))
  f(all(x, p(x) imp q(_1528)))
apply rule to f(all(x, p(x) imp q(_1528)))
  f(p(sk1) imp q(_1528))
apply rule to f(p(sk1) imp q(_1528))
  t(p(sk1))
```

```
  f(q(_1528))
apply rule to t(all(x, exist(y, p(x) imp q(y))))
  t(exist(y, p(sk1) imp q(y)))
apply rule to t(exist(y, p(sk1) imp q(y)))
  t(p(sk1) imp q(sk2))
apply rule to t(p(sk1) imp q(sk2))
branch [1,1]
  f(p(sk1))
closure by f(p(sk1))
branch [1,2]
  t(q(sk2))
apply rule to f(exist(y, all(x, p(x) imp q(y))))
  f(all(x, p(x) imp q(sk2)))
apply rule to f(all(x, p(x) imp q(sk2)))
  f(p(sk3) imp q(sk2))
apply rule to f(p(sk3) imp q(sk2))
  t(p(sk3))
  f(q(sk2))
closure by f(q(sk2))
branch [2]
  t(q(sk0))
apply rule to f(exist(y, all(x, p(x) imp q(y))))
  f(all(x, p(x) imp q(sk0)))
apply rule to f(all(x, p(x) imp q(sk0)))
  f(p(sk1) imp q(sk0))
apply rule to f(p(sk1) imp q(sk0))
  t(p(sk1))
  f(q(sk0))
closure by f(q(sk0))
```

(Expressions like _528 and _1528 are the representations of uninstantiated Prolog variables.)

In fact, using the alternative representation of the initial sequent,

```
[f(exist(y, all(x, p(x) imp q(y)))),
                              t(all(x, exist(y, p(x) imp q(y))))],
```

we obtain a much shorter proof which is already at a quantifier depth 3. Applying rules in a different order causes different restrictions and leads to different proofs.

CLAUSAL FORMS IN TABLEAUX

In §26 we already discussed the example of deducing $\exists y \forall x[P(x) \supset Q(y)]$ from $\forall x \exists y[P(x) \supset Q(y)]$ by means of resolution. If we let $A = \forall x \exists y[P(x) \supset Q(y)]$ and $B = \exists y \forall x[P(x) \supset Q(y)]$, then $C(A) = \forall x[P(x) \supset Q(f(x))]$ and $C(\neg B) = \forall y[P(g(y))]$ & $\forall z[\neg Q(z)]$.

To find a tableau proof from these clauses, we start with an initial sequent in which the clauses are all signed true. In Prolog this means we evaluate

```
  :- tableau([t(all(x, p(x) imp q(f(x)))),
              t(all(y, p(g(y))) and all(z, not q(z)))], 3).
```

yielding the following output.

```
 The initial sequent is:
   t(all(x, p(x) imp q(f(x))))
   t(all(y, p(g(y))) and all(z, not q(z)))
 apply rule to t(all(x, p(x) imp q(f(x))))
   t(p(g(Y1)) imp q(f(g(Y1))))
 apply rule to t(p(g(Y1)) imp q(f(g(Y1))))
 branch [1]
   f(p(g(Y1)))
 apply rule to t(all(y, p(g(y))) and all(z, not q(z)))
   t(all(y, p(g(y))))
   t(all(z, not q(z)))
 apply rule to t(all(y, p(g(y))))
   t(p(g(Y1)))
 closure by t(p(g(Y1)))
 branch [2]
   t(q(f(g(Y1))))
 apply rule to t(all(y, p(g(y))) and all(z, not q(z)))
   t(all(y, p(g(y))))
   t(all(z, not q(z)))
 apply rule to t(all(y, p(g(y))))
   t(p(g(Y2)))
 apply rule to t(all(z, not q(z)))
   t(not q(f(g(Y1))))
 apply rule to t(not q(f(g(Y1))))
   f(q(f(g(Y1))))
 closure by f(q(f(g(Y1))))
```

initial	1	$T\ \forall x[P(x) \supset Q(f(x))]$
sequent	2	$T\ \forall y[P(g(y))] \& \forall z[\neg Q(z)]$
$T\forall(1)$	3	$T\ P(X_1) \supset Q(f(X_1))$
$T\supset(3)$	4.1	$FP(X_1)$
$T\,\&(2)$	5.1	$T\ \forall y[P(g(y))]$
	6.1	$T\ \forall z[\neg Q(z)]$
$T\forall(5.1)$	7.1	$TP(g(Y_1))$
4.1 and 7.1 cause closure if $X_1 = g(Y_1)$		
	4.2	$TQ(f(X_1))$
$T\,\&(2)$	5.2	$T\ \forall y[P(g(y))]$
	6.2	$T\ \forall z[\neg Q(z)]$
$T\forall(5.2)$	7.2	$TP(g(Y_2))$
$T\forall(6.2)$	8.2	$T\neg Q(Z_1)$
$T\neg(8.2)$	9.2	$FQ(Z_1)$
4.2 and 9.2 cause closure if $Z_1 = f(X_1) = f(g(Y_1))$		

Figure 2.

Here we changed by hand the Prolog variable representations like `_1263` to the more readable `Y1` and `Y2`. If we trace how this proof was found, one sees a great similarity between the resolution proof (which is the last example of §26) and this tableau proof (see Figure 2).

We find that bringing formulas in Skolem normal form means that we never need to apply a $T\exists$ or an $F\forall$ rule, and maintaining the restriction list is then superfluous. This makes the tableau method more efficient, and also partially explains the efficiency of the resolution method.

IMPROVING EFFICIENCY

In writing the program presented in the last subsections, we tried to keep it as simple and understandable as possible. The program may be changed to make the evaluation more efficient, or to make it more robust.

Changes to make the program more robust might include a procedure to check whether the initial sequent is syntactically correct, which will produce an error message if it is not or a new `replace` procedure that will only replace free occurrences of a variable rather than all occurrences of that variable.

To improve the efficiency of the program, more !'s can be used. The reader is invited to find out where these can be put without affecting the completeness of the program.

In addition, alternative strategies for applying the tableau rules can be considered. In our program we chose to develop the leftmost branch of the tableau first, and to apply a rule to the leftmost formula in a sequent. This is not prescribed, however, and one might want to try other strategies. Below is an example of an alternative strategy of selecting the formula in a sequent to which we are going to apply a rule. This strategy is a good heuristics for finding shorter proofs, although it is certainly not an unfailing one.

The idea is to delay applications of some rules if there are formulas in the sequent to which 'simpler' rules can be applied. This depends on what we call a 'simple' rule, of course. We shall argue that the closure rule is simpler than any other rule, that negation, alpha and delta rules are simpler than beta rules, which are in turn simpler than gamma rules. Thus, we defend a strategy in which a gamma rule is applied only if no other rule can be applied.

It seems obvious that the closure rule is the simplest rule because, if it succeeds, we need not develop a branch further. Obviously, being eager to close a branch should be a good heuristics for finding a short deduction.

It is wise to delay an application of a beta rule until all possibilities for closure have been tried because, otherwise, after the branch has been split, the same possibilities for closure must be tried twice, namely, once in both resulting branches.

For the same reason, one should not apply a beta rule if it is also possible to apply a negation rule, an alpha rule or a delta rule. For instance, if one can choose between an alpha rule and a beta rule and the beta rule is applied first, then this same alpha rule probably has to be applied twice, namely, in the two resulting branches.

Application of gamma rules will be delayed until no more beta rules can be applied. Note that the opposite order would require applying the gamma rule Q times, where Q is the maximum quantifier depth, before any beta rule is applied. Suppose that a beta rule is then applied to $T\ \forall x[P(x)] \lor G$, then we cannot apply a gamma rule to $T\ \forall x[P(x)]$. Delaying gamma rules has another advantage in our setting. Consider the following tableaux, both with the initial sequent $\{F\ P(a)\ \&\ P(b),\ T\ \forall x[P(x)]\}$. If we apply the gamma rule first, we obtain the following tableau.

initial	1	$F\ P(a)\ \&\ P(b)$
sequent	2	$T\ \forall x[P(x)]$
$T\forall(2)$	3	$TP(X_1)$

Left branch	Right branch
$F\&(1)$ 4.1 $FP(a)$	4.2 $FP(b)$
	$T\forall(2)$ 5.2 $TP(X_2)$
3 and 4.1 cause closure if $X_1 = a$	4.2 and 5.2 cause closure if $X_2 = b$

For this tableau we need a quantifier depth of at least 2 as the gamma rule is applied twice in the second branch. In the following tableau, where application of the gamma rule is delayed, the gamma rule is applied only once in both branches.

initial	1	$F\ P(a)\ \&\ P(b)$
sequent	2	$T\ \forall x[P(x)]$

Left branch	Right branch
$F\&(1)$ 3.1 $FP(a)$	3.2 $FP(b)$
$T\forall(2)$ 4.1 $TP(X_1)$	$T\forall(2)$ 4.2 $TP(X_2)$
3.1 and 4.1 cause closure if $X_1 = a$	3.2 and 4.2 cause closure if $X_2 = b$

These examples illustrate that applying gamma rules after beta rules, as compared to the opposite order, introduces additional different meta-variables, such that more different instantiations are possible at the same quantifier depth.

The ideas mentioned above have been incorporated in the tableaux-based theorem prover developed by W.M.J. Ophelders in his thesis, in which an efficiency comparison is also made with other theorem provers, both resolution-based and tableaux-based ones. See also W.M.J. Ophelders and H.C.M. de Swart, *Tableaux versus Resolution; a Comparison.*

In the program we have given, we check whether the restrictions have been violated after every unification. A more efficient alternative is to *prevent* the restrictions from being violated during the unification. We shall sketch how to write such a program.

Recall that the instantiation of meta-variables is controlled by the predicate `unify`, and the only actual instantiations take place in the second and third clause defining `unify`. One of these clauses is called if one argument is a Prolog variable X and the other argument is a non-variable Prolog term t. X is then instantiated to t if t does not contain X, i.e., if `does_not_occur_in(`X`,`t`)` succeeds.

Assume that the restrictions are not violated before this instantiation takes place. Further assume that t contains a parameter `sk`n. We can then split the list *Rstr* of restrictions into two parts: The part *Left* containing all elements of *Rstr* to the left of `par(sk`n`)`, and the part *Right* containing all elements to the right of `par(sk`n`)`. Obviously, the restrictions will be violated after the instantiation of X by t if there is an element `var(`s`)` in *Right* such that X occurs in s. In that case, after the instantiation, `sk`n will occur in the instantiation s' of s, while `var(`s'`)` occurs to the right of `par(sk`n`)` in the restriction list. So, we must require that X does not occur in any of the instantiations in *Right*, which is precisely the case if `not_in_instantiations(`X`,`*Right*`)` succeeds. If that is the case, then *Right* does not embody any restrictions with respect to X, so, in checking that the other parameters in t will not cause violations of the restrictions, it is no use searching for occurrences of X in this part of the restriction list. That is, for these further checks, we can simply take *Left* to be the restriction list.

Another approach to this problem is used by M. Fitting in his book *First-Order Logic and Automated Theorem Proving*. Instead of using `sk`n, he represents parameters by constructions like `sk(`n`,`L`)`, where L is a list of all (instantiations of) meta-variables that were introduced along the branch so far. It has the advantage that `sk(`n`,`L`)` will never occur in an instantiation of a variable that occurs in L because of the occur check. So, Fitting does not need any separate procedure to check whether the restrictions are violated; it is all done by the occur check. It has the disadvantage of introducing such constructs in the proof, where parameters are easier to read.

ANSWERS TO EXERCISES

25.1. Extend the program in Example 25.1. as follows.

(8) male(bob).
(9) male(tom).
(10) male(jim).
(11) female(pam).
(12) female(liz).
(13) female(pat).
(14) female(ann).
(15) offspring(X, Y) :- parent(Y, X).
(16) father(X, Y) :- parent(X, Y), male(X).
(17) mother(X, Y) :- parent(X, Y), female(X).
(18) sister(X, Y) :- parent(Z, X), parent(Z, Y), female(X).
(19) brother(X, Y) :- parent(Z, X), parent(Z, Y), male(X).

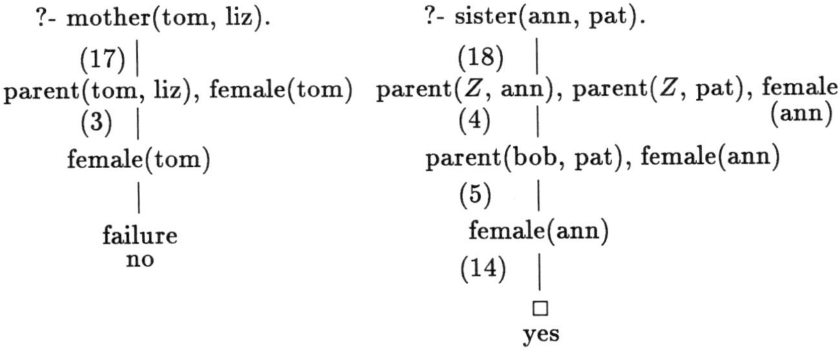

25.2.

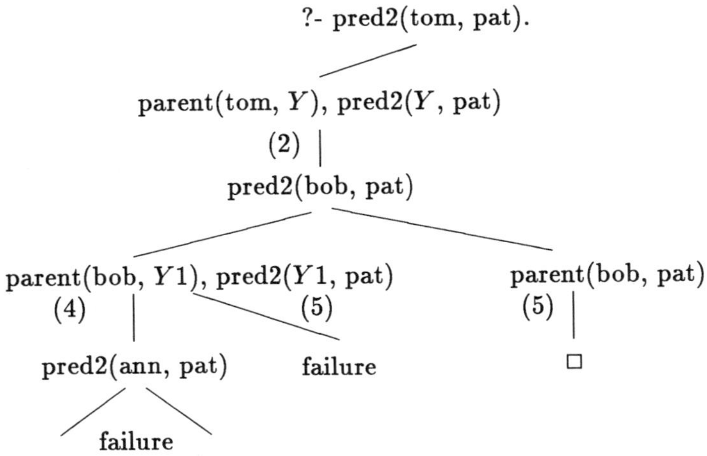

The system has to backtrack many times before it finds a successful branch in the search tree.

25.3. i) Replace Y by $f(X)$; Y does not occur in $f(X)$; result: $p(f(X), Z)$ and $p(f(X), c)$. Next replace Z by c. Consequently, $p(f(X), Z)$ and $p(Y, c)$ can be matched and unified. Result: $p(f(X), c)$.

ii) X and $f(X)$ can be matched, but not unified: replace X by $f(X)$; but X does occur in $f(X)$.

iii) Replace Y by $f(X)$. Result: $p(f(X), c)$ and $p(f(X), f(Z))$. c and $f(Z)$ cannot be matched.

25.4. conc([ ], $L2, L2$).

conc($[X \mid L1], L2, [X \mid L3]$) :- conc($L1, L2, L3$).

25.5. del($X, [X \mid L], L$).

del($X, [Y \mid L], [Y \mid L1]$) :- del($X, L, L1$).

25.6. length([ ], 0).

length($[X \mid L], N$) :- length(L, M), N is $M + 1$.

25.7.

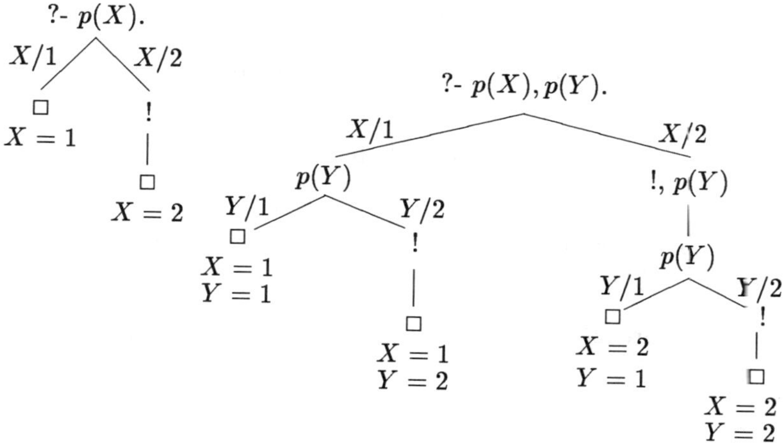

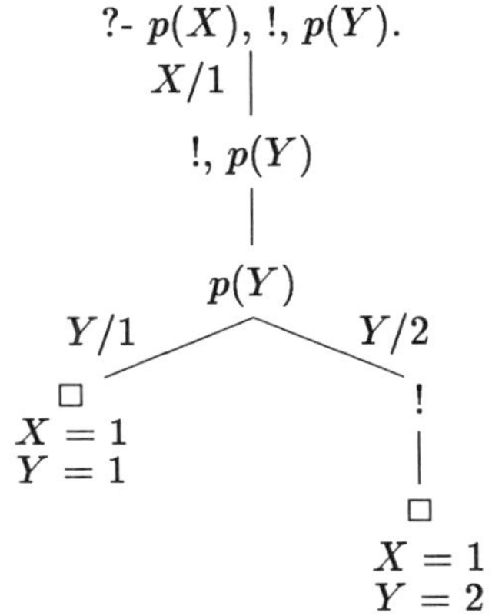

25.8.

```
p(X,0) :- X < 1, ! .
p(X,1) :- X >= 1, X < 2, ! .
p(X,2) :- X >= 2 .
```

25.9.

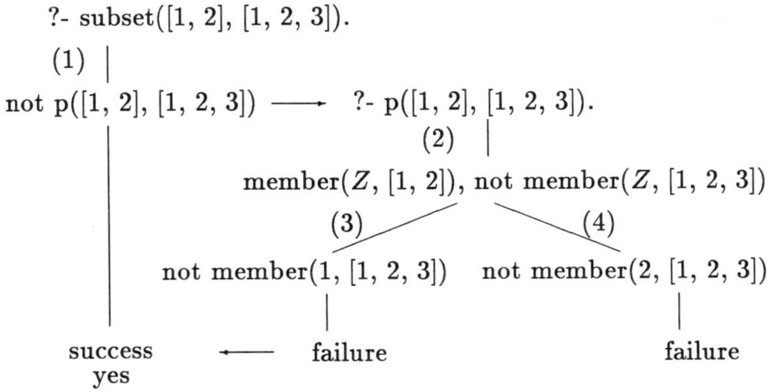

25.10.

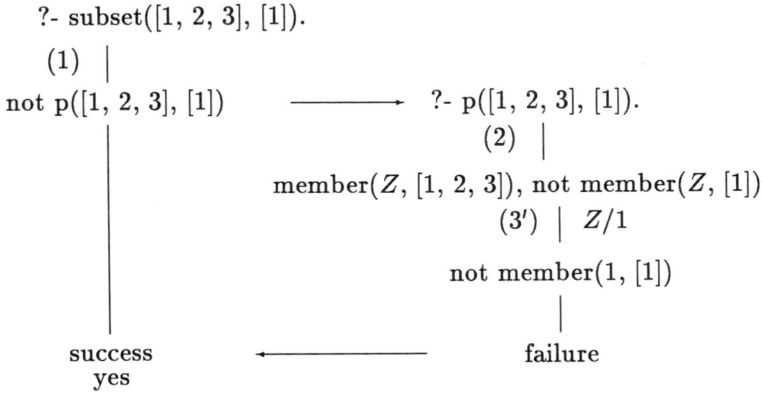

25.11. husband(X) :- family(X, _, _).
wife(X) :- family(_, X, _).
child(X) :- family(_, _, L), member(X, L).
exists(X) :- husband(X); wife(X); child(X).

i) ?- exists(person(N, S, _, _)).
ii) ?- child(person(N, S, date(_, _, 1973), _)).
iii) ?- wife(person(N, S, _, works(_, _))).
iv) ?- exists(person(N, S, date(_, _, Y), unemployed)), $Y < 1960$.
v) ?- exists(person(N, S, date(_, _, Y), works(_, Sal))), $Y < 1960$, Sal > 10000.
vi) ?- family(person(_, S, _, _),_, [_, _ | _]).
vii) ?- family(person(_, S, _, _), _, [ ]).

25.12. i) The reading $(a + b) * c$ has the following structure:

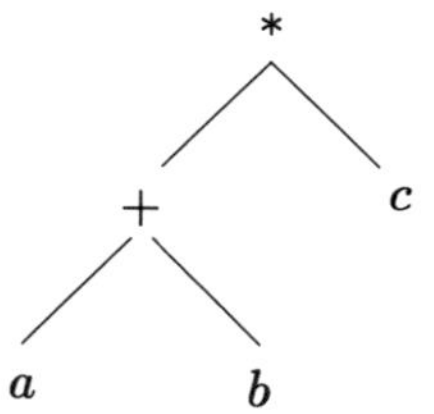

Now the precedence of $a + b$ is 500, which is greater than the precedence of $*$. Therefore this reading is rejected.

ii) The reading $a - (b - c)$ has the following structure:

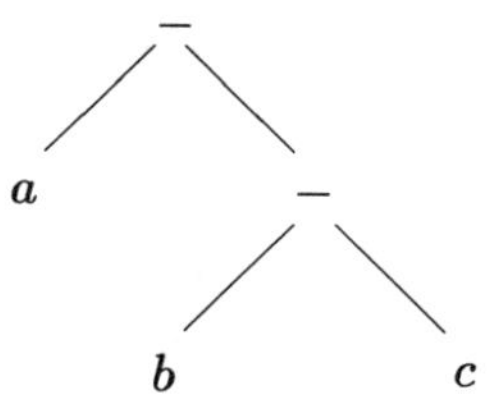

The precedence of $b - c$ is 500, which is not strictly smaller than that of the operator $-$. Since the operator $-$ has been defined to be of type yfx, this reading is impossible

iii) Since "has" has been defined as an infix operator and the arguments "peter" and "information" have precedence 0, which is strictly smaller than the precedence 600 of "has", "peter has information" will be read as "has(peter, information)".

26.1. $U_P = \{a, b, c\}$. $B_P = \{p(a,a), p(a,b), p(a,c), p(b,a), p(b,b), p(b,c), p(c,a), p(c,b), p(c,c)\}$. $M_P = B_P$.

26.2. (a) By definition of ε.

(b) Let $\theta = \{u_1/s_1, \ldots, u_m/s_m\}$ and $\sigma = \{v_1/t_1, \ldots, v_n/t_n\}$. Suppose $t = x$. If $x \notin \{u_1, \ldots, u_m, v_1, \ldots, v_n\}$, then $(x\theta)\sigma = x = x(\theta\sigma)$. If $x = u_i$, then $(x\theta)\sigma = s_i\sigma = x(\theta\sigma)$. If $x = v_j$ and $x \notin \{u_1, \ldots, u_m\}$, then $(x\theta)\sigma = t_j = x(\theta\sigma)$. This result for a variable x can easily be generalized to arbitrary terms and literals.

(c) It suffices to prove that $x((\theta\sigma)\gamma) = x(\theta(\sigma\gamma))$ for a variable x. $x((\theta\sigma)\gamma) = (x(\theta\sigma))\gamma = ((x\theta)\sigma)\gamma = (x\theta)(\sigma\gamma) = x(\theta(\sigma\gamma))$, by (b).

26.3. i) The premisses are (1) $p \vee q$, (2) $\neg p \vee q$ and (3) $\neg q \vee p$. The negation of the (putative) conclusion is (4) $\neg p \vee \neg q$.

(5) q, from (1) and (2);
(6) $\neg p$, from (4) and (5);
(7) $\neg q$, from (3) and (6);
(8) $\Box$, from (5) and (7).

ii) The negation of $p \supset (q \supset p)$ in clausal form yields (1) p, (2) q and (3) $\neg p$. From (1) and (3): $\Box$.

iii) The premisses are (1) $\neg p \vee r$, (2) $\neg q \vee r$. The negation of the (putative) conclusion in clausal form yields (3) $p \vee q$ and (4) $\neg r$.
(5) $q \vee r$, from (1) and (3);
(6) r, from (2) and (5);
(7) $\Box$, from (4) and (6).

iv) The clausal form of the premiss is (1) $p(x, c_1)$. The clausal form of the negation of the conclusion is (2) $\neg p(c_2, y)$. From (1) and (2) follows $\Box$ via the substitution $\{x/c_2, y/c_1\}$.

26.4. i) $\models p \vee \neg p$. Hence $P \vdash r$.

ii)

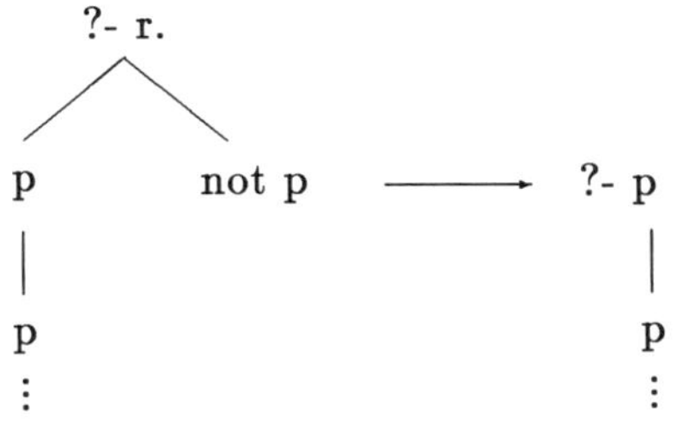

All branches in the search tree for ?- r. are infinite.

27.1.

i)
```
SELECT  t.snm, t.sadr,
        t.sres
FROM    SP t
WHERE   t.wnr = 9
AND     t.nbd > 2
```

ii)
```
SELECT  s.rnr
FROM    R s
WHERE   s.rnr NOT IN
   (SELECT  t.rnr
    FROM    ADM t
    WHERE   t.indat ≤ 800518
    AND     t.indat ≥ 800509
    AND     t.pnr IN
      (SELECT u.pnr
       FROM   P u
       WHERE  u.pres = "Princeton"))
```

iii)
```
SELECT  t.pnr, t.pnm, t.padr, t.pres
FROM    P t
WHERE   t.pnr IN
        (SELECT   u.pnr
         FROM     ADM u
         WHERE    u.snr IN
                  (SELECT  s.snr
                   FROM    SP s
                   WHERE   s.wnr = 9))
```

iv)
```
SELECT  t.snr, t.snm
FROM    SP t
WHERE   t.snr IN  (SELECT  u.snr
                   FROM    ADM u
                   WHERE   u.indat = 800303
                   AND     u.reas = "informaritis")
```

27.2. $D_1 \bowtie D_2$ is the empty set since no tuples agree on the common attribute "name".

$D_3 \bowtie D_2$

nr	wnm	sal	sex	anr	name	man
8	Johnson	2200	male	1	production	9
7	Johnson	3100	female	2	planning	7
9	Kiviat	2900	male	1	production	9

$D_1 \bowtie D_4$

nr	name	sal	sex	dept	anr	anm
7	Johnson	3100	female	2	2	planning
9	Kiviat	2900	male	1	1	production

$D_3 \bowtie D_4$

nr	wnm	sal	sex	anr	anm
7	Johnson	3100	female	2	planning
9	Kiviat	2900	male	1	production

27.3. The patient with number 500 is from Cranbury and has been hospitalized in the period in question in room number 11. So, 11 should not occur in the set of all room-numbers in which no patients from Cranbury were hospitalized between August 11 and 17, 1977. However, 11 is an element of the indicated set. For consider $s1$; $s1(\text{rnr}) = 11$. Then there is $t \in \text{ADM}$, namely $t1$, such that $t(\text{rnr}) = s1(\text{rnr})$ and $770811 \leq t(\text{indat}) \leq 770817$ and $\neg\exists u \in \text{P}[u(\text{pnr}) = t(\text{pnr}) = 400 \;\&\; u(\text{pres}) = \text{"Cranbury"}]$.

28.1. i)The following procedure determines if an integer i, greater than one, is prime.

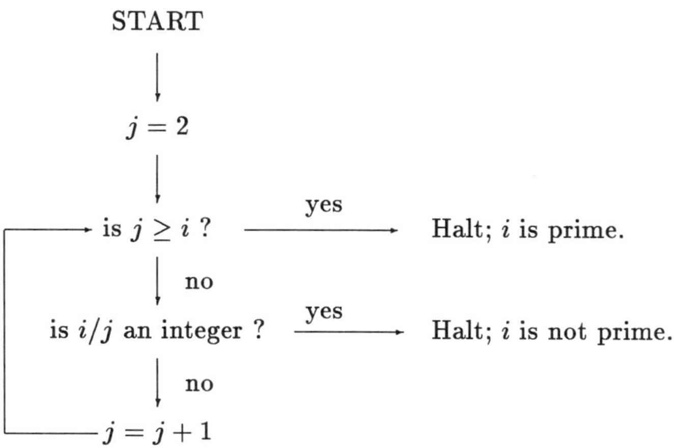

This procedure will terminate for all values of i: either a value of j will be reached which divides i, or j will eventually become equal to or greater than i. In either case the procedure terminates and answers the question whether i is prime. So this procedure is an *algorithm* for determining if an integer greater than one is prime.

ii) The following procedure determines, given a prime number p_n, whether there is a prime number greater than p_n.

1. Set $j = p_n + 1$.
2. If j is prime, then halt. There is a prime number greater than p_n.
3. If j is not prime, set $j = j + 1$.
4. Go back to instruction 2.

At first sight it may seem that this procedure may not terminate. However, since there are infinitely many prime numbers, this procedure will terminate for any input p_n and determine a prime number greater than p_n.

28.2. A procedure for determining if there is a perfect number greater than i is given in the figure below.

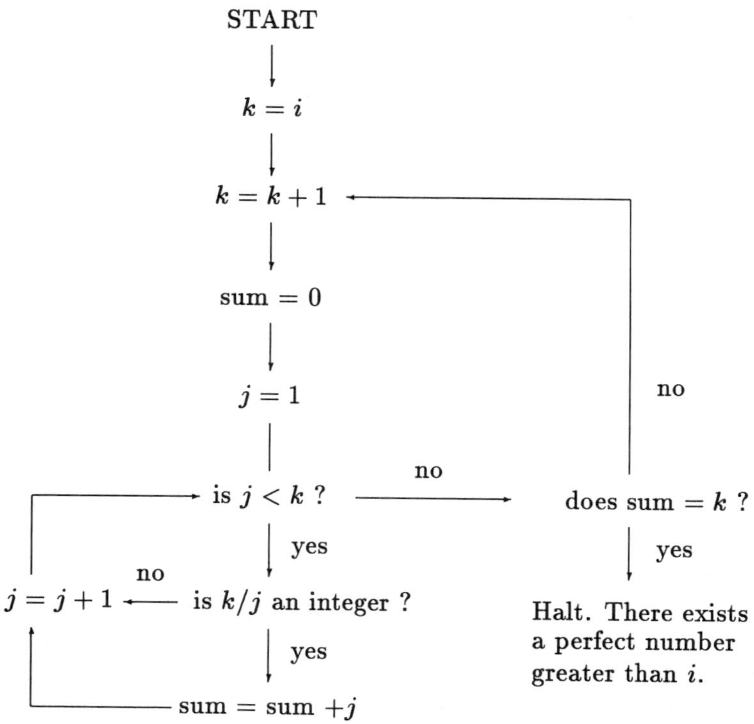

In case there is a perfect number greater than i the procedure, given input i, will find it. But in case there is no perfect number greater than i the procedure, given input i, does not terminate. Since we do not know whether there are infinitely many perfect numbers, we cannot prove that the procedure is an algorithm. It is an algorithm if and only if there are infinitely many perfect numbers. However, the latter statement is an unsolved problem.

28.3. Since no (uniform) method is known to decide for any natural number n whether there is a sequence of n consecutive 7's in the decimal expansion of π, we may not be able to compute $g(n)$ for certain n. In case there is no sequence of n consecutive 7's in the decimal expansion of π, we may never be aware of that. If so, we continue to generate successive digits in the decimal expansion of π looking for such a sequence without ever finding one.

28.4. $T = \langle\{q_0, q_1, q_2\},\ \{\#, 1\},\ q_0,\ \delta\rangle$, where
$\delta(q_0, \#) = (q_1, \#, R)$, $\delta(q_1, 1) = (q_2, \#, R)$, $\delta(q_2, \#) = (h, 1, R)$.
$\delta(q_1, \#) = (h, \#, O)$, $\delta(q_2, 1) = (q_1, \#, R)$,

28.5. $T = \langle\{q_0, q_1\},\ \{\#, 1\},\ q_0,\ \delta\rangle$, where $\delta(q_0, \#) = (q_1, \#, R)$, $\delta(q_1, 1) = (q_1, \#, R)$ and $\delta(q_1, \#) = (h, \#, O)$.

28.6. $T = \langle\{q_0, q_1, q_2, q_3, q_4\},\ \{\#, 1\},\ q_0,\ \delta\rangle$, where

$$\begin{array}{llllll} \delta(q_0,\#) = (q_1,\#,R), & \delta(q_2,1) = (q_2,1,R), & \delta(q_3,\#) = (q_4,\#,L), \\ \delta(q_1,1) = (q_1,\#,R), & \delta(q_2,\#) = (q_3,\#,R), & \delta(q_4,\#) = (q_4,\#,L), \\ \delta(q_1,\#) = (q_2,\#,R), & \delta(q_3,1) = (q_3,\#,R), & \delta(q_4,1) = (h,1,R). \end{array}$$

In order to ensure that this Turing machine handles the number 0 correctly, one should represent a natural number n by a sequence of $n+1$ 1's.

28.7. $T = \langle\{q_0, q_1\},\ \{\#, 1\},\ q_0,\ \delta\rangle$, where $\delta(q_0, \#) = (q_1, \#, R)$, $\delta(q_1, 1) = (q_1, 1, R)$ and $\delta(q_1, \#) = (h, 1, R)$.

28.8. $T = \langle\{q_0, q_1, q_2\},\ \{\#, 0, 1, Y, N\},\ q_0,\ \delta\rangle$, where

$$\begin{array}{lll} \delta(q_0,\#) = (q_1,\#,R), & \delta(q_1,\#) = (h,N,R), & \delta(q_2,0) = (q_2,\#,R), \\ & \delta(q_1,0) = (q_2,\#,R), & \delta(q_2,1) = (q_2,\#,R), \\ & \delta(q_1,1) = (q_1,\#,R), & \delta(q_2,\#) = (h,Y,R). \end{array}$$

Let $L := \{w \in \{0,1\}^* \mid w \text{ contains at least one } 0\}$.
If $w \in L$, then $(q_0, \underline{\#}w\#) \vdash_T^* (h, \#Y\underline{\#})$.
If $w \notin L$, then $(q_0, \underline{\#}w\#) \vdash_T^* (h, \#N\underline{\#})$.

28.9. Let $T = \langle K,\ \Sigma,\ q_0,\ \delta\rangle$ be the Turing machine of Example 28.11. Let $\delta' = \delta$, except $\delta'(q_2, \#) = (q_2, \#, O)$. Then $T' = \langle K,\ \Sigma,\ q_0,\ \delta'\rangle$ accepts $\{a^n \mid n \text{ is even}\}$.

28.10. Let $T = \langle K,\ \Sigma,\ q_0,\ \delta\rangle$ be the Turing machine of Example 28.11. Let $\delta' = \delta$, except $\delta'(q_1, \#) = (h, N, R)$ and $\delta'(q_2, \#) = (h, Y, R)$.
Then $T' = \langle K,\ \Sigma,\ q_0,\ \delta'\rangle$ decides the complement of $\{a^n \mid n \text{ is even}\}$.

28.11. T computes the function $f : \mathbb{N} \to \mathbb{N}$, defined by $f(n) = 0$ if $n \leq 2$ and $f(n) = n - 2$ if $n \geq 2$.

28.12. $T = \langle\{q_0, q_1, q_2, q_3\},\ \{\#, 1\},\ q_0,\ \delta\rangle$, where

$$\begin{array}{llllll} \delta(q_0,\#) &= (q_1,\#,R), & \delta(q_1,\#) &= (q_2,1,R), & \delta(q_2,\#) &= (q_3,\#,L),\\ \delta(q_1,1) &= (q_1,1,R), & \delta(q_2,1) &= (q_2,1,R), & \delta(q_3,1) &= (h,\#,O). \end{array}$$

28.13. i) $T_1 := \langle\{q_0, q_1, q_2, q_3, q_4\},\ \{0, 1, \#\},\ q_0,\ \delta_1\rangle$, where

$$\begin{array}{llllll} \delta_1(q_0,\#) &= (q_1,\#,R), & & & & \\ \delta_1(q_1,0) &= (q_2,0,R), & \delta_1(q_1,1) &= (q_4,1,L), & \delta_1(q_1,\#) &= (q_4,\#,L),\\ \delta_1(q_2,1) &= (q_3,1,R), & \delta_1(q_2,0) &= (q_4,0,L), & \delta_1(q_2,\#) &= (q_4,\#,L),\\ \delta_1(q_3,0) &= (h,0,R), & \delta_1(q_3,1) &= (q_4,1,L), & \delta_1(q_3,\#) &= (q_4,\#,L),\\ \delta_1(q_4,0) &= (q_4,0,O), & \delta_1(q_4,1) &= (q_4,1,O), & \delta_1(q_4,\#) &= (q_4,\#,O). \end{array}$$

T_1 recognizes $\{0\ 1\ 0\}$.
ii) In a similar way one can build a Turing machine $T_2 = \langle\{q_0', q_5, q_6, q_7, q_8\},\ \{0, 1, \#\},\ q_0',\ \delta_2\rangle$ that recognizes $\{1\ 0\ 1\}$.
iii) $T_1 \cup T_2 := \langle\{q_0, q_1, \ldots, q_4, q_0', q_5, \ldots, q_8\},\ \{0, 1, \#\},\ q_0,\ \delta\rangle$, where $\delta = \delta_1 \cup \delta_2$, except that $\delta(q_4, 0) = (q_4, 0, L)$, $\delta(q_4, 1) = (q_4, 1, L)$, $\delta(q_4, \#) = (q_0', \#, O)$. $T_1 \cup T_2$ recognizes $\{0\ 1\ 0\} \cup \{1\ 0\ 1\}$.

28.14. Note that $n = <a_1, \ldots, a_{k+1}>$ is even iff $a_{k+1} = 0$.
Let $T = \langle\{q_0, q_1, q_2, q_3, q_4\},\ \{0, 1, \#, Y, N\},\ q_0,\ \delta\rangle$ with

$$\begin{array}{llllll} \delta(q_0,\#) &= (q_1,\#,R), & \delta(q_2,0) &= (q_3,\#,L), & \delta(q_3,\#) &= (h,Y,R),\\ \delta(q_1,0) &= (q_1,0,R), & \delta(q_2,1) &= (q_4,\#,L), & \delta(q_4,0) &= (q_4,\#,L),\\ \delta(q_1,1) &= (q_1,1,R), & \delta(q_3,0) &= (q_3,\#,L), & \delta(q_4,1) &= (q_4,\#,L),\\ \delta(q_1,\#) &= (q_2,\#,L), & \delta(q_3,1) &= (q_3,\#,L), & \delta(q_4,\#) &= (h,N,R). \end{array}$$

28.15. $T = \langle \{q_0, \ldots, q_9\},\ \{1, \#\},\ q_0,\ \delta \rangle$ with $\delta(q_0, \#) = (q_1, \#, R)$,

$$\begin{array}{lll}
\delta(q_1,1) = (q_1,1,R), & \delta(q_4,1) = (q_4,1,L), & \delta(q_7,1) = (q_7,1,R), \\
\delta(q_1,\#) = (q_2,\#,R), & \delta(q_4,\#) = (q_5,\#,L), & \delta(q_7,\#) = (q_8,\#,R), \\
\delta(q_2,1) = (q_2,1,R), & \delta(q_5,1) = (q_5,1,L), & \delta(q_8,1) = (q_8,1,R), \\
\delta(q_2,\#) = (q_3,\#,L), & \delta(q_5,\#) = (q_6,\#,R), & \delta(q_8,\#) = (q_3,\#,L), \\
\delta(q_3,\#) = (h,\#,O), & \delta(q_6,1) = (q_7,\#,R), & \delta(q_9,1) = (q_9,\#,R), \\
\delta(q_3,1) = (q_4,\#,L), & \delta(q_6,\#) = (q_9,\#,R), & \delta(q_9,\#) = (h,\#,O).
\end{array}$$

Check that

$$\begin{array}{rl}
(q_0,\ \underline{\#}1\ 1\ 1\ 1\ 1\#1\ 1\#) & \vdash_T^* (q_4, \#1\ 1\ 1\ 1\ 1\#\underline{1}\#\#), \\
& \vdash_T^* (q_7, \#\#\underline{1}\ 1\ 1\ 1\#1\#\#), \\
& \vdash_T^* (q_4, \#\#1\ 1\ 1\ 1\underline{\#}\#\#\#), \\
& \vdash_T^* (q_7, \#\#\#\underline{1}\ 1\ 1\#\#\#\#), \\
& \vdash_T^* (h, \#\#\#1\ 1\ 1\underline{\#}\#\#\#),
\end{array}$$

and $(q_0,\ \underline{\#}1\ 1\#1\ 1\ 1\ 1\#) \vdash_T^* (h, \#\#\#\#\#\underline{\#}\#\#\#)$.

28.16. $T = \langle \{q_0, \ldots, q_7\},\ \{1, \#\},\ q_0,\ \delta \rangle$ with

$$\begin{array}{lll}
\delta(q_0,\#) = (q_1,\#,R), & \delta(q_3,\#) = (q_4,1,R), & \delta(q_6,1) = (q_6,1,L), \\
\delta(q_1,1) = (q_2,\#,R), & \delta(q_3,1) = (q_3,1,R), & \delta(q_6,\#) = (q_1,\#,R), \\
\delta(q_1,\#) = (q_7,\#,R), & \delta(q_4,\#) = (q_5,1,L), & \delta(q_7,1) = (q_7,1,R), \\
\delta(q_2,1) = (q_2,1,R), & \delta(q_5,1) = (q_5,1,L), & \delta(q_7,\#) = (h,\#,O). \\
\delta(q_2,\#) = (q_3,\#,R), & \delta(q_5,\#) = (q_6,\#,L), &
\end{array}$$

Check that

$$\begin{array}{rl}
(q_0, \underline{\#}1\ 1\#) & \vdash_T^* (q_5, \#\#1\#\underline{1}\ 1), \\
& \vdash_T^* (q_1, \#\#\underline{1}\#1\ 1), \\
& \vdash_T^* (q_5, \#\#\#\#1\ 1\ \underline{1}\ 1), \\
& \vdash_T^* (q_7, \#\#\#\#\underline{1}\ 1\ 1\ 1), \\
& \vdash_T^* (h, \#\#\#\#1\ 1\ 1\ 1\underline{\#}).
\end{array}$$

28.17. $T = \langle \{q_0, q_1, q_2, q_3, q_4\},\ \{a, b, \#\},\ q_0,\ \delta \rangle$ with

$$\begin{array}{lll}
\delta(q_0,\#) = (q_1,\#,R), & \delta(q_2,a) = (q_2,a,R), & \delta(q_3,b) = (q_4,\#,L), \\
\delta(q_1,a) = (q_2,\#,R), & \delta(q_2,b) = (q_2,b,R), & \delta(q_3,\#) = (q_3,\#,O), \\
\delta(q_1,b) = (q_1,b,O), & \delta(q_2,\#) = (q_3,\#,L), & \delta(q_4,a) = (q_4,a,L), \\
\delta(q_1,\#) = (h,\#,O), & \delta(q_3,a) = (q_3,a,O), & \delta(q_4,b) = (q_4,b,L), \\
& & \delta(q_4,\#) = (q_1,\#,R).
\end{array}$$

Check that

$$\begin{array}{rl}
(q_0,\ \underline{\#}a\ a\ b\ b\#) & \vdash_T^* (h, \#\#\#\underline{\#}\#\#), \\
(q_0,\ \underline{\#}a\ b\ b\#) & \vdash_T^* (q_1, \#\#\underline{b}\#\#),
\end{array}$$

and $(q_0,\ \underline{\#}a\ a\ b\#) \vdash_T^* (q_3, \#\#\underline{\#}\#\#)$.

28.18. $T = \langle\{q_0, q_1, q_2\},\ \{a, b, \#\},\ q_0,\ \delta\rangle$ with

$$\begin{array}{llll} \delta(q_0,\#) = (q_1,\#,R), & \delta(q_2,a) = (q_2,a,O), \\ \delta(q_1,a) = (q_1,a,R), & \delta(q_2,b) = (q_2,b,R), \\ \delta(q_1,b) = (q_2,b,R), & \delta(q_2,\#) = (h,\#,O). \end{array}$$

28.19. $T = \langle\{q_0, q_1, q_2\},\ \{a, b, \#\},\ q_0,\ \delta\rangle$ with $\delta(q_0,\#) = (q_1,\#,R)$,

$$\begin{array}{ll} \delta(q_1,a) = (q_2,a,R), & \delta(q_2,a) = (q_2,a,O), \\ \delta(q_1,b) = (q_1,b,R), & \delta(q_2,b) = (q_1,b,R), \\ \delta(q_1,\#) = (h,\#,O), & \delta(q_2,\#) = (h,\#,O). \end{array}$$

28.20. $T = \langle\{q_0, q_1, q_2\},\ \{a, b, \#\},\ q_0,\ \delta\rangle$ with $\delta(q_0,\#) = (q_1,\#,R)$.

$$\begin{array}{ll} \delta(q_1,a) = (q_2,a,R), & \delta(q_2,a) = (q_2,a,O), \\ \delta(q_1,b) = (q_1,b,O), & \delta(q_2,b) = (q_1,b,R), \\ \delta(q_1,\#) = (h,\#,O), & \delta(q_2,\#) = (q_2,\#,O). \end{array}$$

28.21. $T = \langle\{q_0, q_1\},\ \{a, b, \#\},\ q_0,\ \delta\rangle$ with

$$\begin{array}{ll} \delta(q_0,\#) = (q_1,\#,R), & \delta(q_1,b) = (q_1,b,R), \\ \delta(q_1,a) = (h,a,R), & \delta(q_1,\#) = (q_1,\#,O). \end{array}$$

29.1. $\rho(T) = c\ 1^2\ c\ c\ 1^2\ c\ 1^4\ c\ 1\ c\ 1^6\ c\ 1^3\ c\ c\ 1^2\ c\ 1^6\ c\ 1^2\ c\ 1^4\ c\ 1^2\ c$
$c\ 1^2\ c\ 1^9\ c\ 1^2\ c\ 1^4\ c\ 1^2\ c\ c.$

29.2. String i) is not the encoding of a Turing machine: it starts with $c\ 1^2\ c$, saying that q_0 would be the initial state and then continues with $c\ 1^2\ c\ 1^2\ c\ 1^2\ c\ 1^4\ c\ 1^2\ c$ which is the encoding of $\delta(q_0, R) = (q_0, s_0, R)$. However, this is not an instruction of a Turing machine.

String ii) is the encoding of the Turing machine $T = \langle\{q_0, q_1\},\ \{s_0, s_1\},\ q_0,\ \delta\rangle$ with $\delta(q_0, s_0) = (q_1, s_1, R)$ and $\delta(q_1, s_1) = (h, s_1, O)$.

29.3. Let T_U be a universal Turing machine accepting $H_0 := \{\rho(T)\rho(w) \mid T \text{ accepts } w\}$; i.e., for any Turing machine T and input w, T accepts w iff T_U accepts $\rho(T)\rho(w)$.
Construct Turing machine T_1 as follows: T_1 transforms its input string $\underline{\#}w\#$ into $\underline{\#}w\rho(w)\#$ and then yields control to T_U.
Then $\rho(T) \in H_1$ iff T accepts $\rho(T)$
iff T_U accepts $\rho(T)\rho(\rho(T))$
iff T_1 accepts $\rho(T)$.

So T_1 accepts H_1.

29.4. By Theorem 29.14. $\{A \mid \models A\}$ is Turing-acceptable. It is not difficult to see that $\{A \mid \models \neg A\}$, i.e., $\{A \mid A \text{ is not satisfiable}\}$, is Turing acceptable too. Now, if $L = \{A \mid A \text{ is a formula of the predicate calculus and } A \text{ is satisfiable}\}$ were Turing-acceptable, then both L and $\bar{L}$, the complement of L, would be Turing-acceptable, and hence, by Theorem 28.14., L would be Turing-decidable, contradicting Corollary 29.13. Consequently, in the case of the predicate calculus, $\{A \mid A \text{ is satisfiable}\}$ is not Turing-acceptable.

29.5. $T = \langle\{q_0, q_1, q_2, q_3, q_4\},\ \{a, b, \#\},\ q_0,\ \Delta\rangle$, where Δ contains the following quintuples:

$$
\begin{array}{llll}
(q_0,\#,q_1,\#,R), & (q_2,a,q_2,a,O), & (q_3,a,q_4,a,R), & (q_4,a,h,a,R), \\
(*)\left\{\begin{array}{l}(q_1,a,q_1,a,R), \\ (q_1,a,q_2,a,R),\end{array}\right. & \begin{array}{l}(q_2,b,q_3,b,R), \\ (q_2,\#,q_2,\#,O),\end{array} & \begin{array}{l}(q_3,b,q_3,b,O), \\ (q_3,\#,q_3,\#,O),\end{array} & \begin{array}{l}(q_4,b,q_4,b,O), \\ (q_4,\#,q_4,\#,O).\end{array} \\
(q_1,\#,q_1,\#,O), & & &
\end{array}
$$

$(*)$ T is nondeterministic since it contains two quintuples of the form $(q_1, a, \ , \ , \ )$.

29.6. $T_2 = \langle\{q_0, q_1, q_2, q_3, q_4\},\ \{a, b, \#\},\ q_0,\ \delta\rangle$ with

$$
\begin{array}{llllll}
\delta(q_0,\#) = (q_1,\#,R), & \delta(q_2,a) = (q_1,a,O), & \delta(q_3,b) = (q_1,b,R), \\
\delta(q_1,a) = (q_2,a,R), & \delta(q_2,b) = (q_3,b,R), & \delta(q_3,\#) = (q_3,\#,O), \\
\delta(q_1,b) = (q_1,b,R), & \delta(q_2,\#) = (q_2,\#,O), & \delta(q_4,a) = (h,a,R), \\
\delta(q_1,\#) = (q_1,\#,R), & \delta(q_3,a) = (q_4,a,R), & \delta(q_4,b) = (q_1,b,L), \\
 & & \delta(q_4,\#) = (q_4,\#,O).
\end{array}
$$

Verify that, for instance, $(q_0,\ \underline{\#}a\ b\ a\ b\ a\ a\#) \vdash_T^* (h, \#a\ b\ a\ b\ a\ a\underline{\#})$.

29.7. The following nondeterministic Turing machine accepts the set of all words over $\{a, b\}$ containing at least two consecutive occurrences of a or at least two consecutive occurrences of b.

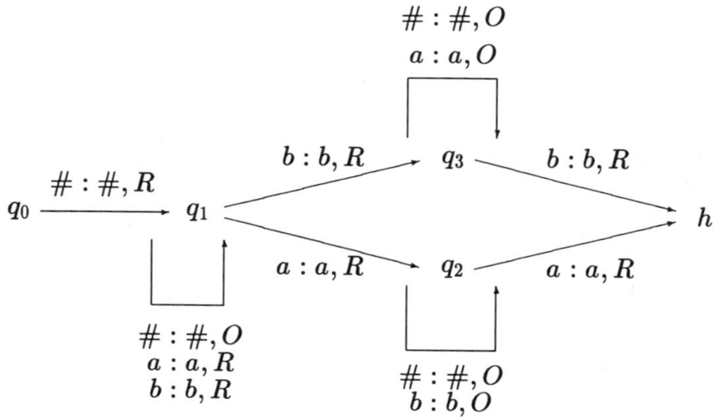

29.8. The following nondeterministic Turing machine accepts the set of all words over $\{a, b\}$ containing at least one occurrence of $a\ b$ or at least one occurrence of $b\ a$.

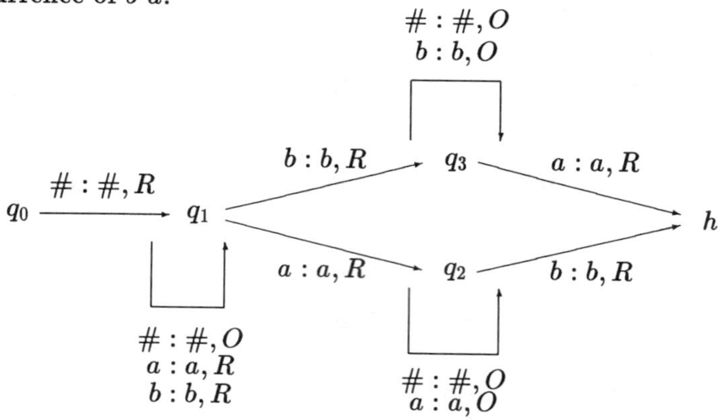

29.9. Suppose T accepts w. Then the intended interpretation is a finite model of $\Delta'_{T,w} \cup \{H\}$.

Conversely, suppose $\Delta'_{T,w} \cup \{H\}$ has a finite model M and suppose T does not accept w. Then for each $n \in \mathbb{N}$, T has not stopped before time n given input w. So, by (**) in the proof of Theorem 29.12,

$\Delta_{T,w}$ implies a description of time n for every $n \in \mathbb{N}$. So for every $n \in \mathbb{N}$, M is a finite model of a description of time n. (1)

Since M is a model of H where $H := \exists t \exists x[H(t, x)]$, there are a time t' and a square number x' in the finite domain of M such that $H^M(t', x')$

holds, H^M being the interpretation of H in M. Since $\Delta'_{T,w}$ contains the formula $\forall x \forall y \forall t[H(t,x) \supset \neg Q_{i_1}(t,y)\ \&\ \ldots\ \&\ \neg Q_{i_n}(t,y)]$, it follows that for all square numbers y' in the finite domain of M,

$$\text{not } Q_{i_1}^M(t',y') \text{ and } \ldots \text{ and not } Q_{i_n}^M(t',y'). \tag{2}$$

But since M is finite, $t' = t'_n$ for some $n \in \mathbb{N}$. Since (by (1)) M is a model of the description of time n, it follows that $Q_i^M(t',x')$ for some $i \in \{i_1,\ldots,i_n\}$ and some square number x' in the finite domain of M. And $Q_i^M(t',x')$ contradicts (2).

30.1. One can design a nondeterministic Turing machine T to accept $L = \{1^n \mid n \text{ is composite}\}$ in polynomial time, by guessing the factors, if there are any. This machine operates as follows, when given input $\underline{\#}\ 1^n\ \#$.

1. Nondeterministically choose two numbers $p, q > 1$ and tranform the input into $\underline{\#}\ 1^n\ \#\ 1^p\ \#\ 1^q\ \#$.
2. Use the multiplication machine to transform the result of Step 1 into $\#\ 1^n\ \#\ 1^{p\cdot q}\ \underline{\#}$.
3. Check whether $1^n = 1^{p\cdot q}$; i.e., check whether the lengths of these strings are equal. If they are equal, then halt; otherwise continue computing forever in some trivial fashion.

30.2. It is trivial to show 1. $f \approx f$ and 2. if $f \approx g$, then $g \approx f$. In order to show the transitivity of $\approx$ (i.e., if $f \approx g$ and $g \approx h$, then $f \approx h$), it suffices to show the transitivity of $\preceq$. So suppose $f \preceq g$ and $g \preceq h$, i.e., there are $c_1, c_2 > 0$ and $n_1, n_2 \in \mathbb{N}$ such that $f(n) \leq c_1 \cdot g(n)$ for all $n \geq n_1$, and $g(n) \leq c_2 \cdot h(n)$ for all $n \geq n_2$. Then $f(n) \leq c_1 \cdot c_2 \cdot h(n)$ for all $n \geq \max(n_1, n_2)$, i.e., $f \preceq h$.

30.3. (a) $a_0 + a_1 n + \ldots + a_k n^k \preceq n^k$ since for all $n \geq 1$, $a_0 + a_1 n + \ldots + a_k n^k \leq (\max(a_0,0) + \max(a_1,0) + \ldots + \max(a_{k-1},0) + a_k) \cdot n^k$. (Notice that possibly $a_j < 0$ for $0 \leq j < k$ and that $a_k > 0$, due to the definition of a polynomial.) Also $n^k \preceq a_0 + a_1 n + \ldots + a_k n^k$ since for all $n \geq 2 \cdot (|a_0| + \ldots + |a_{k-1}|)$, $n^k \leq 2 \cdot (a_0 + a_1 n + \ldots + a_k n^k)$.
(b) It is obvious that $n^k \preceq n^{k+1}$. Now suppose that $n^{k+1} \preceq n^k$, i.e., there is a constant $c > 0$ such that for all $n \geq n_0$, $n^{k+1} \leq c \cdot n^k$. However, this inequality fails for $n = \max(\lceil c \rceil + 1,\ n_0)$. Therefore, not $n^{k+1} \preceq n^k$ and not $n^k \approx n^{k+1}$. (By $\lceil c \rceil$ we mean the greatest integer $\leq c$.)

30.4. The number of tape squares visited by the read-write head cannot be more than the number of steps in the computation. So, for all n, $s(n) \leq t(n)$. Hence, if t is a polynomial, one can take s to be a polynomial too.

31.1. 1)

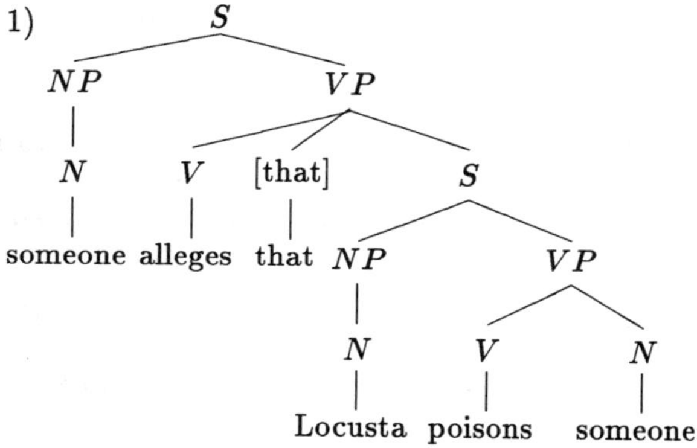

2)

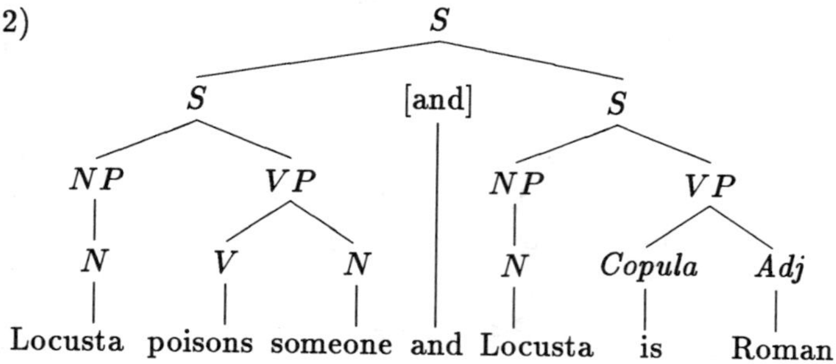

3)

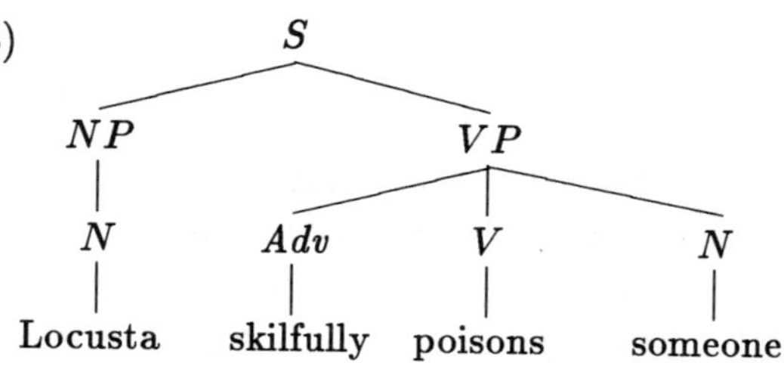

Phrase-marker for the surface structure of the sentences in 1, 2 and 3:

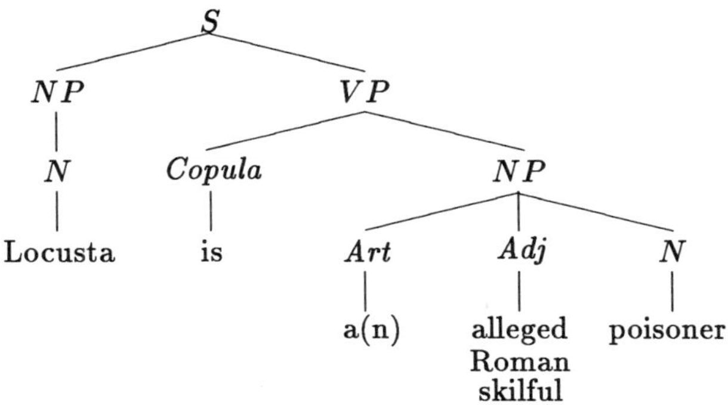

31.2.

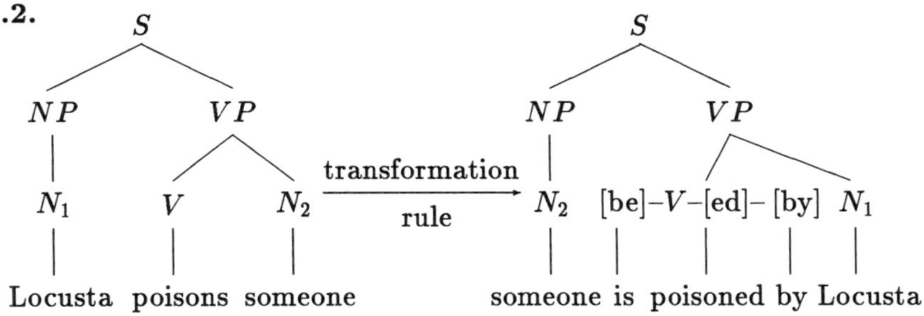

31.3. i) $T = \langle\{q_0, q_1, q_2\},\ \{0, 1, \#\},\ q_0,\ \delta\rangle$ with $\delta(q_0, \#) = (q_1, \#, R)$,

$$\begin{array}{lll} \delta(q_1,1) = (q_1,1,R), & & \delta(q_2,1) = (q_2,1,R), \\ \delta(q_1,0) = (q_2,0,R), & & \delta(q_2,0) = (q_2,0,R), \\ \delta(q_1,\#) = (q_1,\#,O), & & \delta(q_2,\#) = (h,\#,O). \end{array}$$

ii)

$$\begin{array}{rclrclrclrcl} (1) & & & (2) & & & & & & (3) & & \\ S & \to & \#X\#, & q_0\# & \to & \#q_1, & & & & h\# & \to & Y\ , \\ X & \to & X1\ , & q_11 & \to & 1q_1, & q_21 & \to & 1q_2, & 1Y & \to & Y1, \\ X & \to & X0\ , & q_10 & \to & 0q_2, & q_20 & \to & 0q_2, & 0Y & \to & Y0, \\ \#X & \to & q_0\#\ . & q_1\# & \to & q_1\#, & q_2\# & \to & h\#. & \#Y & \to & e. \end{array}$$

$G = \langle V_N,\ V_T,\ P,\ S\rangle$, where $V_N = \{q_0, q_1, q_2, S, X, Y, h, \#\}$, $V_T = \{0, 1\}$, and P consists of the productions in (1), (2) and (3).

31.4. $G = \langle V_N,\ V_T,\ P,\ S\rangle$, where $V_T = \{Q,\ ',\ \sim,\ \supset,\ \&,\ \vee,\ \neg,\ (,\ )\}$, $V_N = \{S, A\}$, and $P = \{S \to A,\ S \to (S \sim S),\ S \to (S \supset S),$ $S \to (S \,\&\, S),\ S \to (S \vee S),\ S \to (\neg S),\ A \to A',\ A \to Q\}$.

31.5. Each string of the form 0^n1^n, $n \geq 1$, is generated by G:

$$S \to 0S1 \to 00S11 \to 0^3S1^3 \to \ldots \to 0^{n-1}S1^{n-1} \to 0^n1^n.$$

Furthermore it is easy to see that these are the only strings in $L(G)$.

31.6. $G = \langle \{S, A\},\ \{0,1\},\ P,\ S\rangle$, where P contains the following productions:

$$\begin{array}{ll} S \to 1 & , A \to 0, \\ S \to 0 & , A \to 0S, \\ S \to 0S & , S \to 1A. \end{array}$$

31.7. $G = \langle \{S, A, B, C\},\ \{a, b, c\},\ P,\ S\rangle$, where P consists of the following productions:

$$\begin{array}{llll} S \to ABC, & A \to a, & AB \to BA, & BC \to CB, \\ S \to SS, & B \to b, & AC \to CA, & CA \to AC, \\ & C \to c, & BA \to AB, & CB \to BC. \end{array}$$

31.8. We have to show that $L - \{e\} = L(G)$, i.e. that $L - \{e\}$ and $L(G)$ have the same elements.

1. So suppose $w \in L - \{e\}$, i.e., w contains an even number of b's and $w \neq e$. Then it is not hard to see that w can be generated by G, i.e., $S \overset{*}{\underset{G}{\Rightarrow}} w$, and hence $w \in L(G)$.
2. Conversely, suppose $w \in L(G)$. Then it follows from the definition of the productions in G that w contains an even number of b's and $w \neq e$, and hence that $w \in L - \{e\}$.

31.9. Similar to 31.8. Note that if an expression u generated by the grammar ($S \overset{*}{\underset{G}{\Rightarrow}} u$) contains three or more consecutive b's, it must also contain the nonterminal D, and hence $u \notin V_T^*$, so $u \notin L(G)$.

31.10. Similar to 31.8.

31.11. Similar to 31.8.

31.12. 1) We show that $S \overset{*}{\underset{G}{\Rightarrow}} a^nb^nc^n$ for each $n \geq 1$.

Using production (1), $S \to aSBC$, $n-1$ times: $S \overset{*}{\underset{G}{\Rightarrow}} a^{n-1}S(BC)^{n-1}$.

Using production (2), $S \to aBC$, once: $S \overset{*}{\underset{G}{\Rightarrow}} a^n(BC)^n$.

Using production (3), $CB \to BC$, repeatedly: $S \overset{*}{\underset{G}{\Rightarrow}} a^nB^nC^n$.

Using production (4), $aB \to ab$, once: $S \overset{*}{\underset{G}{\Rightarrow}} a^nbB^{n-1}C^n$.

Using production (5), $bB \to bb$, $n-1$ times: $S \overset{*}{\underset{G}{\Rightarrow}} a^nb^nC^n$.

Using production (6), $bC \to bc$, once: $S \overset{*}{\underset{G}{\Rightarrow}} a^nb^ncC^{n-1}$.

Using production (7), $cC \to cc$, $n-1$ times: $S \overset{*}{\underset{G}{\Rightarrow}} a^nb^nc^n$.

2) It takes some more trouble to see that the words $a^nb^nc^n$ for $n \geq 1$ are the only terminal strings in $L(G)$. See, for instance, J.E. Hopcroft and J.D. Ullman, *Formal Languages and their relation to Automata*, Example 2.2.

31.13. Similar to 31.12. The grammar is context-free.

31.14.
$$\begin{aligned}
L(G_1) &= \{a^nba^n \mid n \geq 0)\}\text{; type 2.}\\
L(G_2) &= \{a^nb^m \mid n,m \geq 0\} - \{e\}\text{; type 3.}\\
L(G_3) &= \{a^nb^nc^n \mid n \geq 1\}\text{; type 1.}\\
L(G_4) &= \{ww^R \mid w \in \{a,b\}^*\} - \{e\}\text{; type 2.}\\
L(G_5) &= \{a^mb^nc^n \mid n,m \geq 1\}\text{; type 2.}\\
L(G_6) &= \{a^nb^m \mid n,m \geq 1\}\text{; type 2.}\\
L(G_7) &= \emptyset\text{; type 2.}
\end{aligned}$$

31.15. i) Let $G_1 = \langle V_{N_1},\ V_{T_1},\ P_1,\ S_1\rangle$ and $G_2 = \langle V_{N_2},\ V_{T_2},\ P_2,\ S_2\rangle$ be context-free grammars. Without loss of generality assume that $V_{N_1} \cap V_{N_2} = \emptyset$. Let $S_u \notin V_{N_1} \cup V_{N_2}$ and $S_u \notin V_{T_1} \cup V_{T_2}$ and define $G_u :=$

$$\langle (V_{N_1} \cup V_{N_2}) \cup \{S_u\},\ V_{T_1} \cup V_{T_2},\ (P_1 \cup P_2) \cup \{S_u \to S_1,\ S_u \to S_2\},\ S_u\rangle.$$

Then for all $w \in (V_{T_1} \cup V_{T_2})^*$, $S_u \overset{*}{\underset{G_u}{\Rightarrow}} w$ iff $S_1 \overset{*}{\underset{G_u}{\Rightarrow}} w$ or $S_2 \overset{*}{\underset{G_u}{\Rightarrow}} w$.
So, because we have assumed that $V_{N_1} \cap V_{N_2} = \emptyset$,

$$S_u \overset{*}{\underset{G_u}{\Rightarrow}} w \text{ iff } S_1 \overset{*}{\underset{G_1}{\Rightarrow}} w \text{ or } S_2 \overset{*}{\underset{G_2}{\Rightarrow}} w.$$

In other words, $w \in L(G_u)$ iff $w \in L_1$ or $w \in L_2$, i.e., $L(G_u) = L_1 \cup L_2$.
ii) Let $G_1 = \langle V_{N_1},\ V_{T_1},\ P_1,\ S_1\rangle$ and $G_2 = \langle V_{N_2},\ V_{T_2},\ P_2,\ S_2\rangle$ be context-free grammars. Without loss of generality assume that $V_{N_1} \cap V_{N_2} = \emptyset$.

Let $S_c \notin V_{N_1} \cup V_{N_2}$ and $S_c \notin V_{T_1} \cup V_{T_2}$ and define
$G_c := \langle (V_{N_1} \cup V_{N_2}) \cup \{S_c\},\ V_{T_1} \cup V_{T_2},\ (P_1 \cup P_2) \cup \{S_c \to S_1S_2\},\ S_c \rangle$.
Then $S_c \overset{*}{\underset{G_c}{\Rightarrow}} w$ iff for some $w_1 \in V_{T_1}^*$ and for some $w_2 \in V_{T_2}^*$, $w = w_1w_2$ and $S_1 \overset{*}{\underset{G_1}{\Rightarrow}} w_1$ and $S_2 \overset{*}{\underset{G_2}{\Rightarrow}} w_2$. So, $w \in L(G_c)$ iff for some $w_1 \in L_1$ and for some $w_2 \in L_2$, $w = w_1w_2$. In other words, $L(G_c) = L_1L_2$.
iii) Suppose $G = \langle V_N,\ V_T,\ P,\ S \rangle$ is a context-free grammar generating L. Let $G^* := \langle V_N \cup \{S_1\},\ V_T,\ P \cup \{S_1 \to S_1S,\ S_1 \to S\},\ S_1 \rangle$.
Then G^* is context-free and $L(G^*) = L^* - \{e\}$.
iv) Let $G = \langle V_N,\ V_T,\ P,\ S \rangle$ be context-free such that $L(G) = L$.
Define $G^R := \langle V_N,\ V_T,\ P^R,\ S \rangle$ to be the grammar with P^R consisting of all productions of the form $A \to \beta^R$ if $A \to \beta$ is a production in P.
Then G^R is context-free too and $w \in L$ iff $w^R \in L(G^R)$.
v) $L(G_1) = \{a^mb^nc^n \mid n, m \geq 1\}$ is a context-free language (see $L(G_5)$ in exercise 31.14.). Define $G_2 := \langle \{S, B, C\},\ \{a, b, c\},\ \{S \to BC,\ B \to aBb,\ B \to ab,\ C \to cC,\ C \to c\},\ S \rangle$. Then G_2 generates the context-free language $L(G_2) = \{a^nb^nc^m \mid n, m \geq 1\}$.
But $L(G_1) \cap L(G_2) = \{a^nb^nc^n \mid n \geq 1\}$ is not context-free (see exercise 31.14. and exercise 31.17.).
vi) Suppose for all context-free language L, if L is context-free, then so is $\bar{L}$. (∗).
Now consider the grammars G_1 and G_2, given in the solution of v). Let $L_1 := L(G_1)$ and $L_2 := L(G_2)$. Since G_1 and G_2 are context-free, by (∗), $\bar{L}_1$ and $\bar{L}_2$ would be context-free too. Then by exercise i) $\bar{L}_1 \cup \bar{L}_2$ would be context-free; so by (∗), $\overline{\bar{L}_1 \cup \bar{L}_2}$ would be context-free.
But $L_1 \cap L_2 = \overline{\bar{L}_1 \cup \bar{L}_2}$. Contradicting the fact shown in v) that $L_1 \cap L_2$ is not context-free.

31.16. i) $G_1 = \langle \{S\},\ \{a\},\ \{S \to a\},\ S \rangle$.
One can show that for any context-free grammar G with $L(G) \neq \emptyset$ there is an equivalent context-free grammar G_1 with no useless symbols. See, for instance, John E. Hopcroft and Jeffrey D. Ullman, *Formal languages and their relation to automata*, Theorem 4.3.
ii) $G_2 = \langle \{S, A, B\},\ \{b\},\ \{S \to BA,\ A \to BB,\ A \to b,\ B \to BB,\ B \to b\},\ S \rangle$. One can show that for any context-free grammar G there is an equivalent context-free grammar G_2 with no unit productions and no useless symbols. See, for instance, J.E. Hopcroft and J.D. Ullman, *op. cit.*, Theorem 4.4.

iii) The productions for the grammar G_3 in Chomsky normal form are shown below:

$$\begin{array}{llll} S \to C_1A & , S \to C_4B & , C_1 \to b & , C_4 \to a\ , \\ A \to C_2S & , B \to C_5S & , C_2 \to a & , C_5 \to b\ , \\ A \to C_3D_1 & , B \to C_6D_2 & , C_3 \to b & , C_6 \to a, \\ D_1 \to AA & , D_2 \to BB & , A \to a & , B \to b. \end{array}$$

One can show that any context-free language can be generated by a grammar in Chomsky normal form. See, for instance, J.E. Hopcroft and J.D. Ullman, *op. cit.*, Theorem 4.5.

31.17. Suppose $\{a^nb^nc^n \mid n \geq 0\} = L(G)$ for some context-free grammar G. Let k be the number of variables in G. Choose n such that $|a^nb^nc^n| > 2^{k-1}$. By the Pumping Theorem $a^nb^nc^n$ can be written as $u\ v\ z\ x\ y$ such that v or x is nonempty and $u\ v^i\ z\ x^i\ y$ is in $L(G)$ for each $i = 0, 1, 2, \ldots$. But this is impossible since $u\ v^i\ z\ x^i\ y$ cannot be of the form $a^mb^mc^m$.

31.18. T passes from state q_0 to q_1 or from q_1 back to q_0 when a b is read, but T remains in its current state when an a is read. Thus T counts b's modulo 2; since q_0 (the initial state) is the sole final state, T accepts a string if and only if the number of b's is even.

For instance, $(q_0, aabba) \vdash_T (q_0, abba)$
$\vdash_T (q_0, bba)$
$\vdash_T (q_1, ba)$
$\vdash_T (q_0, a)$
$\vdash_T (q_0, e)$.

Therefore, $(q_0, aabba) \vdash_T^* (q_0, e)$ and so $aabba$ is accepted by T.

31.19. As long as three consecutive b's have not been read, T is in state q_i (where i is 0, 1 or 2) immediately after reading a run of i consecutive b's that either began the input string or was preceded by an a. Whenever an a is read and T is in state q_0, q_1 or q_2, T returns to its initial state q_0. States q_0, q_1 and q_2 are all final, so any input string not containing three consecutive b's will be accepted. However, a run of three b's will force T to enter state q_3 which is not final, and T will then remain in this state regardless of the symbols in the rest of the input string.

32.1. $s \equiv (\lambda xy.yxy)(\lambda z.z)(\lambda pq.p)$ and $t \equiv \lambda xyz.xzz(y(xz))$.

32.2. $BV(s) = \{z, x, y\}$, $FV(s) = \{x, y\}$. $BV(t) = \{x, y\}$, $FV(t) = \{z\}$.

32.3. Suppose that $tu =_{\beta\eta} su$ for all λ-terms u, then also $tx =_{\beta\eta} sx$ for some variable x that does not occur free in s or t. Then $\lambda x.tx =_{\beta\eta} \lambda x.sx$ by (ξ) and $t =_{\beta\eta} s$ by (η).

32.4. If the equality $\lambda xy.x =_\beta \lambda x.x$ is added to $\boldsymbol{\lambda}_\beta$, then for any λ-term s, $s =_\beta \mathsf{I}s =_\beta (\lambda x.x)\mathsf{I}s =_\beta (\lambda xy.x)\mathsf{I}s =_\beta \mathsf{I}$. Thus, all terms are equal to I: the extension of $\boldsymbol{\lambda}_\beta$ with $\lambda xy.x =_\beta \lambda x.x$ is inconsistent.

32.5. We are done by constructing a term F that satisfies

$F =_\beta \lambda xy.xFy$, i.e., $F =_\beta (\lambda fxy.xfy)F$. Now take $F := \mathsf{Y}(\lambda fxy.xfy)$.

32.6. If Z is a fixed point of SI, then $Z =_{\beta\eta} \mathsf{SI}Z$ and so for every λ-term t, $Zt =_{\beta\eta} \mathsf{SI}Zt =_{\beta\eta} \mathsf{I}t(Zt) =_{\beta\eta} t(Zt)$, so Zt is a fixed point of t and Z is a fixed point combinator. The other way around, if Z is a fixed point combinator, then $Zt =_{\beta\eta} t(Zt)$ for any λ-term t, so $Zx =_{\beta\eta} x(Zx) =_{\beta\eta} \mathsf{SI}Zx$. Hence $\lambda x.Zx =_{\beta\eta} \lambda x.\mathsf{SI}Zx$ and by the η-rule $Z =_{\beta\eta} \mathsf{SI}Z$, so Z is a fixed point of SI.

32.7. $\mathbf{S}(\mathbf{K}(\mathbf{S}(\mathbf{SKK})))(\mathbf{S}(\mathbf{KK})(\mathbf{SKK}))pq =_w qp$ in **CL**.

32.8. The first is immediate by induction on the structure of s. The second is by induction on u, distinguishing the three cases of Definition 32.16. We treat the last case, so say that $u \equiv st$. Then $(\lambda^* x.u)v \equiv \mathbf{S}(\lambda^* x.s)(\lambda^* x.t)v =_w (\lambda^* x.s)v((\lambda^* x.t)v)$. This last term is by induction hypothesis equal to $s[x := v]t[x := v]$, which is equal to $(st)[x := v]$.

32.9. (i) If $\mathbf{CL} \vdash \mathbf{K}y =_w \mathbf{S}(\mathbf{K}(\mathbf{K}y))\mathbf{I}$, then – because of the Church-Rosser property – $\mathbf{K}y$ and $\mathbf{S}(\mathbf{K}(\mathbf{K}y))\mathbf{I}$ have a common reduct under $\twoheadrightarrow_w$. However, $\mathbf{K}y$ and $\mathbf{S}(\mathbf{K}(\mathbf{K}y))\mathbf{I}$ are in normal form and, hence, they have no common reduct. Contradiction.

(ii) Take $s := \mathbf{K}$. Then $((\mathbf{K})_\lambda)_{\mathbf{CL}} := \mathbf{S}(\mathbf{KK})(\mathbf{SKK})$ and we are done if we show that $\mathbf{S}(\mathbf{KK})(\mathbf{SKK}) \neq_w \mathbf{K}$ in **CL**. By the Church-Rosser property for $\twoheadrightarrow_w$ and the fact that $=_w$ is the least equivalence relation generated by $\twoheadrightarrow_w$, we know that if $\mathbf{S}(\mathbf{KK})(\mathbf{SKK}) =_w \mathbf{K}$, then these two terms have a common $\twoheadrightarrow_w$-reduct. However, both $\mathbf{S}(\mathbf{KK})(\mathbf{SKK})$ and $\mathbf{K}$ are in $\rightarrow_w$-normal form, so they have no common $\twoheadrightarrow_w$-reduct and so $\mathbf{S}(\mathbf{KK})(\mathbf{SKK}) \neq_w \mathbf{K}$ in **CL**.

32.10. Suppose that such an H exists. Then for x and y arbitrary, distinct, variables, $(\lambda x.Hxy)y =_\beta (\lambda x.\mathsf{F})y =_\beta \mathsf{F}$; on the other hand, $(\lambda x.Hxy)y =_\beta Hyy =_\beta \mathsf{T}$, contradicting the Church-Rosser Theorem. So, such a term H cannot exist.

32.11. To show that the Church numerals form a numeral system, we have to define a successor function and a test for zero on c_n. The successor function is defined in Example 32.11. The test for zero can be defined by $\mathsf{Z?}_c := \lambda x.x(\lambda y.\mathsf{F})\mathsf{T}$. The predecessor function P_c^- can be defined by $\mathsf{P}_c^- := \lambda x.T^{-1}(\mathsf{P}^-(Tx))$, where T and T^{-1} are the terms of the Lemma 32.53. Another term that does the job and which is in normal form is $\mathsf{P}_c^- := \lambda xyz.x(\lambda pq.q(py))(\mathsf{K}z)\mathsf{I}$. That the first term defines the predecessor function on c_n is immediate by Lemma 32.53. The second requires a proof by induction.

32.12. For M one can take $\mathsf{M} := \lambda xyz.x(yz)$. That this term λ-defines multiplication is proved by showing that $(c_m z)^n(x) =_\beta z^{m*n}(x)$, by induction on n. For E one can take

$$\mathsf{E} := \mathsf{Y}(\lambda fxy.\ \text{If } \mathsf{Z?}_c y \text{ then } c_1 \text{ else } \mathsf{M}(fx(\mathsf{P}_c^- y))x).$$

Then $\mathsf{E}c_n c_m =_\beta c_{(n^m)}$. A shorter term for E, given by Rosser, is $\mathsf{E} := \lambda xy.yx$. However, Rosser started counting at 1 and this E does not work properly for c_0: $\mathsf{E}c_m c_0 =_\beta c_0 c_m =_\beta \mathsf{I}$ and not c_1.

32.13. First note how a term $x \cdots x$, consisting of n times an x, is constructed from c_n: $\mathsf{P}_c^- c_n(\lambda y.yx)x =_\beta x \cdots x$ (n times), for $n > 0$. Now we have to find a term F such that $Fc_n =_\beta \lambda x.F(x \cdots x)c_{n+1}$ (n times an x). This is accomplished by taking an F such that
$F =_\beta \lambda zx.$ If $\mathsf{Z?}_c z$ then Fc_1 else $F(\mathsf{P}_c^- z(\lambda y.yx)x)(\mathsf{S}^+ z)$.
So, take $F := \mathsf{Y}(\lambda fzx.$ If $\mathsf{Z?}_c z$ then fc_1 else $f(\mathsf{P}_c^- z(\lambda y.yx)x)(\mathsf{S}^+ z))$.

32.14. $F := \mathsf{Y}(\lambda fx.$ If $\mathsf{Z?}_c x$ then Gx else $\mathsf{A}(f(\mathsf{P}_c^- x)(Gx)))$ does the job.

32.15. $\mathsf{I}(\mathsf{II}) \rightarrow_\beta \mathsf{II}$ in two possible ways. A one-step reduction is always standard, so there are two standard reductions from $\mathsf{I}(\mathsf{II})$ to II.

32.16. We write s for the term $\Theta(\lambda f.f(\mathsf{SSS}))$. Now note that $s \twoheadrightarrow_l s(\mathsf{SSS})$ and so $s \twoheadrightarrow_l s(\mathsf{SSS}) \twoheadrightarrow_l s(\mathsf{SSS})(\mathsf{SSS}) \twoheadrightarrow_l \cdots$. The leftmost reduction strategy does not terminate and, hence, s has no normal form.

32.17. If there is an infinite β-reduction path starting from s or t, then there is also an infinite reduction path starting from st. So, by contraposition, if st is strongly normalizing, then also s and t are strongly normalizing. Take $s := \lambda x.x\mathsf{K}\Omega$ and $t := \lambda y.y\mathsf{K}\Omega\mathsf{I}$. Then s and t are both not normalizing, but st is normalizing.

32.18. Let $M = \langle D, [D \to D], F, G\rangle$ be a set-theoretic λ-model with $G \circ F = id_D$. Furthermore, let ρ be a valuation and let s be an arbitrary λ-term with $x \notin FV(s)$. Then $[\![\lambda x.sx]\!]_\rho^D = G(d \mapsto F([\![s]\!]_\rho^D)(d))$. Now, $d \mapsto F([\![s]\!]_\rho^D)(d)$ is the same function as $F([\![s]\!]_\rho^D)$, so $[\![\lambda x.sx]\!]_\rho^D = G(F([\![s]\!]_\rho^D)) = [\![s]\!]_\rho^D$ (by the fact that $G \circ F = id_D$) and we are done.

32.19. The η-rule is not valid in $P\omega$. If it were valid, $[\![\lambda x.x]\!]_\rho$ and $[\![\lambda xy.xy]\!]_\rho$ would be the same in $P\omega$, which is not the case:

$[\![\lambda x.x]\!]_\rho = \{<n, m> \mid m \in e_n\}$ and

$[\![\lambda xy.xy]\!]_\rho = \{<q, <n, m>> \mid \exists e_p \subseteq e_n[<p, m> \in e_q]\}$.

These sets are not the same.

32.20. If $\boldsymbol{\lambda}_\beta \vdash s =_\beta t$, then $P\omega \models s = t$ by Soundness of the set-theoretic λ-models (Theorem 32.67.). Now, $[\![\mathsf{I}]\!]_\rho = \{<n, m> \mid m \in e_n\}$ and $[\![\mathsf{K}]\!]_\rho = \{<n, <p, q>> \mid q \in e_n\}$. These two sets are clearly not the same (and independent of ρ), so $P\omega \not\models \mathsf{I} = \mathsf{K}$ and hence $\boldsymbol{\lambda}_\beta \not\vdash \mathsf{I} =_\beta \mathsf{K}$.

33.1. $\lambda xyz.xy : (\alpha \to \gamma) \to \alpha \to \beta \to \gamma$ in $\lambda{\to}_{Cu}$ and $\lambda x : \alpha \to \gamma \,.\, \lambda y : \alpha \,.\, \lambda z : \beta \,.\, xy : (\alpha \to \gamma) \to \alpha \to \beta \to \gamma$ in $\lambda{\to}_{Ch}$.

33.2. $\lambda xy.y(\lambda z.x) : \beta \to ((\alpha \to \beta) \to \gamma) \to \gamma$ in $\lambda{\to}_{Cu}$ and

$\lambda x:\beta\,.\,\lambda y:(\alpha\to\beta)\to\gamma\,.\,y(\lambda z:\alpha.x):\beta\to((\alpha\to\beta)\to\gamma)\to\gamma$ in $\lambda{\to}_{Ch}$.

33.3. Look at the terms $s := \mathsf{SK}$ and $t := \mathsf{K}_*\ (\equiv \lambda xy.y)$. Then $s \twoheadrightarrow_\beta t$ and the principal type scheme of t is $\alpha\to\beta\to\beta$. The principal type scheme of S is $(\gamma\to\delta\to\epsilon)\to(\gamma\to\delta)\to\gamma\to\epsilon$ and hence the principal type scheme of SK is $(\gamma\to\delta)\to\gamma\to\gamma$, and the equation $(\gamma\to\delta)\to\gamma\to\gamma=\alpha\to\beta\to\beta$ has no solution, so SK cannot be typed with the principal type scheme of $\mathsf{K}_*\ (\equiv\lambda xy.y)$. (Note that to show that SK cannot have the type $\alpha\to\beta\to\beta$, it is not necessary to first compute the principal type scheme.) Now, let $pts(s)$ be σ and $s\twoheadrightarrow_\beta t$. Then $\vdash_{Cu} s:\sigma$ and hence, by Subject Reduction, $\vdash_{Cu} t:\sigma$.

33.4. (a) Take $p_1 := \lambda z.z\mathsf{K} : (\sigma\ \&\ \tau)\to\sigma$, $p_2 := \lambda z.z\mathsf{K}_* : (\sigma\ \&\ \tau)\to\tau$ and $\mathsf{pair} := \lambda xy.\lambda f.fxy : \sigma\to\tau\to\sigma\ \&\ \tau$. Then $p_1(\mathsf{pair}st) = p_1(\lambda f.fst) = (\lambda f.fst)\mathsf{K} = s$ and similarly, $p_2(\mathrm{pair}st) = t$.
(b) Take $\mathsf{in}_1 := \lambda x.\lambda fg.fx : \sigma\to(\sigma\vee\tau)$, $\mathsf{in}_2 := \lambda y.\lambda fg.gy : \tau\to(\sigma\vee\tau)$ and $\mathsf{case} := \lambda fg.\lambda z.zfg : \forall\alpha.(\sigma\to\alpha)\to(\tau\to\alpha)\to(\sigma\vee\tau)\to\alpha$. Then $\mathsf{case}st(\mathsf{in}_1u) = \mathsf{in}_1ust = su$ and similarly $\mathsf{case}st(\mathsf{in}_2u) = tu$.

33.5. (b) Note that $z:N,\ y:N,\ f:o\to o,\ x:o\vdash yf:o\to o$ and that $z:N,\ y:N,\ f:o\to o,\ x:o\vdash zfx:o$.
Hence $z:N,\ y:N,\ f:o\to o,\ x:o\vdash yf(zfx):o$ and we can conclude that $\mathsf{add} := \lambda zyfx.yf(zfx) : N\to N\to N$. That add behaves as addition is easily seen by showing that $\mathsf{add}\ c_nc_m = c_{n+m}$ holds, by induction on m.
(c) For $\lambda xy.yx$ to be of type $N\to N\to N$ in $\lambda{\to}_{Cu}$, it is required that y can be applied to x. This means, by Lemma 33.6, that $y:\rho\to\tau$ and $x:\rho$ for some types ρ and τ. Now $y,x:N\ (\equiv(o\to o)\to o\to o)$, so such types ρ and τ cannot exist.

33.6. (a) By the fact that $\mathsf{add} : N\to N\to N$, one shows that mult is of the required type. (Note that $\mathsf{add}z$ is of type $N\to N$. Furthermore, $y:\forall\alpha.(\alpha\to\alpha)\to\alpha\to\alpha$, and so $y:(N\to N)\to N\to N$. Hence $y(\mathsf{add}z)0$ is of type N.) To prove that mult behaves as multiplication, use the fact that add behaves as addition (see the previous exercise) and show that $\mathsf{mult}\ c_nc_m = c_{m*n}$ by induction on n.

(b) One way to show that exponentiation is definable in $\lambda 2_{Cu}$ is to define it by $\mathsf{exp} := \lambda zy.z(\mathsf{mult}\ y)c_1$. That exp defines exponentiation is proved by showing $\mathsf{exp}\ c_n c_m = c_{m^n}$ by induction on n.

33.7. head nil =

$$(\lambda l{:}L_\sigma.l\sigma t(\lambda z{:}\sigma.\lambda x{:}\sigma.x))(\lambda\alpha.\lambda x{:}\alpha.\lambda f : \alpha\to\sigma\to\alpha\,.\,x) =$$
$$(\lambda\alpha.\lambda x{:}\alpha.\lambda f : \alpha\to\sigma\to\alpha\,.\,x)\sigma t(\lambda z{:}\sigma.\lambda x{:}\sigma.x) = t.$$

head (cons l t') = cons l $t'\sigma t(\lambda z{:}\sigma.\lambda x{:}\sigma.x)) =$

$$(\lambda z{:}\sigma.\lambda x{:}\sigma.x)(l\ \sigma\ t\ (\lambda z{:}\sigma.\lambda x{:}\sigma.x))t' = t'.$$

33.8. A proof in natural deduction style is

$$\cfrac{\cfrac{\cfrac{\cfrac{\forall x,y{:}\sigma\,.\,Rxy\supset Ryx\supset A}{Rxx\supset Rxx\supset A}\quad [Rxx]^1}{Rxx\to A}\quad [Rxx]^1}{\cfrac{A}{Rxx\supset A}\,(1)}}{\forall x{:}\sigma\,.\,Rxx\supset A}$$

A λ-term that denotes this deduction is

$$\lambda p{:}(\Pi x,y{:}\sigma\,.\,Rxy\to Ryx\to A).\lambda x{:}\sigma.\lambda q{:}Rxx\,.\,pxxqq.$$

33.9. Note that in the context $\Gamma, x, y : N$ one can derive $a_2(Sx)y : (Sx)+y = y+(Sx)$ and $a_5yx : y+(Sx) = S(y+x)$, so by transitivity (a_3)
$a_3((Sx)+y)(y+(Sx))(S(y+x))(a_2(Sx)y)(a_5yx) : (Sx)+y = S(y+x)$. (*)
Then $a_2yx : y+x = x+y$ and $a_6S(y+x)(x+y) : (y+x = x+y) \to (S(y+x) = S(x+y))$, so $a_6S(y+x)(x+y)(a_2yx) : S(y+x) = S(x+y)$. (**)
Using transitivity we derive from (*) and (**) that
$\Gamma \vdash \lambda x,y : N\ .\ a_3((Sx)+y)(S(y+x))(S(x+y))(a_3((Sx)+y)(y+(Sx))(S(y+x))(a_2(Sx)y)(a_5yx))(a_6S(y+x)(x+y)(a_2yx))$:

$$\Pi x,y{:}N.(Sx)+y = S(x+y).$$

33.10. The construction is totally similar to the one in Exercise 33.9. One finds that $\Gamma \vdash \lambda x, y{:}N\,.\,a_3((Sx)+y)(S(y+x))(S(x+y))(a_3((Sx)+y)(y+(Sx))(S(y+x))(a_2(Sx)y)(a_5yx))(a_6S(y+x)(x+y)(a_2yx)) :$
$$\Pi x, y{:}N\,.\,T((Sx)+y = S(x+y)).$$

34.1. We have to show that $\neg[K_i]A \supset [K_i]\neg[K_i]A$ is valid. To this end, we take a model $M = \langle W,\ R_1,\ldots,R_m,\ \models\ \rangle$ and an arbitrary $w \in W$. We now have to show that $M \models_w \neg[K_i]A \supset [K_i]\neg[K_i]A$ for any $w \in W$. So suppose $M \models_w \neg[K_i]A$. This means that there is some w' with wR_iw' such that $M \models_{w'} \neg A$. We have to show that $M \models_w [K_i]\neg[K_i]A$, i.e., for all u with wR_iu it holds that $M \models_u \neg[K_i]A$. Consider an arbitrary u such that wR_iu. Since we already know that wR_iw', and R_i is an equivalence relation, we obtain that uR_iw' as well. Since it holds that $M \models_{w'} \neg A$, we may conclude that $M \models_u \neg[K_i]A$.

34.2. (a) is not valid: take an $S5$-model with two worlds, of which one satisfies P and the other does not. Then in either of the two states the formula does not hold.

(b) is valid, since in any $S5$-Kripke model, either there is a world in which P holds or there is one in which P does not hold.

(c) is valid: note that this formula is equivalent with $P \supset [K]\neg[K]\neg P$. This holds in any world in any $S5$-model, since for any world w in which P holds, it holds that for any successor world u of w there is some successor of u in which P holds, viz. the original world w.

(d) is valid, since it is easy to check that $[K]Q \supset [K](P \supset Q)$ is valid.

(e) is not valid; it is even not satisfiable. To illustrate the use of the proof system and its completeness, we show the equivalent statement that the formula is inconsistent, i.e.,
$\vdash \neg(P \,\&\, [K]\neg[K]Q \,\&\, ([K]\neg[K]\neg P \supset [K]Q))$. (We do this in a rather loose way providing the general idea without too much formal humdrum.) So assume $P \,\&\, [K]\neg[K]Q \,\&\, ([K]\neg[K]\neg P \supset [K]Q)$.
From $[K]\neg[K]Q$ we obtain $\neg[K]Q$, by (A3). From $[K]\neg[K]\neg P \supset [K]Q$, we obtain by contraposition that $\neg[K]Q \supset \neg[K]\neg[K]\neg P$. Hence, we get $\neg[K]\neg[K]\neg P$. On the other hand, by (c) we know that $P \supset [K]\neg[K]\neg P$ is valid, and thus (by completeness) provable. (Check this directly for yourself.) Since we have P, we thus have $[K]\neg[K]\neg P$, which together with $\neg[K]\neg[K]\neg P$ yields a contradiction.

Hence we have $\vdash \neg(P \ \& \ [K]\neg[K]Q \ \& \ ([K]\neg[K]\neg P \supset [K]Q))$.

34.3. We have to show that $[C](A \supset [E]A) \supset (A \supset [C]A)$ is valid. To this end, we take a model $M = \langle W, R_1, \ldots, R_m, \models\rangle$ and an arbitrary $w \in W$. Now suppose that $M \models_w [C](A \supset [E]A)$. This means that $M \models_{w'} A \supset [E]A$ for all w' with wR_Cw'. Now suppose that $M \models_w A$. We have to show that $M \models_w [C]A$, i.e., that $M \models_{w'} A$ for all w' with wR_Cw'. For this it is sufficient to show by induction on k that for all $k \geq 0$ it holds that $M \models_{w'} A$ for all w' with $w = w_0$, $w_0R_Ew_1$, $w_1R_Ew_2$, ..., $w_{k-1}R_Ew_k$, $w_k = w'$. The basic case that $k = 0$ is obvious, since we know that $M \models_w A$. Next, assume that the assertion holds for all $k < n$. We now prove the assertion for $k = n$: consider a state w' such that $w = w_0$, $w_0R_Ew_1$, $w_1R_Ew_2$, ..., $w_{n-1}R_Ew_n$, $w_n = w'$. Since $w = w_0$, $w_0R_Ew_1$, $w_1R_Ew_2$, ..., $w_{n-2}R_Ew_{n-1}$, we have, for $u = w_{n-1}$, that $M \models_u A$ by the induction hypothesis. Since we also have that wR_Cu, it holds also that $M \models_u A \supset [E]A$. Thus we have that $M \models_u [E]A$, i.e., $M \models_v A$ for all v with uR_Ev. In particular, we have that $M \models_{w'} A$, which we had to show.

34.4. (a) By (A8), we have $[C]A \supset [E][C]A$. Using (R3) we obtain $[C]([C]A \supset [E][C]A)$. (A10) gives us:
$[C]([C]A \supset [E][C]A) \supset ([C]A \supset [C][C]A)$. Hence, $[C]A \supset [C][C]A$.
(b) This is more difficult than one might expect at first sight. We prove the assertion in four parts:

(i) $[C]A \sim [K_i][C]A$.
(ii) $\neg[C]A \supset [K_i]\neg[C]A$.
(iii) $\neg[C]A \supset [E]\neg[C]A$.
(iv) $\neg[C]A \supset [C]\neg[C]A$.

Proof of **(i)**. By (A8), $[C]A \supset [E][C]A$. By (A6), $[E][C]A \supset [K_i][C]A$. Hence, we have $[C]A \supset [K_i][C]A$.
Since by (A3) we also have $[K_i][C]A \supset [C]A$, we obtain $[C]A \sim [K_i][C]A$.
(ii) By (A3), $[K_i][C]A \supset [C]A$. By contraposition, $\neg[C]A \supset \neg[K_i][C]A$. By (A5), $\neg[K_i][C]A \supset [K_i]\neg[K_i][C]A$. Using (i) we now can infer that $[K_i]\neg[K_i][C]A \sim [K_i]\neg[C]A$. Hence, $\neg[C]A \supset [K_i]\neg[C]A$.
(iii) From (ii) we know that $\neg[C]A \supset [K_i]\neg[C]A$. This holds for all i. Hence, $\neg[C]A \supset ([K_1]\neg[C]A \ \& \ \ldots \ \& \ [K_m]\neg[C]A)$. Consequently, $\neg[C]A \supset [E]\neg[C]A$ by (A6).
(iv) From (iii) we have $\neg[C]A \supset [E]\neg[C]A$. Using (R3) we obtain $[C](\neg[C]A \supset [E]\neg[C]A)$.
(A10) gives us $[C](\neg[C]A \supset [E]\neg[C]A) \supset (\neg[C]A \supset [C]\neg[C]A)$.

Consequently we have that $\neg[C]A \supset [C]\neg[C]A$.

34.5. Instead of providing common knowledge about the situation to his professors the dean now merely gives E-knowledge. It is easy to see by inspecting the proof that in this case the professors cannot ever determine the colour of their hats. So the dean can forget about his conference (and all future ones, for that matter), since he will have to return time and again indefinitely.

34.6. (a) First we note that $M \models_w [C]A$ iff $M \models_{w'} A$ for all $w' \in K_w$. Now suppose that $M \models_w [C]A$. So $M \models_{w'} A$ for all $w' \in K_w$ (*). Take some state $u \in K_w$, so wR_Cu. Let $K_u = \{u' \in S \mid uR_Cu'\}$. (Thus K_u is the upward cone of u.) Clearly, if $v \in K_u$, then $v \in K_w$, since then we have that wR_Cu and uR_Cv, and hence wR_Cv. Thus K_u is a 'subcone' of $K_w : K_u \subseteq K_w$. So, by (*), we obtain that $M \models_{u'} A$ for every $u' \in K_u$. Consequently, we have that $M \models_u [C]A$.
(b) Let $s = (s_1, s_2, \ldots, s_m)$ and $s' = (s'_1, s'_2, \ldots, s'_m)$ be two global states. We have to show that sR_Cs'. Since $m > 1$, we can argue as follows: $(s_1, s_2, \ldots, s_m)R_1(s_1, s'_2, \ldots, s'_m)$ and $(s_1, s'_2, \ldots, s'_m)R_2(s'_1, s'_2, \ldots, s'_m)$. Thus, sR_Cs'. Note that this argument only works if all global states are available.

34.7. We have to show that $[K_i]A \supset [I]A$ is valid.
Take a model $M = \langle W,\ R_1, \ldots, R_m,\ \models\rangle$ and an arbitrary $w \in W$.
Now suppose that $M \models_w [K_i]A$. This means that $M \models_{w'} A$ for all w' with wR_iw'. Since $R_I = R_1 \cap \ldots \cap R_m \subseteq R_i$, we therefore have that $M \models_{w'} A$ for all w' with wR_Iw', i.e., $M \models_w [I]A$.

34.8. Suppose that $\vdash (A_1 \ \& \ \ldots \ \& \ A_m) \supset B$. Then, by I-necessitation we get $\vdash [I]((A_1 \ \& \ \ldots \ \& \ A_m) \supset B)$. By the I-version of (A2) and Modus Ponens (R1), $\vdash [I](A_1 \ \& \ \ldots \ \& \ A_m) \supset [I]B$. By (A1), the I-version of (A2) again and (R1), $\vdash ([I]A_1 \ \& \ \ldots \ \& \ [I]A_m) \supset [I]B$. By (A11) and propositional reasoning we obtain $\vdash ([K_1]A_1 \ \& \ \ldots \ \& \ [K_m]A_m) \supset [I]B$.

34.9. i) $wR_iw' :=$ for all f in A_i, $f(w) = f(w')$. Hence, evidently, wR_iw, if $w_1R_iw_2$ and $w_2R_iw_3$, then $w_1R_iw_3$, and if wR_iw', then $w'R_iw$.
ii) $\alpha R_1\beta$ and $\gamma R_1\delta$; $\alpha R_2\beta$ and $\beta R_2\gamma$.

$\mathcal{E}_1 = \{\{\alpha, \beta\}, \{\gamma, \delta\}\}$ and $\mathcal{E}_2 = \{\{\alpha, \beta, \gamma\}, \{\delta\}\}$.

34.10.

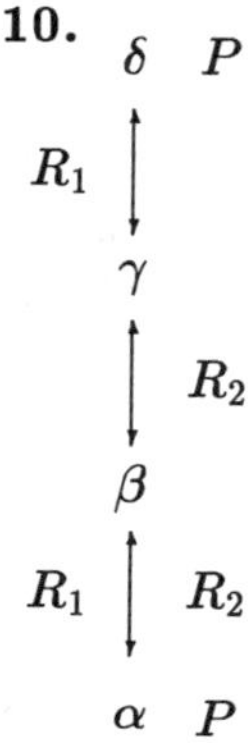

i) In world α agent 1 considers two worlds as possible, namely α and β, and only in one of them P holds. So $M \models_\alpha p_1(P) = 1/2$.

ii) In world α agent 2 considers three worlds as possible, namely α, β and γ, and only in one of them P holds. Hence, $M \models_\alpha p_2(P) = 1/3$.

34.11. i) a) Suppose $w' \in R_i(w)$ for some $w \in R_C(w_0)$, i.e., wR_iw' for some w such that w_0R_Cw. Then w_0R_Cw', i.e., $w' \in R_C(w_0)$. Conversely, suppose $w' \in R_C(w_0)$, i.e., w_0R_Cw'. Because R_i is reflexive, we have $w'R_iw'$ and hence $w' \in R_i(w')$.

b) Suppose $R_i(w_1) \neq R_i(w_2)$, say there is some w' such that $w' \in R_i(w_1)$ and $w' \notin R_i(w_2)$. In other words, w_1R_iw' and not w_2R_iw'. (1)

Next suppose that $R_i(w_1) \cap R_i(w_2) \neq \emptyset$, say $w \in R_i(w_1) \cap R_i(w_2)$, i.e., w_1R_iw and w_2R_iw. (2)

From (2) and the symmetry of R_i it follows that wR_iw_1. (3)

From (1), (3) and the transitivity of R_i it follows that wR_iw'. (4)

From (2), (4) and the transitivity of R_i it follows that w_2R_iw'. (5)

(1) and (5) are contradictory. Therefore, if $R_i(w_1) \neq R_i(w_2)$, then $R_i(w_1) \cap R_i(w_2) = \emptyset$.

ii) Suppose $M \models_w C(p_1(A) = q_1)$, i.e., for all $w' \in W$ with wR_Cw', $M \models_{w'} p_1(A) = q_1$; in other words, for all $w' \in W$ with wR_Cw',

$\mu(X_A \cap R_1(w')) = q_1 \cdot \mu(R_1(w'))$. (***)

$$\begin{aligned}
\mu(X_A \cap R_C(w)) &=_{\text{i)}} \mu(X_A \cap (R_1(w_0) \cup \ldots \cup R_1(w_k))) \\
&= \mu((X_A \cap R_1(w_0)) \cup \ldots \cup (X_A \cap R_1(w_k))) \\
&=_{(**)} \mu(X_A \cap R_1(w_0)) + \ldots + \mu(X_A \cap R_1(w_k)). \\
&=_{(***)} q_1 \cdot \mu(R_1(w_0)) + \ldots + q_1 \cdot \mu(R_1(w_k)) \\
&= q_1 \cdot (\mu(R_1(w_0)) + \ldots + \mu(R_1(w_k))) \\
&=_{(**)} q_1 \cdot \mu(R_1(w_0) \cup \ldots \cup R_1(w_k)) \\
&=_{\text{i)}} q_1 \cdot \mu(R_C(w)). \qquad \text{(I)}
\end{aligned}$$

Next suppose that $M \models_w C(p_2(A) = q_2)$. Then it follows in a similar way that $\mu(X_A \cap R_C(w)) = q_2 \cdot \mu(R_C(w))$. (II)
And from (I) and (II) it follows that $q_1 = q_2$.

34.12. 1) By definition, $K_i(X) \subseteq X$. 2) $K_i(X) := \cup\{V \mid V \in \mathcal{D}_i$ and $V \subseteq X\}$. Now suppose $w \in K_i(X)$. Because $K_i(X) \in \mathcal{D}_i$ and $K_i(X) \subseteq X$ it follows that $w \in K_i(K_i(X))$. 3) If $(W, \mathcal{D}_i)$ is a topological space given by a partition, then every set in $\mathcal{D}_i$ is both open and closed. $W - K_i(W - X)$ is the complement of an open set; so by definition, it is closed and hence open. Therefore, $K_i(W - K_i(W - X)) = W - K_i(W - X)$.
$K_1(\{\alpha\}) = \emptyset$, $K_1(\{\alpha, \beta\}) = \{\alpha, \beta\}$, $K_1(\{\alpha, \beta, \gamma\}) = \{\alpha, \beta\}$,
$K_1(\{\alpha, \beta, \gamma, \delta\}) = \{\alpha, \beta, \gamma, \delta\}$; $K_2(\{\alpha\}) = \emptyset$, $K_2(\{\alpha, \beta\}) = \emptyset$,
$K_2(\{\alpha, \beta, \gamma\}) = \{\alpha, \beta, \gamma\}$, $K_2(\{\delta\}) = \{\delta\}$, $K_2(\{\alpha, \beta, \gamma, \delta\}) = \{\alpha, \beta, \gamma, \delta\}$.
For instance, $K_1(\{\alpha, \beta, \gamma\}) = \{\alpha, \beta\}$ is the set of all worlds or objects which are classified by agent 1 *without any doubt* as elements of $\{\alpha, \beta, \gamma\}$; γ does not belong to this set because agent 1 cannot distinguish between γ and δ.

35.1. The reachable states of the algorithm MUTEX:
$\{n_1, n_2, s_1\}, \{n_1, t_2, s_2\}, \{n_1, c_2, s_2\}, \{n_1, n_2, s_2\}, \{t_1, n_2, s_1\}, \{c_1, n_2, s_1\}$,
$\{t_1, t_2, s_1\}, \{t_1, t_2, s_2\}, \{t_1, c_2, s_1\}, \{c_1, t_2, s_2\}$.

35.2. Let $M = \langle W, R, V\rangle$ be a modal model with R serial and suppose $M, w \models \Box\psi$, i.e., for all $w' \in W$ if $w\, R\, w'$, then $M, w' \models \psi$. Since R is serial, there is a $w' \in W$ such that $w\, R\, w'$. So, there is a $w' \in W$ such that $w\, R\, w'$ and $M, w' \models \psi$. I.e., $M, w \models \Diamond\psi$. For the proof of the other statements see the answer to Exercise 13.10 in Volume I.

35.3. We show only that $G(\phi \supset X\phi) \supset (\phi \supset G\phi)$ is valid and that the rule "if $\vdash \phi$, then $\vdash G\phi$ " preserves validity. Let $M = \langle p, V\rangle$ be a propositional LTL model and $M, w_0 \models G(\phi \supset X\phi)\ \&\ \phi$. Therefore, for each $i \geq 0$: $M, w_i \models \phi$. By the semantics of G, it follows that $M, w_0 \models G\phi$.
Let ϕ be a valid formula. Then ϕ is true in each model. Let $M = \langle p, V\rangle$ be a model. Thus, $M, w_0 \models \phi$. Since each $\langle p_i, V\lceil p_i\rangle$ is a model as well, $M, w_i \models \phi$ for all $i \geq 0$. Therefore, $M \models G\phi$.

35.4. Let $M = \langle p, D, V \rangle$ be a LTL model and $M, w_0 \models F(\phi \ \& \ \psi)$. Therefore, there is $i \geq 0$ such that $M, w_i \models (\phi \ \& \ \psi)$. By the semantics of F, it follows that $M, w_0 \models F\phi \ \& \ F\psi$.

35.5. Let $M = \langle p, D, V \rangle$ be a LTL model such that $V(w_i, q_1) = true$ iff i is even, and $V(w_i, q_2) = true$ iff i is odd. Then, $M, w_0 \models Fq_1 \ \& \ Fq_2$ but $M, w_0 \not\models F(q_1 \ \& \ q_2)$.

35.6. • The live semantics of the program in Example 35.9:

$$\chi_{PL} := z = 0 \ \& \ G(\exists u[z = u \ \& \ X(z = s(u))]).$$

• The just semantics of the program in Example 35.9:

$$\chi_{PJ} := z = 0 \ \& \ G(\exists u[z = u \ \& \ X(z = s(u))] \oplus \exists u[z = u \ \& \ X(z = u)]) \ \& \ GF(\exists u[z = u \ \& \ X(z = s(u))]).$$

35.7. Notice that the following model is a model for χ_{PJ}, but it is not a model for N7: $M = \langle p, D, V \rangle$, where $V_{Flex}(w_{2j})(z) = s^j(0)$ and $V_{Flex}(w_{2j+1})(z) = V_{Flex}(w_{2j})(z)$ for $j \geq 0$. The model M corresponds to the behaviour of the program P such that the assignment is executed only at the states with even indices.

35.8. Specification:

• A non-miraculous (but lossy) buffer: $\forall m[G\neg(out = m) \lor (F(in = m \ \& \ XF(out = m)) \ \& \ G(out = m \supset XG\neg(out = m)))]$. The first disjunct covers messages that were never put into the buffer or got lost. The second disjunct covers messages that have been put into the buffer and appear in the output – it rules out duplication of messages.

• Adding liveness: $G(\exists m[in = m \ \& \ \neg F(out = m)] \supset F\neg(out = \emptyset))$.
This clause ensures that as long as there are messages in the buffer, output will appear. It does not guarantee that in the case of infinite input, no messages will be delayed forever.

• Adding faithfulness: $\forall m[G(in = m \supset XF(out = m))]$.
This clause ensures that no messages get lost.

• Adding order:
$\forall m, m'[G(in = m \ \& \ XF(in = m') \supset F(out = m \ \& \ XF(out = m')))]$.
This clause ensures that the order between messages is preserved.

The respective clauses are written in a manner to be easily understood intuitively. As they strengthen the specification incrementally, weakening is quite possible. For example, because of the exclusion of miracles, faithfulness could also be obtained through the weaker requirement $\forall m[F(in = m) \supset F(out = m)]$.

35.9. Past time operators that can be viewed as counterparts to the future time LTL operators are S (since), Y (last time - "yesterday"!) and the derived ones like H (has always been in the past) and P (previously). Formulas, possibly in combination with future time operators, are constructed in the obvious manner. The same models as in the case of future time LTL are used for interpretating the past time operators.

- $M, w_i \models Y\phi$ iff $i > 0$ and $M, w_{i-1} \models \phi$.
- $M, w_i \models (\phi S \psi)$ iff there is k, $0 \leq k \leq i$, such that $M, w_k \models \psi$ and for all j, $k \leq j \leq i$, $M, w_j \models \phi$.
- Eventual and always are derived operators: $P\phi = (true S \phi)$,

  $H\phi = \neg P \neg \phi$.

As an example of an obvious gain in clarity of expression by the use of past time operators consider non-miraculousness:
$\forall m[G(out = m \supset YP(in = m))\ \&\ (out = m \supset XG\neg(out = m))]$. The first conjunct states that if a message appears in the output, it must have been inputted sometime before. The second conjunct rules out duplication of messages.

35.10. Only fair executions need to be considered now, i.e., executions for which the LTL formula $\bigwedge_{t \in F} GFen(t) \supset GFexe(t)$ holds, where $exe(t)$ is an assertion saying that t was executed at a state and F is a set of fair transitions. No transition of F can be enabled an infinite number of times and never be executed.

• Weaken the access conditions to only $\neg c_2$ for the process P_1 and $\neg c_1$ for the process P_2, i.e., s_1 and s_2 are no longer used. Make F equal to $\{t_1c_1, c_1n_1, t_2c_2, c_2n_2\}$. Eventual access cannot be guaranteed under justice only, for whichever justice set, because of the following scenario. Let P_1 be in its trying section t_1, let only P_2 make moves, entering and leaving its critical section an infinite number of times. Then t_1c_1, although enabled infinitely many times, never becomes continuously enabled, so justice does not apply .

• The adaptation of the rules to reflect the new scheduling regime is quite straightforward. It consists of replacing the justice set J by a set of fair transitions F and requiring eventual enabledness of a fair transition rather than continuous enabledness of a just one.

	RESP′		CHAIN′
R1	$G(\phi \supset (\psi \vee \gamma))$	C1	$G(\phi \supset (\bigvee_i \gamma_i)),\ i \geq 0$
R2	$\{\gamma\}T\{\psi \vee \gamma\}$	C2	$\{\gamma_i\}T\{\bigvee_{j\leq i} \gamma_j\},\ i > 0$
R3′	$\{\gamma\}t_h\{\psi\}$, for some $t_h \in F$	C3′	$\{\gamma_i\}h_i\{\bigvee_{j<i} \gamma_j\}, h_i \in F, i > 0$
R4′	$G(\gamma \supset Fen(t_h))$	C4′	$G(\gamma_i \supset Fen(h_i)),\ i > 0$
	$G(\phi \supset F\psi)$		$G(\phi \supset F\psi)$

Note that although the fourth premiss R4′ seems to make the rule RESP′ circular, this is in fact not the case, as the program over which the premiss has to be proven is smaller than the original one by the transition t_h. A similar remark applies to the chain-version of the rule.

• The adaptation of the proof is straightforward and left to the reader.

36.1. First, we show that two models are CTL modally equivalent iff they are CTL* modally equivalent. To prove this, it is sufficient to show that two models are CTL* modally equivalent iff they are bisimilar. The proof can be carried out by easy induction on the complexity of a formula. As CTL* is an extension of LTL, each two models which are not LTL modally equivalent are not CTL* modally equivalent. And therefore, they are not CTL modally equivalent.

36.2. Let $M = \langle W, R, V\rangle$ with root v_0 be an anchored CTL model such that $M, v_0 \models EG\psi \ \&\ AF\phi$. Therefore, M contains a full path with ψ and each full path in M contains a world with ϕ. So, M contains a full path with a world satisfying both ψ and ϕ. Thus, $M, v_0 \models EF(\psi \ \&\ \phi)$.

36.3. Let $M = \langle W, R, V\rangle$ with root v_0 be the following CTL model, in which the relation R is denoted by arrows and only propositions that are true at states are written:

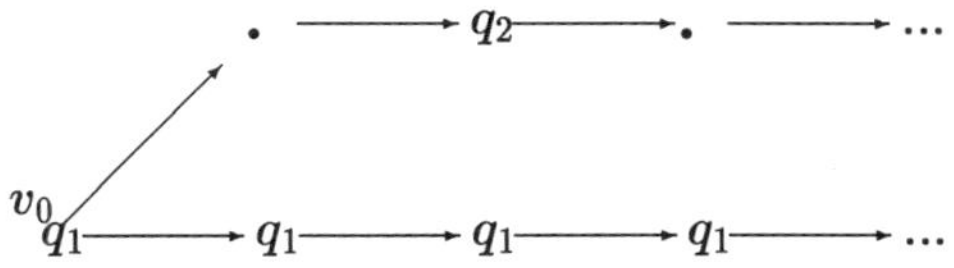

Notice that $M \models EGq_1 \;\&\; EFq_2$, but that $M \not\models EF(q_1 \;\&\; q_2)$.

36.4. The live CTL semantics for MUTEX is as follows. For readability, we use a simplification that can be applied in a concrete case like this. In clause (1), enabled transitions are represented by just writing down a (reachable) state and selecting the corresponding enabled transitions. Two transitions are written out in full; all others can be obtained analogously. The clause (2) evaluates to *true* as there is an enabled transition at each reachable state.

$$
\begin{array}{rcl}
\chi_{MUTEX} & = & \bigwedge \overline{\{n_1, n_2, s_1\}} \;\&\; AG(\\
 & & \{\bigwedge \overline{\{n_1, n_2, s_1\}} \supset \\
(1) & & [EX(\bigwedge \overline{\{t_1, n_2, s_1\}}) \;\&\; EX(\bigwedge \overline{\{n_1, t_2, s_2\}}) \;\&\; \\
 & & AX(\bigwedge \overline{\{t_1, n_2, s_1\}} \vee \bigwedge \overline{\{n_1, t_2, s_2\}}) \;\&\; \\
(2) & & true]\} \\
\& & & \\
 & & \{\bigwedge \overline{\{t_1, n_2, s_1\}} \supset \\
(1) & & [EX(\bigwedge \overline{\{c_1, n_2, s_1\}}) \;\&\; EX(\bigwedge \overline{\{t_1, t_2, s_2\}}) \;\&\; \\
 & & AX(\bigwedge \overline{\{c_1, n_2, s_1\}} \vee \bigwedge \overline{\{t_1, t_2, s_2\}}) \;\&\; \\
(2) & & true]\} \\
\& & & \\
(1) & & \{\ldots \\
 & & \;\& \\
(2) & & \ldots\})
\end{array}
$$

36.5. Model checking of the formula AFc_1 for MUTEX.

The formula is equivalent to the formula $\neg EG\neg c_1$ which does not contain the derived operators.

• The worlds $\{n_1, n_2, s_1\}, \{n_1, t_2, s_2\}, \{n_1, c_2, s_2\}, \{n_1, n_2, s_2\}, \{t_1, n_2, s_1\}, \{t_1, t_2, s_1\}, \{t_1, t_2, s_2\}, \{t_1, c_2, s_1\}$ are labelled $\neg c_1$.

• The worlds $\{n_1, n_2, s_1\}, \{n_1, t_2, s_2\}, \{n_1, c_2, s_2\}$, $\{n_1, n_2, s_2\}$ are labelled $EG\neg c_1$.

• The world $\{n_1, n_2, s_1\}$ is labelled $EG\neg c_1$.

Therefore AFc_1 does not hold.

36.6. The model for MUTEX with an idle transition at each world.

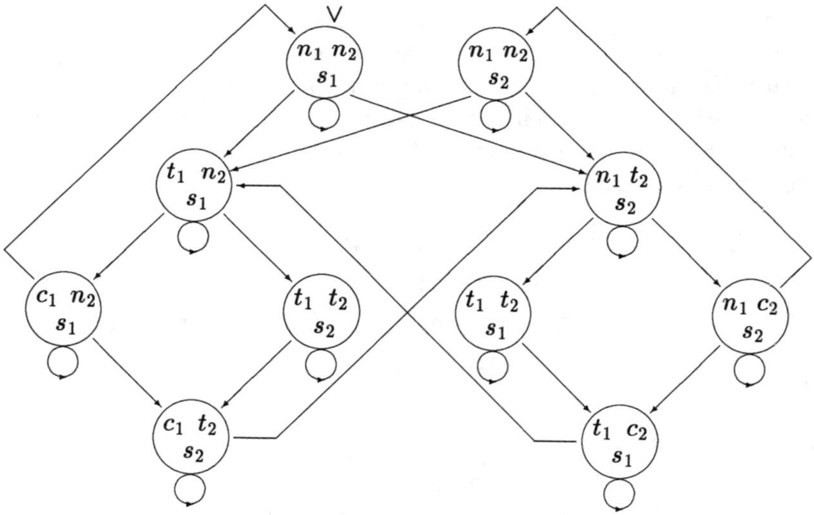

The diagram for model checking MUTEX with idling steps

Model checking for $AG\neg(c_1 \& c_2)$: The formula is equivalent to the formula $\neg E(trueU(c_1 \& c_2))$ which does not contain the derived operators.

- There are no worlds labelled c_1 & c_2.
- There are no worlds labelled $E(trueU(c_1 \& c_2))$.
- All the worlds are labelled $\neg E(trueU(c_1 \& c_2))$.

Therefore, $AG\neg(c_1 \& c_2)$ is true.

Model checking for $AG(t_1 \supset AFc_1)$: The formula is equivalent to the formula $\neg E(trueU(t_1 \& EG\neg c_1))$ which does not contain the derived operators.

- The worlds $\{n_1, n_2, s_1\}, \{n_1, t_2, s_2\}, \{n_1, c_2, s_2\}, \{n_1, n_2, s_2\}, \{t_1, n_2, s_1\}, \{t_1, t_2, s_1\}, \{t_1, t_2, s_2\}, \{t_1, c_2, s_1\}$ are labelled $\neg c_1$.
- The worlds $\{n_1, n_2, s_1\}, \{n_1, t_2, s_2\}, \{n_1, c_2, s_2\}, \{n_1, n_2, s_2\}, \{t_1, n_2, s_1\}, \{t_1, t_2, s_2\}, \{t_1, t_2, s_1\}, \{t_1, c_2, s_1\}$ are labelled $EG\neg c_1$.
- The worlds $\{t_1, n_2, s_1\}, \{t_1, t_2, s_2\}, \{t_1, t_2, s_1\}, \{t_1, c_2, s_1\}$ are labelled $t_1 \& EG\neg c_1$.
- All the worlds are labelled $E(trueU(t_1 \& EG\neg c_1))$.

Therefore, $AG(t_1 \supset AFc_1)$ is not true.

36.7. Modification is only necessary for checking formulas of the form $EG\psi$. As before, the algorithm finds maximal strongly connected components in $W' = \{w \mid M, w \models \psi\}$. Then a check is made to determine whether there is a maximal strongly connected component SC that is just, i.e., for each $t \in J$, if t is enabled at all the states of SC, then t is executed between some two states of SC. Those worlds which belong to just maximal components of size greater than 1 or with a self-loop are selected. Then, as before, the algorithm traverses the graph backwards and finds all worlds which can be reached by a path in which each world is labelled ψ.

36.8. The formula $AG\neg(c_1 \ \& \ c_2)$ is *true* in the model for MUTEX in Exercise 36.6 over just paths since the worlds in the just paths are the same as the worlds in all the paths. $AG\neg(c_1 \ \& \ c_2)$ was *true* over all the paths of the model. Model checking for $AG(t_1 \supset AFc_1)$ over just paths with $J = \{t_1c_1, c_1n_1, t_2c_2, c_2n_2\}$:
– The formula is equivalent to the formula $\neg E(trueU(t_1 \ \& \ EG\neg c_1))$ which does not contain the derived operators.
– The worlds $\{n_1, n_2, s_1\}, \{n_1, t_2, s_2\}, \{n_1, c_2, s_2\}, \{n_1, n_2, s_2\}, \{t_1, n_2, s_1\}, \{t_1, t_2, s_1\}, \{t_1, t_2, s_2\}, \{t_1, c_2, s_1\}$ are labelled $\neg c_1$.
– The worlds $\{n_1, n_2, s_1\}, \{n_1, t_2, s_2\}, \{n_1, c_2, s_2\}, \{n_1, n_2, s_2\}$ are labelled $EG\neg c_1$.

– There are no worlds labelled t_1 & $EG\neg c_1$.
– There are no worlds labelled $E(trueU(t_1 \,\&\, EG\neg c_1))$.
– All the worlds are labelled $\neg E(trueU(t_1 \,\&\, EG\neg c_1))$.
Therefore, $AG(t_1 \supset AFc_1)$ is true.

36.9. CTL formulas will be checked over "fair" paths, i.e., the paths satisfying the LTL formula $\bigwedge_{t \in F} GFen(t) \supset GFexe(t)$, where $exe(t)$ is an assertion saying that t was executed at a state and F is a set of fair transitions. No transition of F can be enabled an infinite number of times and never be executed.

• Weaken the access conditions to only $\neg c_2$ for the process P_1 and $\neg c_1$ for the process P_2, i.e., s_1 and s_2 are no longer used. Make F equal to $\{t_1c_1, c_1n_1, t_2c_2, c_2n_2\}$. For a discussion of eventual access under justice only, see the answer to exercise 35.10.

• The modification of the model checking algorithm only concerns checking formulas of the form $EG\psi$. As before, the algorithm finds maximal strongly connected components in $W' = \{w \mid M, w \models \psi\}$. Then a check is made to determine whether there is a maximal strongly connected component that is fair, i.e., containing a strongly connected subcomponent SC such that for each $t \in F$, if t is enabled at some world of SC, then t is executed between some two worlds of SC. Those worlds are selected which belong to fair maximal components of size greater than 1 or with a self-loop. Then, as before, the algorithm traverses the graph backwards and finds all worlds which can be reached by a path in which each world is labelled ψ.

• Model checking of the formula $AF(t_1 \supset AFc_1)$ for the new version of MUTEX over fair paths with $F = \{t_1c_1, c_1n_1, t_2c_2, c_2n_2\}$.

The figure below shows the model for the new version of MUTEX with an idle transition at each world.

– The formula is equivalent to the formula $\neg E(trueU(t_1 \,\&\, EG\neg c_1))$, which does not contain the derived operators.
– The worlds $\{n_1, n_2\}, \{n_1, t_2\}, \{n_1, c_2\}, \{t_1, n_2\}, \{t_1, t_2\}, \{t_1, c_2\}$ are labelled $\neg c_1$.
– The worlds $\{n_1, n_2\}, \{n_1, t_2\}, \{n_1, c_2\}$ are labelled $EG\neg c_1$.
Notice that these worlds constitute a maximal strongly connected component of worlds satisfying $\neg c_1$ which is fair.
– There are no worlds labelled t_1 & $EG\neg c_1$.

– There are no worlds labelled $E(trueU(t_1 \,\&\, EG\neg c_1))$.
– All the worlds are labelled $\neg E(trueU(t_1 \,\&\, EG\neg c_1))$.
Therefore, $AG(t_1 \supset AFc_1)$ is true.

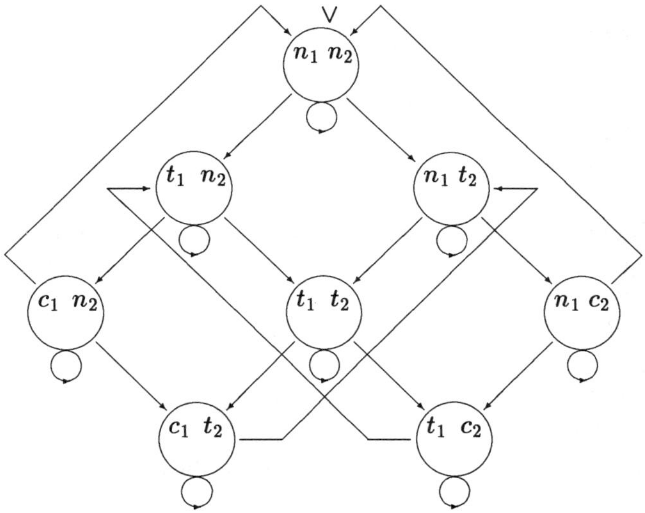

The diagram for the new version of MUTEX with idling steps

LITERATURE

Allen, J.P.B., and Buren, P. van, *Chomsky: Selected Readings.* Oxford University Press, London, 1975.

Apt, K.R., Ten years of Hoare's logic, a survey, part I. *ACM Trans. Programming Lang. Syst.* 3, pp. 431-483, 1981.

Barendregt, H.P., *The lambda calculus: its syntax and semantics*, revised edition. Studies in Logic and the Foundations of Mathematics, Vol. 103, North Holland, 1984.

Barendregt, H.P., Typed lambda calculi. In: *Handbook of Logic in Computer Science*, eds. Abramski et al., Oxford Univ. Press, 1992.

Böhm, C., and Berarducci, A., Automatic synthesis of typed Λ-programs on term algebras. *Theoretical Computer Science*, 39, pp. 135-154, 1985.

Boolos, G., and Jeffrey, R., *Computability and Logic.* Cambridge University Press, London, 1974, 1980, 1982.

Börger, E., *Trends in Theoretical Computer Science.* Computer Science Press, Rockville, 1988.

Bratko, I., *Prolog; Programming for Artificial Intelligence.* Addison-Wesley, 1986, 1990.

Brock, E.O. de, *The Foundations of Semantic Databases.* Prentice Hall, 1993, to appear.

Brookshear, J. Glenn, *Theory of Computation; Formal Languages, Automata and Complexity.* The Benjamin/Cummings Publishing Company, New York, 1989.

Browne, M.C., Clarke, E.M., Grumberg, O., Characterizing finite Kripke structures in propositional temporal logics. *Theoretical Computer Science*, 59 (1, 2), pp. 115-131, 1988.

Bruijn, N.G. de, A survey of the project Automath. In *To H.B. Curry: Essays on Combinatory Logic, Lambda Calculus and Formalism,* eds. J.P. Seldin, J.R. Hindley. Academic Press, New York, pp. 580-606, 1980.

Burstall, M., Program proving as hand simulation with a little induction. *Proc. of IFIP*, pp. 308-312, 1974.

Chomsky, N., *Syntactic Structures.* Mouton, the Hague, 1957.

Chomsky, N., *Aspects of the Theory of Syntax.* M.I.T. Press, Cambridge, Mass., 1965.

Chomsky, N., *Language and Mind.* Harcourt, Brace & World, New York, 1968.

Clarke, E.M., Emerson, E.A. and Sistla, A.P., Automatic verification of finite state concurrent systems using temporal logic specifications: A practical approach. *ACM Transactions on Programming Languages and Systems*, 8(2), pp. 244–263, 1986.

Constable, R.L., et al., *Implementing Mathematics with the Nuprl Proof Development System.* Prentice-Hall, 1986.

Cook, S.A., The Complexity of Theorem-Proving Procedures. *Proceedings 3rd Annual ACM Symposium on the Theory of Computing*, 1971, pp. 151-158.

Coquand, Th., Metamathematical investigations of a calculus of constructions. In: *Logic and Computer Science*, ed. P.G. Odifreddi, APIC series, vol. 31, Academic Press, pp. 91-122, 1990.

Coquand, Th. and Huet, G., The calculus of constructions. *Information and Computation*, 76, pp. 95-120, 1988.

Curry, H.B. and Feys, R., *Combinatory Logic, Volume 1.* North-Holland, 1958.

Cutland, N., *Computability, An introduction to recursive function theory.* Cambridge University Press, Cambridge, 1980.

Davis, M., *Computability & Unsolvability.* McGraw-Hill, New York, 1958.

Davis, M., (ed.), *The Undecidable.* Basic papers on undecidable propositions, unsolvable problems and computable functions. Raven Press, Hewlett, New York, 1965.

Dowek, G., Felty, A., Herbelin, H., Huet, G., Paulin-Mohring, Ch., Werner, B., *The Coq proof assistent*, version 5.6, user's guide. INRIA Rocquencourt-CNRS ENS, Lyon, 1991.

Emerson, E.A. , In: J. van Leeuwen, editor, *Handbook of Theoretical Computer Science.* Elsevier Science Publishers B.V., 1990.

Fitting, M., *First-order Logic and Automated Theorem Proving.* Springer-Verlag, New York, 1990.

Gabbay, D., *Investigations in Modal and Tense Logics with Applications to Problems in Philosophy and Linguistics.* Reidel, Dordrecht, 1976.

Galton, A., *Logic for Information Technology.* John Wiley and Sons, Chichester, 1990.

Genesereth, M.R. and Nilsson, N.J., *Logical Foundations of Artificial*

Intelligence. Morgan Kaufmann, Los Altos, 1987.

Geuvers, J.H., *Logics and Type Systems.* PhD. thesis, Nijmegen University, The Netherlands, 1993.

Girard, J.-Y., Interprétation fonctionelle et élimination des coupures dans l'arithmétique d'ordre supérieur. PhD. thesis, Université Paris VII, France, 1972.

Girard, J.-Y., Lafont, Y. and Taylor, P., *Proofs and Types.* Cambridge Tracts in Theoretical Computer Science 7, Cambridge University Press, 1989.

Goldblatt, R., *Logics of Time and Computation.* CSLI Lecture Notes No. 7, 1992.

Harman, G., (ed.), *On Noam Chomsky: Critical Essays.* Anchor Books, New York, 1974.

Harper, R., Honsell, F. and Plotkin, G., A framework for defining logics. *Proceedings Second Symposium on Logic in Computer Science,* (Ithaca, N.Y.), IEEE, Washington DC, pp. 194-204, 1987.

Herken, R. (ed.), *The Universal Turing Machine. A Half-Century Survey.* Oxford University Press, Oxford, 1988.

Hindley, J.R. and Seldin, J.P., *Introduction to Combinators and λ-Calculus.* London Mathematical Society Student Texts 1, Cambridge University Press, 1986.

Hopcroft, J.E. and Ullman, J.D., *Formal Languages and their Relation to Automata.* Addison-Wesley, Reading, Mass., 1969.

Hopcroft, J.E. and Ullman, J.D., *Introduction to Automata Theory, Languages and Computation.* Addison-Wesley, Reading, Mass., 1979.

Howard, W.A., The formulas-as-types notion of construction. In: *To H.B. Curry: Essays on Combinatory Logic, Lambda Calculus and Formalism,* eds. J.P. Seldin, J.R. Hindley. Academic Press, New York, pp. 479–490, 1980.

Hughes, G.E. and M.J. Cresswell, *An Introduction to Modal Logic.* Methuen, London and New York, 1968.

Hughes, G.E. and M.J. Cresswell, *A Companion to Modal Logic.* Methuen, London and New York, 1984.

Karp, R.M., Reducibility among combinatorial problems. In: R.E. Miller and J.W. Thatcher (eds.), *Complexity of Computer Computations.* Plenum Press, New York, pp. 85-103, 1972.

Kleene, S.C., *Introduction to Metamathematics.* Wolters-Noordhoff Publ.

and North-Holland Publ. Co., Amsterdam, 1952-1957-1959-1962-1964-1967-1971-1974-1980.

Kleene, S.C., *Mathematical Logic.* John Wiley & Sons, Inc., New York, 1967.

Kröger, F., Formalization of algorithmic reasoning. *Proc. of MFCS*, pp. 287-293, 1975.

Lamport, L., What good is temporal logic ? *Information Processing* '83, pp. 657-668, 1983.

Lamport, L., The Temporal Logic of Actions. *Transactions of Programming Languages and Systems.* To appear.

Lewis, H.R. and Papadimitriou, C.H., *Elements of the Theory of Computation.* Prentice-Hall, Englewood Cliffs, New Jersey, 1981.

Lloyd, J.W., *Foundations of Logic Programming.* Springer Verlag, Berlin, 1987.

Luo, Z. and Pollack, R., *Lego proof development system*: User's manual. Dept. of Computer Science, University of Edinburgh, 1992.

Maier, D., *The theory of relational databases.* Computer Science Press, Rockville, Md, 1983.

Manna, Z. and Pnueli, A., *The Temporal Logic of Reactive and Concurrent Systems Specification.* Springer, 1991.

Manna, Z. and Pnueli, A., Completing the temporal picture. *Theoretical Computer Science*, 83(1), pp. 97–139, 1991.

Manna, Z. and Pnueli, A., A Temporal Proof Methodology for Reactive Systems. In: M. Broy (ed.), *Program Design Calculi.* Springer, 1993.

Martin-Löf, P., An intuitionistic theory of types: predicative part. *Logic Colloquium '73*, North-Holland 1975, pp 73-118, 1975.

Martin-Löf, P., *Intuitionistic Type Theory*; Studies in Proof theory. Bibliopolis, Napoli, 1984.

McCune, W.W., *OTTER 2.0 Users Guide.* Argonne National Laboratory, ANL-90/9, March 1990.

Meyer, J.-J.Ch. and Hoek, W. van der, *Epistemic Logic for AI and Computer Science.* Cambridge University Press, 1994.

Mohring, Ch., Algorithm development in the calculus of constructions. In *Proceedings of the first symposium on Logic in Computer Science.* Cambridge, Mass., IEEE, pp. 84-91, 1986.

Nederpelt, R.P., Type systems – basic ideas and applications. In: A.J.

van de Goor, ed., *Proceedings of CSN '90*, Stichting Mathematisch Centrum, 1990.

Nordström, B., Petersson, K., Smith, J.M., *Programming in Martin-Löf's Type Theory*. Oxford University Press, 1990.

Odifreddi, P., *Classical Recursion Theory*. Studies in Logic and the Foundations of Mathematics, Vol 125, North Holland, 1989.

Ophelders, W.M.J., *Automated Theorem Proving based upon a Tableau-method with Unification under restrictions: Theory, Implementation and Empirical Results*. PhD. Thesis, Tilburg University, the Netherlands, 1992.

Ophelders, W.M.J. and de Swart, H.C.M., Tableaux versus Resolution; a comparison. *Fundamenta Informaticae*, Vol. 18, 1993, pp. 109-127.

Peterson, G.L., A New Solution to Lamport's Concurrent Programming Problem. *ACM Trans. Prog. Lang. Sys.*, 5(1), pp. 56-65, 1983.

Pnueli, A., The temporal logic of programs. *Proc. of MFCS*, pp. 45-57, 1977.

Prior, A., *Time and Modality*. Clarendon Press, Oxford, 1957.

Rauszer, C.M., Rough Logic for Multi-Agent Systems. *Proceedings* of the Conference *Logic at Work*, Amsterdam, 1992. To be published by Springer Verlag.

Reynolds, J.C., Towards a theory of type structure. *Proceedings, Colloque sur la Programmation*, LNCS 19, pp. 408-425, 1974.

Salomaa, A., *Computation and Automata*. Cambridge University Press, Cambridge, 1985.

Scott , D.S., The language LAMBDA (abstract), *Journal of Symbolic Logic* 39, pp. 425-427, 1974.

Scott , D.S., Lambda calculus: Some models, some philosophy. In: *The Kleene Symposium*. Studies in Logic 101, North-Holland, eds. J.K. Barwise, H.J. Keisler and K. Kunen, pp. 223-266, 1980.

Searle, J.R., Chomsky's Revolution in Linguistics. *New York Review of Books*, June 29, 1972. Reprinted in Harman, G. (ed.), *On Noam Chomsky: Critical Essays*. Anchor Books, New York, 1974, pp. 2-33.

Sommerhalder, R. and van Westrhenen, S.C., *The Theory of Computability; Programs, Machines, Effectiveness, and Feasibility*. Addison-Wesley, New York, 1988.

Stanczyk, S., *Theory and Practice of Relational Databases*. Pitman, London, 1990.

Sterling, L. and Shapiro, E., *The Art of Prolog: advanced programming techniques.* MIT Press, 1986.

Stirling, C., Comparing linear and branching time temporal logics, *Lecture Notes in Computer Science*, 398, pp. 1-20, Springer, 1987.

Swart, H.C.M. de, Linking Kripke semantics and game theory; a logical approach to Aumann's theorem. *Proceedings Computing Science in the Netherlands*, 1992, (ed. J.L.G. Dietz), Stichting Mathematisch Centrum, pp. 316-326.

Swart, H.C.M. de, and Ophelders, W.M.J., Tableaux, Resolution and Complexity of Formulas. *Methods of Logic in Computer Science*, 1994, to appear.

Szalas, A., Concerning the Semantic Consequence Relation in First-Order Temporal Logic. *Theoretical Computer Science*, 47, pp. 329-334, 1986.

Szalas, A., A complete axiomatic characterization of first-order temporal logic of linear time. *Theoretical Computer Science*, 54, pp. 199-214, 1987.

Takahashi, M., Parallel reductions in λ-calculus, *Journal of Symbolic Computation* 7, pp. 113–123, 1989.

Turing, A.M., *Mechanical Intelligence*, ed. by D.C. Ince. Elsevier North-Holland, Amsterdam, 1992.

Turing, A.M., On computable numbers, with an application to the Entscheidungsproblem, 1936-1937. Reprinted in Davis, M., *The Undecidable*, pp. 115-154.

Ullman, J.D., *Principles of Database and Knowledge Base Systems*, Vol. I. Computer Science Press, Rockville, Maryland, 1988.

Ullman, J.D., *Principles of Database and Knowledge Base Systems*, Vol. II. Computer Science Press, Rockville, Maryland, 1989.

Wood, D., *Theory of Computation.* John Wiley & Sons, New York, 1987.

Yang, Chao-Chih, *Relational Databases*, Prentice Hall, 1986.

INDEX